Exploring Mathematics

An Engaging Introduction to Proof

Exploring Mathematics gives students experience with doing mathematics – interrogating mathematical claims, exploring definitions, forming conjectures, attempting proofs, and presenting results – and engages them with examples, exercises, and projects that pique their curiosity. Written with a minimal number of pre-requisites, this text can be used by college students in their first and second years of study and by independent readers who want an accessible introduction to theoretical mathematics. Core topics include proof techniques, sets, functions, relations, and cardinality, with selected additional topics that provide many possibilities for further exploration.

With a problem-based approach to investigating the material, students develop interesting examples and theorems through numerous exercises and projects. In-text exercises, with complete solutions or robust hints included in an appendix, help students explore and master the topics being presented. The end-of-chapter exercises and projects provide students opportunities to confirm their understanding of core material, learn new concepts, and develop mathematical creativity.

John Meier is the David M. '70 and Linda Roth Professor of Mathematics at Lafayette College, where he also served as Dean of the Curriculum. His research focuses on geometric group theory and involves algorithmic, combinatorial, geometric, and topological issues that arise in the study of infinite groups. In addition to teaching awards from Cornell University and Lafayette College, Professor Meier is the proud recipient of the James Crawford Teaching Prize from the Eastern Pennsylvania and Delaware section of the Mathematical Association of America.

Derek Smith is Associate Professor of Mathematics at Lafayette College. His research focuses on algebra, combinatorics, and geometry. He has taught a wide variety of undergraduate courses in mathematics and other subjects in both the United States and Europe. He is the recipient of multiple teaching awards at Lafayette, and his work has been supported by the Mathematical Association of America and the National Science Foundation. Professor Smith is a former editor of the problem section of *Math Horizons*.

CAMBRIDGE MATHEMATICAL TEXTBOOKS

Cambridge Mathematical Textbooks is a program of undergraduate and beginning graduate level textbooks for core courses, new courses, and interdisciplinary courses in pure and applied mathematics. These texts provide motivation with plenty of exercises of varying difficulty, interesting examples, modern applications, and unique approaches to the material.

A complete list of books in the series can be found at www.cambridge.org/mathematics
Recent titles include the following:

Chance, Strategy, and Choice: An Introduction to the Mathematics of Games and Elections, S. B. Smith
Set Theory: A First Course, D. W. Cunningham
Chaotic Dynamics: Fractals, Tilings, and Substitutions, G. R. Goodson
Introduction to Experimental Mathematics, S. Eilers & R. Johansen
A Second Course in Linear Algebra, S. R. Garcia & R. A. Horn
Exploring Mathematics: An Engaging Introduction to Proof, J. Meier & D. Smith
A First Course in Analysis, J. B. Conway

Exploring Mathematics

An Engaging Introduction to Proof

JOHN MEIER
Lafayette College, PA, USA

DEREK SMITH
Lafayette College, PA, USA

CAMBRIDGE
UNIVERSITY PRESS

University Printing House, Cambridge CB2 8BS, United Kingdom

One Liberty Plaza, 20th Floor, New York, NY 10006, USA

477 Williamstown Road, Port Melbourne, VIC 3207, Australia

4843/24, 2nd Floor, Ansari Road, Daryaganj, Delhi – 110002, India

79 Anson Road, #06–04/06, Singapore 079906

Cambridge University Press is part of the University of Cambridge.

It furthers the University's mission by disseminating knowledge in the pursuit of education, learning, and research at the highest international levels of excellence.

www.cambridge.org
Information on this title: www.cambridge.org/9781107128989

First published 2017

Printed in the United States of America by Sheridan Books, Inc.

A catalogue record for this publication is available from the British Library.

ISBN 978-1-107-12898-9 Hardback

For Noah and Robert

Contents

Preface

Mathematics is a fascinating discipline that calls for creativity, imagination, and the mastery of rigorous standards of proof. This book introduces students to these facets of the field in a problem-focused setting. For over a decade, we and many others have used draft chapters of *Exploring Mathematics* as the primary text for Lafayette's *Transition to Theoretical Mathematics* course. Our collective experience shows that this approach assists students in their transition from primarily computational classes toward more advanced mathematics, and it encourages them to continue along this path by demonstrating that while mathematics can at times be challenging, it is also very enjoyable.

Here are some of the key features of *Exploring Mathematics.*

- The sections are short, and core topics are covered in chapters that present important material with minimal pre-requisites. This structure provides flexibility to the instructor in terms of pacing and coverage.
- Mathematical maturity requires both a facility with writing proofs and comfort with abstraction and creativity. We help students develop these abilities throughout the book, beginning with the initial chapters.
- A student does not learn mathematics by passive reading. It is through the creation of examples, questioning if results can be extended, and other such in-the-margin activities that a student learns the subject. We encourage this behavior by including frequent in-text exercises that serve not only to check understanding, but also to develop material.
- We construct many mathematical objects that are elementary in their definition and commonly referenced in upper-level classes. These are woven throughout the text, with related exercises providing numerous opportunities for independent investigations of their important properties.
- Each chapter concludes with a robust mixture of exercises ranging from the routine to rather challenging problems, and the book concludes with a collection of projects: guided explorations that students can work on individually or in groups.

These and other fundamental aspects of *Exploring Mathematics* are described in greater detail below, where we also indicate different ways an instructor can map out the material that can be covered in a single term.

An Active Approach

Our experience teaching courses that introduce students to mathematical proofs shows that the spirit of mathematics is effectively taught with a focus on problem-solving. It is in doing mathematics – by exploring definitions, forming conjectures, and working on the writing of proofs – that students begin to understand the discipline. We have packed into this book well over 600 exercises that engage students in these important activities.

Each section has in-text exercises, and we expect the reader to work on all of these, preferably at the moment they are first encountered. A good reader of mathematics needs to develop the skill of frequently interrogating the definitions, examples, and arguments presented on the page. The in-text exercises model this behavior, and we rely on the earnest efforts of the reader as important material is developed in many of them. Most in-text exercises are checks of the reader's understanding or a vehicle to introduce examples, but some are challenging. Complete solutions, or at the very least some good hints, are provided in an appendix for those who are stuck. This structure makes it possible for someone to work through every chapter in this book, whether they are doing so in the context of a class or as independent reading. The solution appendix also serves as an additional resource for a student seeking additional examples of mathematical arguments.

Chapters conclude with a collection of exercises, segregated by the section or sections they are most closely aligned with. Many of these exercises are straightforward, but there are quite a few challenging problems as well. We conclude each chapter's exercises with 'More Exercises!', which are related to the topics of the chapter but which are wider ranging in scope or do not neatly align with any specific section.

The book concludes with a collection of projects in Chapter 11. These give students the opportunity to tackle questions that require multiple steps and the exploration of definitions and examples. They are written as guided research projects, with the amount of difficulty, direction, and independence varying project-to-project. At the start of each project we indicate the background material that is required for that particular project. We have had great success assigning these as essay projects, where the ultimate goal is for a small group of students to work together and produce a mathematical document, with definitions, theorems, and proofs presented in the standard style of mathematical exposition.

A Flexible Structure

Figure 1 outlines the topics covered by this text and the chapter dependence. In this chart we highlight the chapters that contain what we believe are the core sections of the book. We recognize, though, that what is considered mandatory material varies by institution and often by the preferences of individual instructors. Drafts of this book have been used successfully by instructors who have widely differing pedagogical styles, research specialties, and priorities for material to be covered.

Chapter 1: *Let's Play* is an introduction to the book, including topics that will reappear throughout the text and a preview of mathematical proofs. The intention is not that students will be able to instantly digest these arguments or be able to immediately construct their own impeccable proofs. Rather, we allow students to gain immediate exposure to theoretical mathematics through intriguing examples and ideas that we begin developing in greater detail in Chapter 2. Chapter 1 should not occupy more than the first two weeks of a semester-long term; one of us covers this chapter in one week. It is possible for an instructor to present some of her own favorite topics to illustrate the nature of theoretical mathematics in class, assigning portions of Chapter 1 as required out-of-class reading.

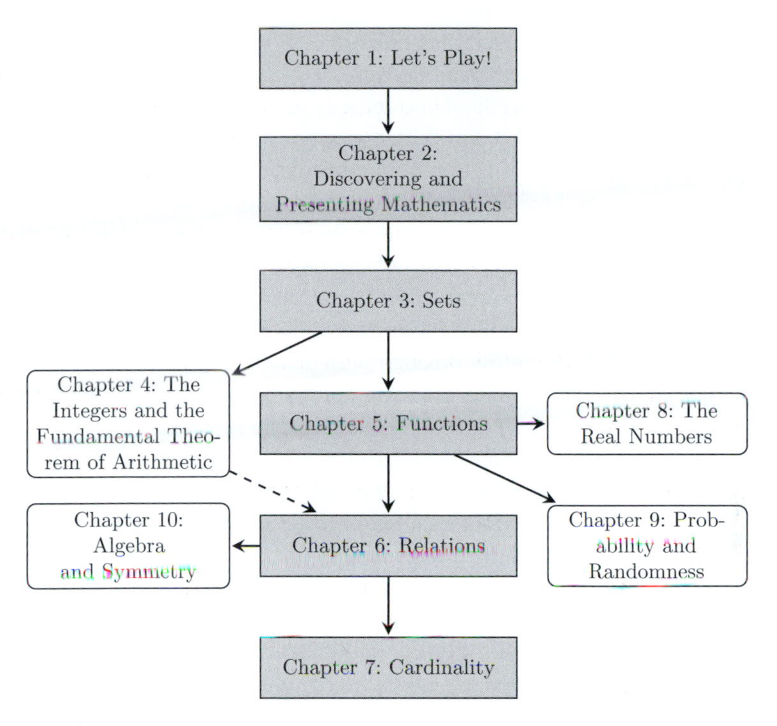

Figure 1. Arrows indicate chapter dependence. The core sections are contained in the shaded chapters. The final two sections of Chapter 6, but not its core sections, depend on some material from Chapter 4.

Chapter 2: *Discovering and Presenting Mathematics* begins by describing the standard techniques of mathematical proof – direct proofs, proofs by contradiction, and proofs by induction – along with a brief presentation of formal logic. We then transition to the creative side of mathematics, discussing how to form and test conjectures and ultimately how to present ideas when breakthroughs happen. This chapter anchors everything that is to follow.

After Chapter 2 gives students a grounding in creating and presenting mathematics, the remaining chapters present the fundamental ideas that have become standard in transition courses, including sets, functions, relations, and cardinality. For us, the core of this book consists of the following 35 sections:

Let's Play!, Sections 1.1, 1.2, and 1.4–1.7
Discovering and Presenting Mathematics, Sections 2.1–2.9
Sets, Sections 3.1–3.7
Functions, Sections 5.1–5.6
Relations, Sections 6.1 and 6.3
Cardinality, Sections 7.1–7.5

Many of these sections are short, and often more than three sections can be discussed in a week. We tend to cover these thirty-five core sections and a couple of related projects in no more than twelve weeks, leaving ample time for additional topics.

If you find that there is indeed room for additional material, you could choose to add depth to the topics already covered by spending time on sections not listed above. Depending on the length of the term, you might choose to cover enumerative combinatorics (Sections 3.8 and 9.3); functions applied to subsets (Sections 5.7 and 5.8); partially ordered sets (Section 6.2 and Project 11.6); modular arithmetic (Sections 6.4, 6.5, and 6.6); or the Schröder–Bernstein Theorem (Section 7.6 and Project 11.8).

Another approach is to add a fuller discussion of the integers and real numbers, which can be accomplished by including sections of Chapter 4: *The Integers and the Fundamental Theorem of Arithmetic* and Chapter 8: *The Real Numbers*, such as the material in Sections 4.1 through 4.3 (the proof of the Fundamental Theorem of Arithmetic) and 8.1 through 8.3 (providing practice with proofs of quantified statements).

The one model we have not seen attempted, and would argue against, is to try and cover all of the sections in this book in a single semester.

Two of our chapters cover material that is not always included in other texts. Chapter 10 is an introduction to abstract algebra, leading quite quickly to the idea of isomorphic groups. If this chapter is to be covered at the end of a single semester, you will need to spend time on Chapter 4 and have experience with modular arithmetic, which is presented in the later sections of Chapter 6. Chapter 9 is an introduction to discrete probability – including such notable objects as Pascal's triangle – with a narrative that focuses on the notion of randomness. Covering either one of these chapters in addition to the core sections can be done in a single semester, assuming one plans for this goal at the beginning and paces the course accordingly.

What We Assume

Our approach demands the reader's active participation. It is not demanding, though, in terms of pre-requisites. We have chosen not to build number systems from scratch, relying instead on students' exposure to standard ways of expressing the rationals and reals via fractions and decimals. An instructor who wants to provide some understanding of these constructions can certainly do so, and in particular should assign Exercise 6.75, where the rationals are constructed from equivalence classes in $\mathbb{Z} \times \mathbb{N}$, and Project 11.9, which develops the real numbers via Cauchy sequences (and so has Chapter 8 as a pre-requisite).

In addition to a naive understanding of the integers, rationals, and reals, we assume that students have been exposed to trigonometric functions, logarithms and exponentials, absolute value, and other standard functions included in basic mathematical knowledge from high school. While we occasionally appeal to material from calculus, these points of connection are never necessary for the development of the material and are included to help students see that what they did in their calculus courses is not fully orthogonal to the material in this book. Curiosity will serve a reader better than a background in calculus.

Conclusion

Exploring Mathematics is balanced toward the active and away from the rote. Students are encouraged and in fact required to explore and conjecture. While we present many proofs, the student who wants to simply mimic model proofs will soon need to step away from that strategy and become more daring.

We hope that everyone finds this book to be an interesting, challenging, and effective text. All readers should discover that they cannot learn this material without some, or a lot, of effort. Engaging with theoretical mathematics, its concepts and methodology, is a transformative process. At the end it should be the case that students realize that mathematics is not just "creative and rigorous" but is in fact delightfully creative and exceptionally rigorous, and that those two features of the discipline produce a fascinating field of study.

Acknowledgments

We thank our students and colleagues at Lafayette, many of whom suffered through some rather poor initial drafts of this text. Your collective feedback, even though often contradictory, has improved this book through its many iterations of use, and your questions have helped us to align the presentation with our long held desire to create a book that rewards an active reader with an understanding that the mathematics they have yet to learn is likely to be more interesting and engaging than the mathematics they know.

The students who provided responses are countable but too numerous to list here by name. We thank all of our colleagues, especially Professors Ethan Berkove, Jonathan Bloom, Justin Corvino, Evan Fisher, Trent Gaugler, Gary Gordon, Chawne Kimber, Elizabeth McMahon, and Lorenzo Traldi for their tremendously helpful feedback. We thank Professor Peter Winkler for his encouragement and a number of timely suggestions. We also greatly appreciate feedback from Jed Mihalisin and two dozen anonymous reviewers. And we are grateful to Jayne Trent for helping us to get drafts into the hands of students over the years.

We thank Kaitlin Leach, Charles Howell, Mairi Sutherland and all the folks at Cambridge University Press for their assistance and willingness to work with authors who are not particularly good at meeting deadlines.

Tracing the origins of all of the exercises in this text is somewhere on the spectrum between daunting and impossible. Almost all of the exercises were initially created by ourselves or others as part of homework sets, and it is certainly the case that we borrowed some ideas for exercises from resources we have forgotten. We know that some of our challenging exercises originated in problem contests like the American Invitational Mathematics Examination, the Putnam Examination, and the Mathematical Olympiads. The reader interested in finding additional challenging problems to work on is encouraged to seek out questions from those contests. Versions of several of our exercises can be found in the following excellent sources for engaging problems: Peter Winkler's *Mathematical Puzzles: A Connoisseur's Collection* [**Win04**]; the problem books created from training materials for the USA IMO Team, such as *102 Combinatorial Problems* [**AF03**]; and Rick Gillman's collection, *A Friendly Mathematics Competition* [**Gil03**].

Finally, we couldn't have finished this long-term project without the steady support of our families over the course of more than a decade. When the to-do lists seemed never-ending, they kept reminding us that this day would arrive.

1 Let's Play!

In this chapter we give you immediate exposure to reading and doing mathematics. In the next chapter we will begin to introduce the basic building blocks of mathematical reasoning, but here we preview the three most important types of mathematical argument: direct proof, proof by mathematical induction, and proof by contradiction. Along the way, we will introduce some basic properties of the integers, to be used in many examples throughout the text.

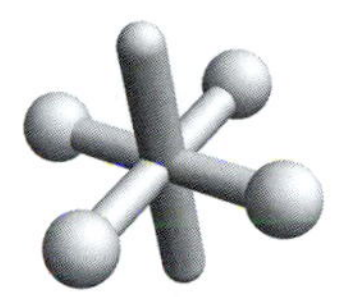

1.1. A Direct Approach

The discipline of mathematics has its own perspective on intellectual inquiry. As this book progresses, we will see that mathematicians often have an extremely high standard for establishing the veracity of statements; they enjoy "playing" with abstract mathematical objects (which are occasionally called "toys"); and they particularly enjoy when there are surprising connections between seemingly unrelated ideas, such as in the example presented in Section 1.3. Mathematics is certainly a serious business, but it is one that involves exploration and creativity, and this aspect of mathematics is given short shrift far too often.

Consider the following "theorem," which is what mathematicians call an important fact that has been proven.

THEOREM 1.1. *The medians of a triangle divide the triangle into six subtriangles of equal areas.*

In order to understand this statement you need to know or recall that a *median* of a triangle is a line segment connecting a vertex to the midpoint of its opposite side. The theorem also assumes that you know that the three medians of a triangle intersect in a single point. This point is called the *centroid* of the triangle.

Figure 1 shows the six subtriangles referenced in the statement of the theorem. They are formed by choosing a vertex and an adjacent midpoint, and always using the centroid as the third vertex of the triangle. If the original triangle is equilateral, then the six subtriangles will be congruent and hence they will have equal areas. The case of an equilateral triangle is, however, quite special. Compare the situation for an equilateral triangle with the triangle illustrated in Figure 1, where no pair of subtriangles is congruent. In this situation, and in fact for any triangle, why should all of the areas A, B, C, D, E, and F be equal? Try to figure this out on a separate sheet of paper *before* you read any further! It might help to recall the formula for the area of a triangle in terms of its height and base length.

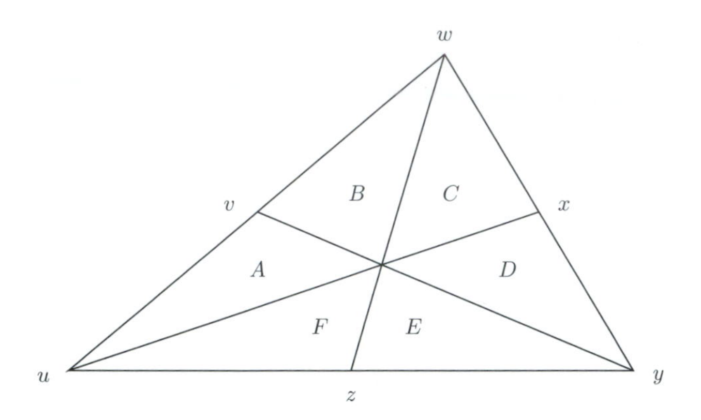

Figure 1. A triangle divided into six smaller triangles by its medians.

PROOF. The area of a triangle with base b and height h is $\frac{1}{2}bh$. Consider the two triangles whose areas are E and F. The bases for these triangles have the same length, since z is the midpoint of the line segment $\overline{uy}$. The height of both triangles is the same, as for both triangles it is the distance from the centroid to the line through u and y. Thus we have $E = F$. The same form of argument shows that $A = B$ and $C = D$.

However, the argument so far does not explain why $A = E$, for example. In order to show this, we consider the larger subtriangles $\triangle uwx$ and $\triangle uxy$. These triangles also have the same area. Since x is the midpoint of $\overline{wy}$, the lengths of $\overline{wx}$ and $\overline{xy}$ are the same. And the height of both triangles is given by the distance from u to the line through w and y. This means that

$$\text{Area of } \triangle uwx = A + B + C = D + E + F = \text{Area of } \triangle uxy,$$

which in light of the previous equalities yields

$$2A + C = C + 2E.$$

From this we can conclude that $A = E$, which immediately implies that $A = B = E = F$.

To show that we can include C and D with these equalities as well, consider the subtriangles $\triangle uvy$ and $\triangle vwy$. One more application of the "equal bases and equal heights" argument shows they have equal areas, hence

$$A + F + E = B + C + D.$$

This reduces to

$$A + 2E = A + 2C,$$

so we conclude that $E = C$ and thus that $A = B = C = D = E = F$. □

Notice what happened on the way from the statement of the theorem to the end of the proof. A figure was drawn. Labels like A and x were introduced. Relevant facts were employed, like the formula for the area of a triangle and the fact that the three medians of a triangle intersect in a common point. And the argument, which proceeded *directly* from the known facts to the desired conclusion, was written in the form of sentences and paragraphs that altogether constitute what's called a *direct proof*. Also notice what wasn't presented: all of the hard work and scribbles required in arriving at the finished product.

An important question arises: what are the "known facts" that you can assume in this book? As we discussed in the Preface, generally you should feel free to use basic mathematical knowledge from high school along with any other facts you have already proven. For instance, in proving the following theorem, known as the arithmetic–geometric mean inequality, you may wish to use this fact about inequalities: if x and y are real numbers with $x \geq y \geq 0$, then $\sqrt{x} \geq \sqrt{y} \geq 0$.

THEOREM 1.2. *If a and b are two non-negative real numbers, then*

$$\frac{a+b}{2} \geq \sqrt{ab}.$$

Exercise 1.1 Try to construct a direct proof of this theorem! Don't expect to come up with a "perfect" proof. Proving a new result and then presenting it well are skills that are continually developed over years, not days.

REMARK 1.3. At the beginning of this section we mentioned that a theorem is an important mathematical statement that has been proven. There are two other common terms that also denote mathematical statements that have proofs. A *proposition* is a proven mathematical statement, but one that is perhaps not so important as to be called a theorem. A *lemma* is also a proven mathematical statement, possibly quite important, whose primary function is as a key step or tool to proving other results. These terms should make more sense by the time you have worked through Section 2.7.

1.2. Fibonacci Numbers and the Golden Ratio

In this section, we introduce two important and seemingly unrelated mathematical objects. In Section 1.3, we will exhibit a surprising connection between them.

1.2.1. The Fibonacci Numbers. In the early thirteenth century, Leonardo of Pisa published a well-received book, the *Liber Abaci*, intended to instruct the growing merchant class in the art of using Hindu–Arabic numerals to do arithmetic. Most of the exercises in his book are mundane calculations of profits, conversions of currency, and so forth. But one question was a real gem:

> How many pairs of rabbits are created by one pair in one year? A certain man had one pair of rabbits together in a certain enclosed place, and one wishes to know how many are created from the pair in one year when it is the nature of them in a single month to bear another pair, and in the second month those born to bear also.[1]

[1] This translation is from Sigler's *Fibonacci's Liber Abaci: A Translation into Modern English of Leonardo Pisano's Book of Calculation* [**Pis02**].

We admit that this is not the easiest statement to comprehend. However, by working through what happens month by month you can start to see a pattern. Let's assume that at the very start the man has no rabbits, but then:

(1) In the first month the man has only one pair of rabbits.
(2) In the second month they are still the only pair, as the last phrase of the question indicates that it takes two months before a pair of rabbits will bear children.
(3) In the third month the original pair of rabbits creates a new pair, bringing the total to two pairs of rabbits.
(4) In the fourth month the original pair creates another pair, while the second pair is maturing, bringing the total to three pairs of rabbits.
(5) In the fifth month the original pair and their first offspring each create offspring, bringing the total number of pairs to five. Note: There is an assumption that not only do pairs of rabbits always breed two children in a litter, they also always breed a male and female pair. And they never die!

Figure 2 displays the growing family of rabbit pairs in the first few months. While the overall pattern may not yet be apparent to you, it is clear that the sequence of number of pairs in each month begins

$$1, 1, 2, 3, 5, \ldots$$

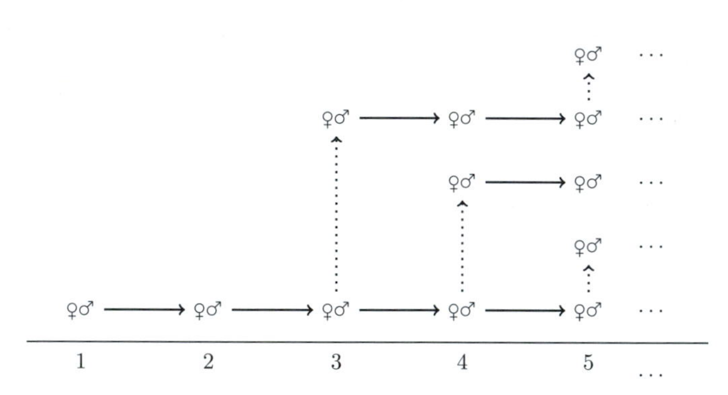

Figure 2. Breeding rabbit pairs over the first five months. The dotted arrows point to offspring.

Exercise 1.2 Compute the next two numbers in the sequence.

On the face of it, this is just an amusing diversion. But there are a number of surprising connections between this sequence and some important mathematics. So let's begin to study this sequence by giving it a proper mathematical definition. The *Fibonacci numbers* are a sequence of integers, where we declare that $f_1 = 1, f_2 = 1$, and the rest of the Fibonacci numbers are defined using the "recursive" formula

$$f_{n+1} = f_n + f_{n-1}$$

for $n \geq 2$. Thus

$$\begin{aligned} f_3 &= f_2 + f_1 = 1 + 1 = 2, \\ f_4 &= f_3 + f_2 = 2 + 1 = 3, \\ f_5 &= 3 + 2 = 5, \quad f_6 = 5 + 3 = 8, \quad f_7 = 8 + 5 = 13, \ldots \end{aligned}$$

and so on.

Exercise 1.3 Compute f_8, f_9, and f_{10} using the recursive definition.

We can also define the Fibonacci sequence to start with f_0, as long as its value is consistent with the values of f_n for positive n. Since the recursive formula requires

$$1 = f_2 = f_1 + f_0 = 1 + f_0,$$

we see that $f_0 = 0$, which would indeed be the number of rabbit pairs before the first month. Exercise 1.24 at the end of the chapter considers the values of f_n for negative n.

Exercise 1.4 Just because two sequences agree on their first five terms doesn't mean that they will always agree. Explain why the definition for f_n is a correct model for the number of rabbit pairs described by Fibonacci. As a hint, consider that every rabbit pair that exists in month $n + 1$ is exactly one of two types: it existed in month n, or it was just born in month $n + 1$.

Exercise 1.5 Here are all of the ways to write the first few positive integers as ordered sums of 1s and 2s, that is, sums of 1s and 2s where the placement of the 1s and 2s matters.

$$\mathbf{1} = 1, \qquad \mathbf{2} = 2 = 1 + 1, \qquad \mathbf{3} = 2 + 1 = 1 + 2 = 1 + 1 + 1,$$

$$\mathbf{4} = 2 + 2 = 2 + 1 + 1 = 1 + 2 + 1 = 1 + 1 + 2 = 1 + 1 + 1 + 1.$$

Can you guess a connection between the number of ordered sums and the Fibonacci numbers?

1.2.2. The Golden Ratio. Proposition VI.30 in Euclid's *Elements* asks how one can divide a line segment AB into two parts, AP and PB, so that the ratio AB/AP equals the ratio AP/PB.[2] This is sometimes referred to as dividing the segment into "extreme and mean ratio." With a clever construction, Euclid shows how to find the point P providing such a division of AB.

Exercise 1.6 Let $A = 0$ on the number line and let $B = 1$. Pick any value you like for P, just as long as the number you pick is between 0 and 1. Did you happen to pick the correct value for P? Should P be equal to, greater than, or less than $1/2$?

You can determine the numerical value of this ratio using a bit of algebra. Because we desire

$$\frac{AB}{AP} = \frac{AP}{PB},$$

and we know that $PB = AB - AP$, we have

$$\frac{AB}{AP} = \frac{AP}{AB - AP},$$

which upon cross-multiplying gives

$$AB^2 - AB \cdot AP = AP^2\,.$$

2 Notice that here we use a term like AB to denote both a line segment and its length. Such notational double-duty is not uncommon in mathematics, often with the explanation that the context should make the particular usage clear. We hope it is in this case!

Since

$$AB^2 - AB \cdot AP - AP^2 = 0\,,$$

dividing by AP^2 gives

$$\left(\frac{AB}{AP}\right)^2 - \left(\frac{AB}{AP}\right) - 1 = 0\,.$$

Using the quadratic formula we find that[3]

$$\left(\frac{AB}{AP}\right) = \frac{1+\sqrt{5}}{2} \quad \text{or} \quad \frac{1-\sqrt{5}}{2}\,.$$

Since the ratio is positive, the correct value is $(1+\sqrt{5})/2$.

This particular number has shown up in so many different contexts (see Exercises 1.25 and 1.26), it has earned the name the *golden ratio*. It is often denoted by the Greek letter "phi":

$$\phi = \frac{1+\sqrt{5}}{2}\,.$$

In our derivation of ϕ we stumbled upon an important fact: ϕ is one of the solutions of the quadratic equation

$$x^2 = x + 1. \tag{1.1}$$

We also learned that the other solution of Equation 1.1 is $(1-\sqrt{5})/2$, the *conjugate* of ϕ.

Exercise 1.7 Show that the conjugate of ϕ is equal to both $1-\phi$ and $-1/\phi$.

1.3. Inductive Reasoning

There is a surprising connection between the Fibonacci numbers (a sequence of integers) and the golden ratio (which involves $\sqrt{5}$) that is given in the following proposition. It will take a bit of work to prove this result, but the effort is worth it, since it provides an excellent example of a proof technique called mathematical induction that will be discussed in greater detail in Section 2.6.

PROPOSITION 1.4. *Let f_n denote the nth Fibonacci number, and let ϕ be the golden ratio. Then*

$$f_n = \frac{\phi^n - (1-\phi)^n}{\sqrt{5}} \tag{1.2}$$

for each non-negative integer n.

3 You will find phrases like "we find that" used frequently in mathematical writing. They are signals for you to do some work: on another sheet of paper, actually use the well-known quadratic formula to verify that the displayed values are indeed the roots of the given quadratic equation. To read mathematics is to work through mathematics; we will discuss this further in Section 2.8. But to amplify this now: while reading the first paragraph of this section, you should have drawn your own line segment labeled with A, B, and P since no figure was provided for you.

Before attempting a proof of this fact, we should verify that it is true for small values of n and keep our eyes open for any helpful patterns or insight. Let's first check Equation 1.2 for the case $n = 0$. Since any non-zero number raised to the zero power is 1,

$$\frac{\phi^0 - (1-\phi)^0}{\sqrt{5}} = \frac{1-1}{\sqrt{5}} = 0 = f_0.$$

While the claimed formula holds for $n = 0$, we have gained little insight.

Moving on to the case $n = 1$, we see that the formula holds since

$$\frac{\phi^1 - (1-\phi)^1}{\sqrt{5}} = \frac{(1+\sqrt{5}) - (1-\sqrt{5})}{2\sqrt{5}} = \frac{2\sqrt{5}}{2\sqrt{5}} = 1 = f_1.$$

Again, we have gained little insight.

We now consider the case $n = 2$. We start with

$$\frac{\phi^2 - (1-\phi)^2}{\sqrt{5}},$$

but instead of first trying to simplify the numerator with algebra, we note that ϕ and $(1 - \phi)$ both satisfy Equation 1.1, so we may replace ϕ^2 and $(1 - \phi)^2$ with $\phi + 1$ and $(1 - \phi) + 1$, respectively. Thus the expression simplifies to

$$\frac{\{\phi + 1\} - \{(1-\phi) + 1\}}{\sqrt{5}} = \frac{2\phi - 1}{\sqrt{5}} = \frac{\sqrt{5}}{\sqrt{5}} = 1 = f_2,$$

as desired.

Exercise 1.8 Verify that $\frac{\phi^2-(1-\phi)^2}{\sqrt{5}} = f_2$ by simplifying the numerator without an appeal to Equation 1.1. Then verify that $\frac{\phi^3-(1-\phi)^3}{\sqrt{5}} = f_3$ by using the equations $\phi^3 = \phi^2 + \phi$ and $(1 - \phi)^3 = (1 - \phi)^2 + (1 - \phi)$, which are direct consequences of Equation 1.1 after multiplying both sides of the equation by x.

At this point you might have gained some insights toward proving Proposition 1.4, including the possible usefulness of Equation 1.1 and the realization that we cannot prove that the theorem is true for *all* non-negative integers n using a case-by-case analysis. If we are going to prove the proposition, we will need some effective way of working with the golden ratio raised to positive integer powers. One of the key facts about ϕ is that it and its conjugate satisfy a relation reminiscent of the Fibonacci recursion.

LEMMA 1.5. *For any positive integer n,*

$$\phi^{n+1} = \phi^n + \phi^{n-1} \quad \textit{and} \quad (1-\phi)^{n+1} = (1-\phi)^n + (1-\phi)^{n-1}.$$

PROOF. Since ϕ and $1 - \phi$ both satisfy Equation 1.1, we know $\phi^2 = \phi + 1$ and $(1 - \phi)^2 = (1 - \phi) + 1$. Multiplying by ϕ^{n-1} and $(1 - \phi)^{n-1}$, respectively, establishes the formulas in the statement of the lemma. □

We are now ready to prove our theorem.

PROOF OF PROPOSITION 1.4. By our prior experimenting, we already know that Equation 1.2 holds for f_0, f_1, and f_2. So let's *assume* that the equation holds for all non-negative integers up to some unspecified integer $n \geq 2$ and explore the next Fibonacci number, f_{n+1}.

By the definition of the Fibonacci numbers, $f_{n+1} = f_n + f_{n-1}$. Since we are assuming that Equation 1.2 holds for f_n and f_{n-1}, we can use

$$f_n = \frac{\phi^n - (1-\phi)^n}{\sqrt{5}} \quad \text{and} \quad f_{n-1} = \frac{\phi^{n-1} - (1-\phi)^{n-1}}{\sqrt{5}}.$$

Combining these facts and rearranging some terms gives us

$$\begin{aligned} f_{n+1} &= f_n + f_{n-1} \\ &= \frac{\phi^n - (1-\phi)^n}{\sqrt{5}} + \frac{\phi^{n-1} - (1-\phi)^{n-1}}{\sqrt{5}} \\ &= \frac{\left[\phi^n + \phi^{n-1}\right] - \left[(1-\phi)^n + (1-\phi)^{n-1}\right]}{\sqrt{5}}. \end{aligned}$$

Lemma 1.5 tells us that $\phi^{n+1} = \phi^n + \phi^{n-1}$ and $(1-\phi)^{n+1} = (1-\phi)^n + (1-\phi)^{n-1}$, so we may substitute these values into the equation above, to establish

$$f_{n+1} = \frac{\phi^{n+1} - (1-\phi)^{n+1}}{\sqrt{5}}.$$

Here's the punchline. Since Equation 1.2 actually does hold for f_2 and f_1, we now know that it holds for f_3. And since the equation holds for f_3 and f_2, it holds for f_4. And since it holds for f_4 and f_3, it holds for f_5. And so on, with the conclusion that it holds for any f_n with $n \geq 0$. □

The proof just given of Proposition 1.4 is an example of a *proof by induction*. Such a proof does not prove the result claimed directly. Rather, it proves that you can prove each case if called upon to do so. We will encounter many more proofs by induction later; they are discussed in earnest in Section 2.6.

Exercise 1.9 Following the argument given in the proof of Proposition 1.4, establish the particular case

$$f_4 = \frac{\phi^4 - (1-\phi)^4}{\sqrt{5}}$$

by assuming similar formulas already hold for f_3 and f_2.

The Fibonacci numbers and the golden ratio are two examples of the sorts of objects mathematicians like to examine. The fact that there is a connection between these two seemingly unrelated things is the sort of revelation that mathematicians enjoy discovering. But in addition to the insight, mathematicians have an expectation that such an insight needs to be buttressed by a rigorous argument that establishes the result. This trio of ideas (interesting objects, insights, and rigorous arguments) underlies all of theoretical mathematics.

Exercises 1.28 and 1.29 at the end of the chapter ask how the connection between f_n and ϕ^n can be used to efficiently compute f_n for large values of n.

1.4. Natural Numbers and Divisibility

In Chapter 2 we discuss mathematical exploration, insight, and proofs. In Chapter 3 we introduce the concept of a "set," a standard building block of mathematics. To enliven those discussions, the next few sections introduce some basic ideas from the study of numbers, leading to two famous proofs by contradiction.

The *natural numbers* are the numbers $1, 2, 3, 4$, and so on; they do not include zero or negative numbers. Together they are commonly denoted by $\mathbb{N}$.

DEFINITION 1.6. If a and b are two natural numbers, we say that *a divides b*, or a is a *divisor* of b, and we write $a|b$, when there is a natural number c such that $b = ac$. Another way to say this is that $a|b$ if and only if b is a positive multiple of a. When a does not divide b, we write $a \nmid b$.

Here are a few examples:

$$3|12, \quad 12|108, \quad \text{and} \quad 7|1001,$$

while

$$3\nmid 4, \quad 8\nmid 4, \quad \text{and} \quad 2\nmid 3^k$$

for any natural number k. Also note that $a|a$ and $1|a$ for any natural number a.

Let's prove the following number-theoretical fact.

PROPOSITION 1.7. *If a, b, and c are any natural numbers such that both $a|b$ and $a|c$, then $a|(b+c)$.*

PROOF. Since $a|b$, there is a natural number m such that $b = am$. Similarly, since $a|c$, there is a natural number n such that $c = an$. This means that

$$b + c = am + an = a(m + n),$$

which shows that $b + c$ is a positive multiple of a. In other words, $a|(b + c)$. □

This short direct proof shares several of the same characteristics of the proof of Theorem 1.1, including the introduction of new variables (m and n) and the use of a known fact (the distributive law). What's different about this proof is the use (three times) of a definition.

In addition to proving that certain claims are true, mathematicians also like to demonstrate that certain claims are false.

NOT-A-THEOREM 1.8. *If a, b, and c are any natural numbers such that both $a|c$ and $b|c$, then $(a + b)|c$.*

PROOF THAT THE STATEMENT IS FALSE. This is a general statement claiming that something is true regardless of the actual values of the natural numbers a, b, and c, so a proof that the statement is false is simply a specific *counterexample*. For instance, if we let $a = 2$, $b = 2$, and $c = 6$, then we have $a|c$ and $b|c$. Since $a + b = 4$ and $4\nmid 6$, we know that $(a + b)\nmid c$. Thus, the general statement above is not true. □

We emphasize that it does not matter that if $a = 3$, $b = 6$, and $c = 18$, then $a|c$, $b|c$, and $(a + b)|c$. The statement above is not true because it's not true for *every* triple of a, b, and c satisfying the conditions in the statement.

Exercise 1.10 Here are several statements that are either true or false. After experimentation with many triples of numbers a, b, and c, your job is to provide an appropriate proof of the truth or falseness of each.

(a) If a, b, and c are any natural numbers such that both $a|c$ and $b|c$, then $ab|c$.
(b) If a, b, and c are any natural numbers such that $a|b$, then $a|bc$.
(c) If a, b, and c are any natural numbers such that $a|b$ and $b|c$, then $a|c$.
(d) If a, b, and c are any natural numbers such that both $a|b$ and $a|c$, then $a|(2b+3c)$.
(e) If a, b, and c are any natural numbers such that both $a|b$ and $a|(b+c)$, then $a|c$.

Exercise 1.11 Here are three statements involving a natural number n that are either true or false. After considering specific examples, either prove the statement or provide a counterexample.

(a) If n is even, then n^2 is even. Or said another way, if $2|n$, then $2|n^2$.
(b) If $2|n^2$, then $2|n$.
(c) If $4|n^2$, then $4|n$.

Exercise 1.12 Prove that if a and b are natural numbers such that $a|b$, then $a \leq b$.

1.5. The Primes

DEFINITION 1.9. A natural number p is *prime* if it has exactly two distinct natural number divisors: 1 and p.

Exercise 1.13 Point out why this definition implies that the number 1 is not prime.

Exercise 1.14 There are eight prime numbers less than 20: they are 2, 3, 5, 7, 11, 13, 17, and 19. How many primes are less than 100?

A natural number greater than 1 that is not prime is called *composite*.

Our goal in this section is to establish that there are infinitely many primes. The argument we present is essentially the proof written down by Euclid, and it requires one of the results from Exercise 1.10.

LEMMA 1.10. *If a, b, and c are any natural numbers such that both $a|b$ and $a|(b+c)$, then $a|c$.*

PROOF. Since $a|b$, there is a natural number m such that $b = am$. Similarly, since $a|(b+c)$, there is a natural number n such that $b + c = an$. Thus

$$c = (b+c) - b = an - am = a(n-m).$$

Since $n > m$, $n - m$ is a natural number, so we conclude that $a|c$. □

We'll also need a result that we'll prove later in Section 2.6: every natural number $n \geq 2$ is either prime or can be written as a product of primes.[4] For example, 84 can be written as $2 \cdot 2 \cdot 3 \cdot 7$. In fact, the only other ways to do this simply permute the prime

4 This is a rare instance when we use a result we will prove later. Our goal is to present a classical example of a certain type of proof, not to prove the result from scratch.

factors, such as $84 = 2 \cdot 3 \cdot 2 \cdot 7 = 2 \cdot 7 \cdot 3 \cdot 2$. The general statement of these ideas is called the Fundamental Theorem of Arithmetic (Theorem 4.11), which is the central result in Chapter 4.

Exercise 1.15 Write out all twelve ways to factor 84 into primes.

Euclid's proof is a type of argument, called *proof by contradiction*, that is widely used in mathematics. Other examples of this technique can be found in the proofs of Theorem 1.12 and Proposition 1.14 on the pages that follow, and in Section 2.3.

THEOREM 1.11 (Euclid). *There are infinitely many prime numbers.*

PROOF. *Assume to the contrary* that there are only finitely many prime numbers, say k of them. We can then list them in increasing order as

$$p_1, p_2, p_3, \ldots, p_k,$$

where $p_1 = 2, p_2 = 3$, and so on. We do not know the exact value of the natural number k, the length of the list; we are only assuming that k exists.

Now let

$$n = p_1 p_2 p_3 \cdots p_k + 1.$$

Since $n \geq 2$, we know that it is prime or can be expressed as a product of primes. Let p be a prime divisor of n. Since p is prime, it is on our complete list of primes, so $p = p_i$ for some index i. We now know two things:

(a) $p \mid n$,
(b) $p \mid p_1 p_2 p_3 \cdots p_k$.

It follows from Lemma 1.10 (using $b = p_1 p_2 p_3 \cdots p_k$ and $c = 1$) that $p|1$, so the result of Exercise 1.12 implies that $p \leq 1$. This contradicts the fact that $p \geq 2$. Thus our initial assumption about there being only finitely many prime numbers must be false. □

1.6. The Integers

The *integers* consist of the natural numbers, their negatives, and zero. They are commonly denoted by $\mathbb{Z}$, for the German word *Zahlen*:

$$\mathbb{Z} = \{\ldots, -2, -1, 0, 1, 2, 3, \ldots\}.$$

In discussing the natural numbers $\mathbb{N}$ and the integers $\mathbb{Z}$, we are beginning to meet some sets. We delay developing set theory to Chapter 3, but one bit of set theory notation will be handy for us to have now. Namely, we express the statement that 1 is a natural number by writing "$1 \in \mathbb{N}$," and similarly we write "$0 \notin \mathbb{N}$" to say that 0 is not a natural number. The symbol $\in$ is read "is an element of" or "is in."

Many of the things we have discussed previously in the context of $\mathbb{N}$ extend seamlessly to $\mathbb{Z}$. For example, we can say that $a|b$ for $a, b \in \mathbb{Z}$ whenever there is a $c \in \mathbb{Z}$ such that $b = ac$. Thus we have $3|12$, and we also have $(-3)|12$ and $3|(-12)$.

Exercise 1.16 Explain why $a|0$ for all $a \in \mathbb{Z}$, but $0 \nmid a$ for all non-zero integers a. Should $0|0$?

Exercise 1.17 State and prove Proposition 1.7 and Lemma 1.10 in the context of the integers (instead of the natural numbers). Lemma 1.10 is slightly easier to prove in the context of integers because you don't have to justify that $n - m > 0$: just use the fact that the difference of two integers must be an integer.

We end this section by proving a very useful fact about the integers. Look for the argument by contradiction employed in the middle of the proof.

THEOREM 1.12 (The Division Algorithm). *Given an integer a and a natural number b, there are unique integers m and r such that*

$$a = mb + r$$

with $0 \leq r < b$.

Before giving the proof, we have a couple of remarks. The theorem asks for two results: that appropriate integers m and r exist, and that no other pair of integers would work. Also, the number r is called the *remainder*. Figure 3 shows $a = 14$ sitting among several multiples of $b = 3$. The equation $14 = 4 \cdot 3 + 2$ shows that the remainder is 2. It is certainly true that $14 = 5 \cdot 3 + (-1)$ and $14 = 3 \cdot 3 + 5$ as well, but in both cases the putative remainders are out of the acceptable range.

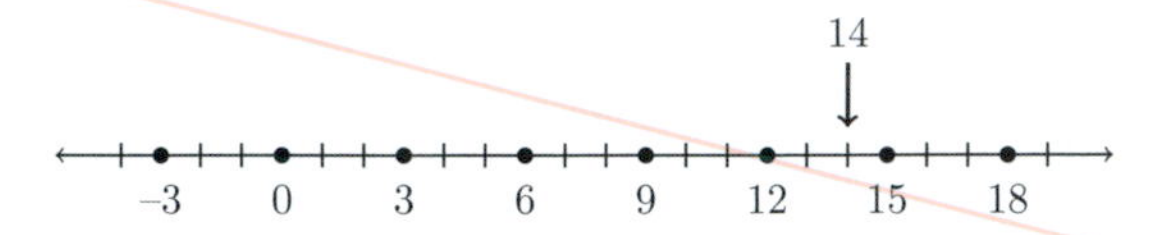

Figure 3. The remainder upon dividing 14 by 3 is $14 - 12 = 2$.

PROOF. Consider all of the integers of the form $a - kb$ with $k \in \mathbb{Z}$. Some of these are non-negative, and one of these is the least non-negative integer you can express as $a - kb$. Let this number be $r = a - mb$. We claim that m and r are the two numbers we want.

We know that $0 \leq r$, but must $r < b$? Assume to the contrary that $r \geq b$. Then

$$0 \leq r - b = (a - mb) - b = a - (m + 1)b,$$

so $r - b$ would be a smaller non-negative integer of the form $a - kb$, contradicting the fact that r was chosen to be the smallest such integer. So we conclude that indeed $r < b$, and therefore $a = mb + r$ with $0 \leq r < b$.

We now address uniqueness: are m and r the only possible choices for the two numbers we want? Let m' and r' be any two integers satisfying $a = m'b + r'$ with $0 \leq r' < b$. Then

$$r' - r = (a - m'b) - (a - mb) = (m - m')b,$$

and thus $b|(r' - r)$. But we know that $-b < r' - r < b$, and the only multiple of b strictly between $-b$ and b is 0. So $r' - r = 0$, which gives $r = r'$; and from $m'b + r' = mb + r$ we conclude that $m = m'$ as well. Therefore, there is only one possibility for each of m and r in the expression $a = mb + r$: they are unique. $\square$

Exercise 1.18 Find the remainder r when $a = 83$ and $b = 11$ by finding the least non-negative integer of the form $a - kb$ with $k \in \mathbb{Z}$. Then do the same for $a = -11$ and $b = 4$.

Exercise 1.19 Use the division algorithm to prove that every integer can be written in the form $4k$, $4k+1$, $4k+2$, or $4k+3$ for some $k \in \mathbb{Z}$.

1.7. The Rationals, the Reals, and the Square Root of 2

Just as the natural numbers can be augmented to give the integers, the integers can be expanded to the rational numbers, which can be expanded to the real numbers. The *rational numbers* are sometimes colloquially referred to as the "fractions," and we denote the set of them by the symbol $\mathbb{Q}$.

DEFINITION 1.13. A *rational number* is a real number that can be expressed in the form m/n with $m \in \mathbb{Z}$ and $n \in \mathbb{N}$.

There is no claim that any rational number's expression as m/n is unique, because it is not. For example:

$$\frac{-6}{14} = \frac{-3}{7} = \frac{-36}{84}.$$

Often we describe a rational number by its *lowest terms* expression, such as $-3/7$ above, where all common integer factors of m and n greater than 1 have been removed.

You should notice that every natural number is an integer, but not every integer is a natural number. Similarly, every integer is a rational number, but not every rational is an integer. The main result of this section is Proposition 1.14, which establishes a similar claim: not every real number is rational.

PROPOSITION 1.14. *The square root of 2 is not a rational number.*

PROOF. Assume for the moment, contrary to the theorem statement, that $\sqrt{2}$ is in fact a rational number. Thus $\sqrt{2}$ can be expressed as $\sqrt{2} = m/n$ in lowest terms. Since $\sqrt{2} > 0$, both m and n will be natural numbers.

Squaring both sides of $\sqrt{2} = m/n$ and cross-multiplying gives

$$m^2 = 2n^2.$$

Since $2n^2$ is even, it must be the case that m^2 is even, and so m is even. This means that $m = 2m'$ for some integer m'. But then we have $4(m')^2 = 2n^2$, and hence

$$n^2 = 2(m')^2.$$

Since $2(m')^2$ is even, by similar reasoning we know that n must be even.

This contradicts m/n being in lowest terms, since now we see that both m and n are even and so have a common factor of 2. Thus our assumption that $\sqrt{2}$ is rational cannot be true. □

Exercise 1.20 Determine if each of the following statements is true or false.

(a) $0 \in \mathbb{N}$
(b) $0 \in \mathbb{Z}$

(c) $\sqrt{2} \in \mathbb{Q}$
(d) $-\sqrt{2} + \sqrt{2} \in \mathbb{Q}$
(e) $1 + \sqrt{2} \in \mathbb{Q}$
(f) $2\sqrt{2} \in \mathbb{Q}$
(g) $\sqrt{2} + \sqrt{2} \in \mathbb{Q}$
(h) $\sqrt{2}\sqrt{2} \in \mathbb{Q}$

The set of real numbers is denoted $\mathbb{R}$, and it is often described as "all the points on the number line" or "all the numbers you can write in decimal notation." Some key properties of the real numbers are developed in Chapter 8, and an argument that real numbers can always be expressed via decimal expansions is the underlying theme of the first part of Project 11.9. We will, however, assume for the time being that you are familiar with the association between real numbers and decimal expansions.

Real numbers that are not rational are *irrational*. We proved above that $\sqrt{2}$ is irrational, and you are asked to establish that other numbers are irrational in Exercises 1.42, 1.43, and 1.44. This is a topic we return to in several places throughout the book.

GOING BEYOND THIS BOOK. Conway and Shipman have a very nice article discussing a number of proofs of the irrationality of $\sqrt{2}$ [**CS13**] to illustrate their notion of an "extreme proof." We recommend you read it, not only to see additional proofs of this core fact, but also to gain some understanding of how these two mathematicians think about the field of mathematics.

1.8. End-of-Chapter Exercises

At the end of each chapter, some additional exercises are recommended after you complete your initial study of certain sections, while other exercises are unclassified.

Also, in this introductory chapter a handful of the end-of-chapter exercises (e.g. Exercises 1.24, 1.45, 1.46, 1.48, 1.49, 1.51) might be best solved using induction-style arguments. While the formal treatment of mathematical induction appears in Section 2.6, we couldn't resist giving you the chance to learn additional things about several of the topics in this chapter and, possibly, to discover the basic inductive ideas used to prove them. Of course, you'll have a chance to return to these exercises after studying Section 2.6.

Exercises you can work on after Sections 1.1 and 1.2

1.21 Prove that $(a+b+c)^3 = a^3 + b^3 + c^3 + 3(a+b)(b+c)(a+c)$.

1.22 Prove that for any three real numbers a, b, and c

$$a^2 + b^2 + c^2 \geq ab + bc + ac\,. \tag{1.3}$$

As a first step, explain why

$$(a-b)^2 + (b-c)^2 + (a-c)^2 \geq 0 \tag{1.4}$$

and then show how Inequality 1.4 directly implies Inequality 1.3.

1.23 A standard calculus problem has you prove that the rectangle of greatest area for a fixed perimeter is a square. Prove this without calculus using the arithmetic–geometric mean inequality.

1.24 What happens if you apply the recursive Fibonacci formula $f_{n+1} = f_n + f_{n-1}$ for $n \leq 0$? What values do you get for f_{-1}, f_{-2}, and f_{-10}? (This is now only in the realm of mathematical theory – do not expect these values to count the number of rabbit pairs in any month!) Guess a connection between f_n and f_{-n} for any positive integer n, and express it as a mathematical equation. Then try to prove that the relationship holds.

1.25 The golden ratio shows up in a number of places in a regular pentagon. The first place is in comparing a diagonal with an edge:

(a) The ratio of a diagonal of a regular pentagon to an edge of the pentagon is ϕ.

The collection of all the diagonals of a regular pentagon forms a regular pentagram, as shown on the right in Figure 4. The intersections of the diagonals then divide each diagonal into a center portion and two side portions. In fact:

(b) The ratio of a full diagonal to a center-plus-side portion is ϕ.
(c) The ratio of a center-plus-side portion to a side portion is ϕ.
(d) The ratio of a side portion to a center portion is ϕ.

Show that all four of these statements are true.

1.26 The golden ratio can be expressed in many ways. For example, it is an infinite nested sum of square roots:

$$\phi = \sqrt{1 + \sqrt{1 + \sqrt{1 + \cdots}}}\,.$$

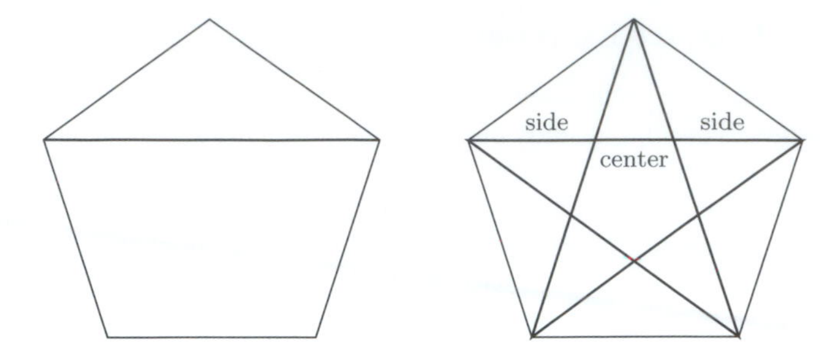

Figure 4. A diagonal of a regular pentagon (left) and the regular pentagram formed by the diagonals of a regular pentagon (right).

It is also a "continued fraction":

$$\phi = 1 + \cfrac{1}{1 + \cfrac{1}{1 + \cfrac{1}{1 + \cdots}}}.$$

Prove that each of these expressions is, indeed, equal to ϕ. You may assume that each expression represents *some* real number x, so that your job is to prove that $x = \phi$. Exercise 8.34 hints at why x must exist.

1.27 This exercise is reminiscent of the first theorem we presented. Prove that a triangle's area is 7 times the area of the triangle framed by side trisectors, as shown in Figure 5.

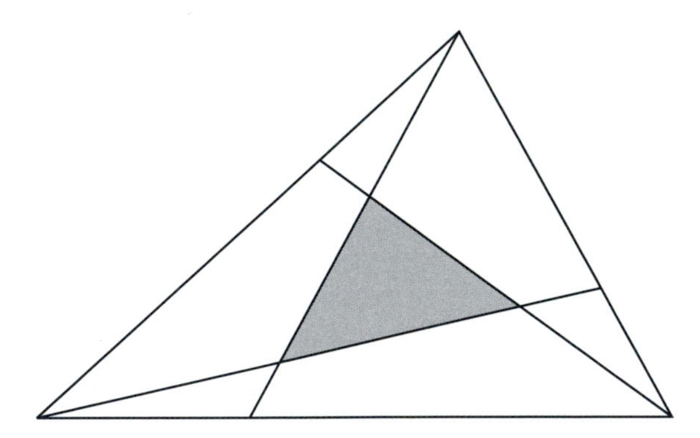

Figure 5. The side trisectors of a triangle frame an inner triangle whose area is $1/7$ the area of the original triangle.

Exercises you can work on after Sections 1.3 and 1.4

1.28 Let round(x) be the nearest integer to a real number x; for example, round(π) $= 3$ and round($\sqrt{3}$) $= 2$. We used a computer to compare the nth Fibonacci number with round($\phi^n/\sqrt{5}$) for n between 1 and 10, and we found something rather surprising:

n	f_n	round$\left(\frac{\phi^n}{\sqrt{5}}\right)$
1	1	1
2	1	1
3	2	2
4	3	3
5	5	5
6	8	8
7	13	13
8	21	21
9	34	34
10	55	55

(a) Use a computer to verify that this pattern continues up to $n = 100$.
(b) Justify the formula

$$f_n = \text{round}\left(\frac{\phi^n}{\sqrt{5}}\right)$$

for positive integers n.

1.29 What might be a fast way for a computer to compute the value, or even an approximate value, of $f_{1\,000\,000}$, the millionth Fibonacci number? Can you think of a method that computes $f_{1\,048\,576}$ more quickly? (1 048 576 is 2 to the 20th power.) Assume that computers can rapidly do things like add, subtract, multiply, and divide numbers, as well as compute square roots and round. If you have additional knowledge about the relative speed of these operations performed by computer, feel free to use it.

1.30 Prove each of the following statements.

(a) When $a = 3n + 1$ for some natural number n, then $3|(a^2 - 1)$.
(b) When $a = 3n + 2$ for some natural number n, then $3|(a^2 - 1)$.
(c) When $a = 3n$ for some natural number n, then $3\nmid(a^2 - 1)$.

1.31 The number 2 appears to divide every third Fibonacci number: based on values we have already computed, we know $2|f_3$, $2|f_6$, and $2|f_9$. In this exercise you will see if there is a similar pattern emerging for the Fibonacci numbers that are divisible by 3.

(a) Does it seem like there is a similar pattern for the Fibonacci numbers that are divisible by 3? Since $f_4 = 3$ we know $3|f_4$. What is the second Fibonacci number that is divisible by 3? Is there a third Fibonacci number that is divisible by 3?
(b) Your work on part (a) might lead you to guess that every fourth Fibonacci number is divisible by 3, that is $3|f_{4n}$ for every natural number n. Test if this claim is true for $n = 1$ up to $n = 10$. We recommend using a computer or a list of Fibonacci numbers for this problem.
(c) Are any other Fibonacci numbers between f_1 and f_{40} divisible by 3? Or does it appear that only the Fibonacci numbers of the form f_{4n} are divisible by 3?

Please note, we are not asking you to construct any proofs in this exercise . . . we leave that for Exercise 2.81.

Exercises you can work on after Section 1.5

1.32 Find a sequence of ten consecutive natural numbers, each of which is composite and less than 1 000 000. As a hint, it is not necessary but it might help to start with a number that is divisible by $2, 3, 5, 7$, and 11.

1.33 Inspired by Euclid's proof of Theorem 1.11, you notice that

$$\begin{aligned} 2+1 &= 3 \text{ is prime,} \\ 2\cdot 3+1 &= 7 \text{ is prime,} \\ 2\cdot 3\cdot 5+1 &= 31 \text{ is prime.} \end{aligned}$$

Is one more than the product of the first k primes always a prime? If so, then Euclid's proof can be simplified.

In order to facilitate working on this question, if n is an integer greater than 1, let $\widehat{n}$ be the product of all primes that are less than or equal to n. For example, $\widehat{3} = 2\cdot 3 = 6$ and $\widehat{6} = \widehat{5} = 2\cdot 3\cdot 5 = 30$. Is $\widehat{n}+1$ always prime?

(a) Verify that $\widehat{7}+1$ and $\widehat{11}+1$ are both prime.
(b) Use a computer to verify that $\widehat{379}+1$ is prime.
(c) Show that $\widehat{n}+1$ is not always prime by considering $\widehat{13}+1$.
(d) Can our method be rescued? Define $a_1 = 2$ and then recursively define

$$a_{i+1} = a_1 a_2 \cdots a_i + 1.$$

Thus $a_2 = 2+1 = 3$, $a_3 = 2\cdot 3+1 = 7$, and $a_4 = 2\cdot 3\cdot 7+1 = 43$. Are all of the a_is prime?
(e) Can *you* devise a simple recursive formula that generates a sequence of distinct primes?

1.34 The previous exercise attempted to find a formula that would output an infinite sequence of distinct prime values. In this exercise, you'll show that no polynomial of a specific form can produce only prime values.

(a) Consider $f(n) = n^2+n+41$. Check that $f(1) = 43$, $f(2) = 47$, and $f(3) = 53$, all primes so far. Does this formula only generate primes when you input natural numbers?
(b) If $a_0 > 1$, explain why no polynomial of the form

$$a_n x^n + a_{n-1}x^{n-1} + \cdots + a_1 x + a_0$$

with each $a_i \in \mathbb{N}$ can generate only prime values when you input natural numbers for x.

1.35 *Twin primes* are two odd primes that are "nearest neighbors"; that is, they are primes that are only two apart. For instance 11 and 13 are twin primes, as are 41 and 43. While it has been known for thousands of years that there are infinitely many primes, it is still unknown if there are infinitely many twin primes.

Here's your exercise: find all of the *triplet primes*, which are three primes of the form p, $p+2$, and $p+4$, and prove that your list is complete.

1.36 Prove that there exist consecutive prime numbers that are arbitrarily far apart. Put another way, show that if k is any positive integer, then there are positive integers $n, n+1, \ldots, n+k-1$, none of which is prime.

Exercises you can work on after Section 1.6

1.37 For each of the following statements, determine if it is true or false. Then either prove the statement or provide a counterexample.

(a) If a, b, and c are integers such that both $a|b$ and $a|c$, then $a|(2b - 3c)$.
(b) If a, b, and c are integers such that both $a|c$ and $b|c$, then $ab|c$.
(c) If a, b, and c are integers such that both $a \nmid b$ and $a \nmid c$, then $a \nmid (b + c)$.
(d) If a, b, and c are integers such that both $a \nmid b$ and $b \nmid c$, then $a \nmid c$.

1.38 For each value of a and b, find the integers m and r such that $a = mb + r$ with $0 \leq r < b$.

(a) $a = 1800$ and $b = 4$
(b) $a = 1743$ and $b = 1826$
(c) $a = 1826$ and $b = 44$
(d) $a = -1826$ and $b = 44$

1.39 Let r_1 be the remainder when $n_1 \in \mathbb{N}$ is divided by 3, and let r_2 be the remainder when $n_2 \in \mathbb{N}$ is divided by 3.

(a) Show by example that it is possible that $r_1 + r_2$ is *not* the remainder when $n_1 + n_2$ is divided by 3.
(b) Show by example that it is possible that $r_1 \cdot r_2$ is *not* the remainder when $n_1 \cdot n_2$ is divided by 3.
(c) Since the remainders upon division by 3 are 0, 1, or 2, the sum $r_1 + r_2$ is an integer between 0 and 4. Copy and complete the following table:

If $r_1 + r_2$ equals	Then the remainder of $n_1 + n_2$ is
0	0
1	?
2	?
3	0
4	?

(d) Since the remainders upon division by 3 are 0, 1, or 2, the product $r_1 \cdot r_2$ is 0, 1, 2 or 4. Copy and complete the following table:

If $r_1 \cdot r_2$ equals	Then the remainder of $n_1 \cdot n_2$ is
0	?
1	?
2	?
4	?

(e) Prove that the entries in your tables are correct.

1.40 A natural number $n \in \mathbb{N}$ is a *perfect square* if $n = m^2$ for some $m \in \mathbb{N}$. The first four perfect squares are 1, 4, 9, and 16.

(a) What are the remainders when you divide the first four perfect squares by 4?
(b) Prove that every perfect square $n = m^2$ leaves a remainder of 0 or 1 when divided by 4.

Exercises you can work on after Section 1.7

1.41 Definition 1.13 states that a rational number is a real number that can be written as the ratio of an integer and a natural number, but in practice we commonly use other forms of expression as well. For example,

$$\frac{-6}{14}, \frac{6}{-14}, \text{ and } -\frac{6}{14}$$

are used to specify the same real number. Could we have instead defined a rational number to be a real number that can be expressed in the form m/n with $m \in \mathbb{Z}$ and $n \in \mathbb{Z}$?

1.42 Prove that $\sqrt[3]{2} \notin \mathbb{Q}$.

1.43 Prove that $\sqrt{3} \notin \mathbb{Q}$.

1.44 John has written up a proof by contradiction showing that $\log_2(3)$ is irrational. For some strange reason, the sentences he wrote have become jumbled into the order below. What is the correct order for these sentences?

(a) We know that $m > 0$ because $\log_2(3) > 0$.
(b) Let $\log_2(3) = m/n$ for some $m \in \mathbb{Z}$ and $n \in \mathbb{N}$.
(c) This is a contradiction, so our assumption that $\log_2(3)$ is rational must be false.
(d) We then have

$$\left(2^{m/n}\right)^n = 3^n,$$

which implies $2^m = 3^n$.
(e) Assume to the contrary that $\log_2(3)$ is a rational number.
(f) By the definition of $\log_2$, we know that $2^{m/n} = 3$.
(g) However, since $m > 0$ we know 2^m is an even number, and 3^n is odd.

More Exercises!

1.45 Let f_n denote the nth Fibonacci number. Prove the following two summation formulas:

(a) $\sum_{i=0}^{n-1} f_{2i+1} = f_{2n}$. For example, when $n = 3$ this formula says that

$$\sum_{i=0}^{2} f_{2i+1} = f_1 + f_3 + f_5 = 1 + 2 + 5 = 8 = f_6.$$

(b) $1 + \sum_{i=0}^{n} f_{2i} = f_{2n+1}$. For example, when $n = 3$ this formula says that

$$1 + \sum_{i=0}^{3} f_{2i} = 1 + f_0 + f_2 + f_4 + f_6 = 1 + 0 + 1 + 3 + 8 = 13 = f_7.$$

In each case, determine all of the values of n for which the formulas hold.

1.46 Prove that every natural number n can be expressed as an increasing sum of distinct Fibonacci numbers. For example, $5 = 5$ and $17 = 1 + 3 + 13$.

1.47 We commonly use a "base-10" or "decimal" system to express integers. For example, when we write $n = 5802$ we are saying

$$n = \underline{5} \cdot 10^3 + \underline{8} \cdot 10^2 + \underline{0} \cdot 10^1 + \underline{2} \cdot 10^0.$$

The numbers $5, 8, 0$, and 2 are the *digits* used in expressing n. More generally, if the decimal expansion for n is

$$n = a_k a_{k-1} \cdots a_1 a_0,$$

then $0 \leq a_i \leq 9$ for each i and

$$n = a_k 10^k + a_{k-1} 10^{k-1} + \cdots + a_2 10^2 + a_1 10^1 + a_0.$$

Prove the following divisibility trick:

Trick! *Let n be a natural number whose decimal expansion is*

$$n = a_k a_{k-1} \cdots a_1 a_0.$$

If the sum $a_k + a_{k-1} + \cdots + a_1 + a_0$ *of the digits of n is divisible by* 3, *then n is divisible by* 3.

For example, $5 + 8 + 0 + 2 = 15$, which is divisible by 3. Therefore $n = 5802$ is also divisible by 3.

Then prove that the same trick works if you replace "3" with "9."

1.48 You can also use a "base-2" or "binary" system to express integers. For example, the number we express as 6 in decimal notation is written as "110" in binary because

$$6 = \underline{1} \cdot 2^2 + \underline{1} \cdot 2^1 + \underline{0} \cdot 2^0.$$

(a) Write down the first ten natural numbers in binary notation.

(b) Show that every natural number can be written in binary notation. That is, every natural number n can be written as:

$$n = a_k 2^k + a_{k-1} 2^{k-1} + \cdots + a_2 2^2 + a_1 2^1 + a_0,$$

where each a_i is either 0 or 1. The terms a_i are often called "bits," short for "binary digits."

1.49 There is nothing special about 2 or 10: any positive integer $b > 1$ can serve as the base of a system for representing integer values. For example, since the underlined coefficients in

$$16 = \underline{1} \cdot 3^2 + \underline{2} \cdot 3^1 + \underline{1} \cdot 3^0$$

are all between 0 and 2, we write "121" as the base-3 notation for the number expressed as 16 in decimal notation. A common shorthand way to express this relationship is $16_{10} = 121_3$.

(a) Write 16_{10} in base-b notation for $b = 4, 5, 6$, and 7.

(b) Show that every natural number can be written in base-b notation for $b > 1$. That is, every natural number n can be written as:

$$n = a_k b^k + a_{k-1} b^{k-1} + \cdots + a_2 b^2 + a_1 b^1 + a_0,$$

where each a_i is an integer such that $0 \leq a_i \leq b - 1$.

(c) The following convention makes base-16 ("hexadecimal") notation manageable:

$$\mathrm{A} = 10 \quad \mathrm{B} = 11 \quad \mathrm{C} = 12 \quad \mathrm{D} = 13 \quad \mathrm{E} = 14 \quad \mathrm{F} = 15$$

For example,

$$1000 = \underline{3} \cdot 16^2 + \underline{14} \cdot 16^1 + \underline{8} \cdot 16^0$$

shows that $1000_{10} = 3\mathrm{E}8_{16}$. Compute hexadecimal expressions for the decimal numbers 107, 511, 618, and 43 962.

(d) Does addition with hexadecimal notation work in the way you would expect, including carrying? Consider your answers to part (c) and the fact that $107 + 511 = 618$.

1.50 This problem refers to binary notation, described in Exercise 1.48. Prove the following trick:

Trick! *Let n be a natural number whose binary expression is*

$$n = a_k a_{k-1} \cdots a_1 a_0.$$

If the alternating sum of the bits equals zero, so that

$$(-1)^k a_k + (-1)^{k-1} a_{k-1} + \cdots - a_1 + a_0 = 0\,,$$

then n is divisible by 3.

For example, the number expressed as 9 in decimal notation is written as 1001 in binary notation. In this case the alternating sum is zero ($-1 + 0 - 0 + 1 = 0$) and the number is indeed divisible by 3.

1.51 If m is any natural number, m *factorial* is defined as

$$m! = m \cdot (m-1) \cdots 2 \cdot 1\,.$$

For example, $4! = 4 \cdot 3 \cdot 2 \cdot 1 = 24$.

We can express natural numbers in terms of factorials. For example,

$$77 = 3 \cdot 4! + 0 \cdot 3! + 2 \cdot 2! + 1 \cdot 1!$$

Prove that every natural number n can be expressed as

$$n = a_k \cdot k! + a_{k-1} \cdot (k-1)! + \cdots + a_2 \cdot 2! + a_1 \cdot 1!,$$

where k is some integer and $0 \leq a_i \leq i$ for each i.

1.52 In previous exercises, you found ways to represent any natural number as a sum whose terms involve Fibonacci numbers, powers of 2 or 10, and factorials. Which, if any, of these types of expressions always give a unique representation? That is, for which of these expressions is it the case that there is one and only one way to write each natural number in the prescribed form? Why?

1.53 Here's another trick.

Trick!
(a) Think of a 5-digit integer with all of its digits distinct.
(b) Permute the digits so that no digit stays in the same spot.
(c) Subtract the smaller of (a) and (b) from the larger.
(d) Multiply the resulting difference in (c) by any natural number.
(e) Then circle any digit of the result in (d) that's not a 0, and compute the sum of the digits you *didn't* circle.

For example, if the integer you pick is 51 637 and you permute the digits to get 17 365, after subtracting you get 34 272. You might then choose to multiply it by 5 to get 171 360, and then you might circle the second 1, so that the sum of the uncircled digits is 17.

Regardless of your choices, if you tell us the sum of your uncircled digits in (e), we'll tell you the digit you circled. How do we know?!

1.54 For each of the following theorems, which approach do you think would be best to try first: a direct proof, a proof by induction, or a proof by contradiction? Please don't attempt to actually prove the results! We just want you to think about what approach you would attempt first.

(a) THEOREM *The maximum number of regions into which n lines can divide the plane is* $\frac{1}{2}\left(n^2+n+2\right)$.
(b) THEOREM *There are no positive integer solutions to the quadratic Diophantine equation* $x^2 - y^2 = 1$.
(c) THEOREM *For a right triangle, the square of the hypotenuse equals the sum of the squares of the other two sides.*

We should mention that what seems like a good approach at first can turn out to not be the best way to construct a proof. So you should not always expect to be able to look at the statement of a theorem and immediately know the best proof technique.

1.55 You break a 3×4 chocolate bar across a horizontal or vertical division to form two smaller bars; Figure 6 shows one of the five possible ways to do this. You then break one of these two pieces, and so on, until you only have the constituent rectangles left. Prove that every such method of dividing the bar into its smallest pieces uses exactly eleven breaks.

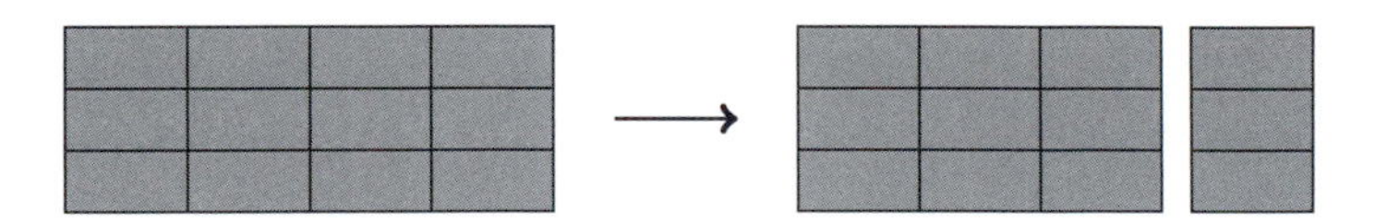

Figure 6. Breaking up a chocolate bar.

If You Have Studied Calculus

If you have already studied some calculus, you may have seen direct proofs, proofs by induction, and proofs by contradiction, either explicitly or implicitly. If you have not yet studied calculus, don't worry: in this book, we will flag the relatively few exercises and examples that require it.

1.56 Find a calculus text and determine what form of argument is used to establish the following results.

(a) The derivative of sine is cosine.
(b) The derivative of x^n is nx^{n-1}.
(c) The Mean Value Theorem.
(d) Exercises like "$x^3 + x = 1$ has at most one solution." These are usually found soon after a presentation of the Mean Value Theorem.
(e) The Divergence Test for infinite series.

2 Discovering and Presenting Mathematics

Underlying all of mathematics is the notion of proof, and the skeleton supporting mathematical proof is formal logic. Our treatment of formal logic is terse. The idea behind this chapter is to provide just a glimpse into the foundations of mathematics and pair that with a broader discussion of doing mathematics.

2.1. Truth, Tabulated

There are five common logical connectives used in mathematics:

English	Symbol
"and"	$\wedge$
"or"	$\vee$
"not"	$\sim$
"if-then" or the "conditional"	$\Rightarrow$
"if and only if" or the "biconditional"	$\Leftrightarrow$

The meaning of these symbols is best expressed via truth tables. For example, the truth table for $\sim$ is

P	$\sim P$
T	F
F	T

This table indicates that if a statement P is true, then not-P is false. Similarly, if P is false, then not-P is true. For an example of the latter, if P is the statement "$\sqrt{5} < \sqrt{3}$," then $\sim P$ can be phrased as "$\sqrt{5} \geq \sqrt{3}$," corresponding to the last line of the table.

The truth tables for $\wedge$ and $\vee$ are larger, since they involve two statements being joined together. The following table indicates whether $P \wedge Q$ is true based on the truth status of P and Q:

P	Q	$P \wedge Q$
T	T	T
T	F	F
F	T	F
F	F	F

Thus "P and Q" is considered to be true only when both P and Q are true. For example,

$$\underbrace{\sqrt{2} \text{ is irrational}}_{P} \text{ and } \underbrace{\text{greater than 1}}_{Q=\text{``}\sqrt{2}>1\text{''}}$$

is a true statement.

EXAMPLE 2.1. In the Land of Knights and Knaves everyone is either a knight or a knave, the distinguishing feature being that knights always tell the truth while knaves always lie. The trouble is, there are no visible signs that distinguish knights from knaves. You have to determine an individual's position by the statements that she and other citizens make.

For example, when you arrive in the Land of Knights and Knaves you meet Alice and Bob. Alice says "Bob is a knave." And Bob says "Alice and I are both knights." Who is a knight? Who is a knave?

Let A be the proposition "Alice is a knight," and let B mean "Bob is a knight." Alice's statement is then expressed by $\sim B$ and Bob's statement is $A \wedge B$. Consider the following truth table:

A	B	$\sim B$	$A \wedge B$
T	T	F	T
T	F	T	F
F	T	F	F
F	F	T	F

The first two columns of the table cover all four possibilities for Alice and Bob to be knights and knaves. Since knights tell the truth and knaves lie, and since the statements A and B are about being knights, we are looking for rows where the first entry matches the third and the second matches the fourth. This only happens in the second row, which means that Alice is a knight and Bob is a knave.

Exercise 2.1 You next meet Cecelia and Desmond. Cecelia says "I'm a knave, but Desmond isn't." What do you know about Cecelia and Desmond?

The truth table for $\vee$ is

P	Q	$P \vee Q$
T	T	T
T	F	T
F	T	T
F	F	F

Notice that in mathematics we use "or" in a non-exclusive manner. For example,

$$\underbrace{\text{11 is prime}}_{P} \text{ or } \underbrace{\text{13 is prime}}_{Q}$$

is a true statement. This can occasionally be confusing, since in English we frequently use "or" in a "but not both" sense:

$$\underbrace{\text{You can get a bonus mug}}_{P} \text{ or } \underbrace{\text{you can get free shipping}}_{Q}.$$

In other words, if we let $\veebar$ denote "exclusive-or," then the truth table for $\veebar$ would look like

P	Q	$P \veebar Q$
T	T	F
T	F	T
F	T	T
F	F	F

This non-standard logical connective, however, can be expressed in terms of $\wedge$, $\vee$, and $\sim$, because the right-hand column above can also be created using the expression

$$(P \vee Q) \wedge (\sim P \vee \sim Q).$$

To verify this claim, we build the truth table for this expression, where we have numbered the columns to help explain the process.

(1)	(2)	(3)	(4)	(5)	(6)	(7)
P	Q	$P \vee Q$	$\sim P$	$\sim Q$	$\sim P \vee \sim Q$	$(P \vee Q) \wedge (\sim P \vee \sim Q)$
T	T	T	F	F	F	F
T	F	T	F	T	T	T
F	T	T	T	F	T	T
F	F	F	T	T	T	F

Column (3) lists the truth values for $P \vee Q$; column (6) gives the truth values for $\sim P \vee \sim Q$; and so we apply the $\wedge$ connective to columns (3) and (6) to produce the final column (7).

Exercise 2.2 If $P =$ "$4|6^2$" and $Q =$ "$4|6$," determine the truth status of $P \vee \sim Q$, and express it as an English sentence.

Exercise 2.3 Make separate truth tables for $P \vee Q$ and $\sim (\sim P \wedge \sim Q)$.

The *conditional*, often expressed in English via the construction "If P, then Q" – as in "If I win the lottery, then I'll loan you ten dollars" – is encapsulated by

P	Q	$P \Rightarrow Q$
T	T	T
T	F	F
F	T	T
F	F	T

Notice that the only way $P \Rightarrow Q$ is false is if P is a true statement while Q is false. As examples, the following conditionals are all considered to be true statements:

"If 2 is prime, then 11 is prime."
"If all even numbers are prime, then 11 is prime."
"If all even numbers are prime, then all odd numbers are prime."

But "If 2 is prime, then all odd numbers are prime" is a false statement.

The *biconditional*, often expressed in English via the construction "if and only if" – as in "I'll loan you ten dollars if and only if I win the lottery" – is encapsulated by

P	Q	$P \Leftrightarrow Q$
T	T	T
T	F	F
F	T	F
F	F	T

Often the phrase "if and only if" is abbreviated to "iff."

Exercise 2.4 Make separate truth tables for $(P \wedge Q) \Rightarrow R$ and $(P \vee Q) \Leftrightarrow R$. Before you begin, determine how many rows each table will have.

REMARK 2.2. Exercise 2.3 shows that the truth table for $\vee$ can be replicated by one in terms of $\sim$ and $\wedge$, as can the tables for $\Rightarrow$ and $\Leftrightarrow$ (see Exercise 2.47). Unlike $\veebar$, we emphasize the connectives $\vee$, $\Rightarrow$, and $\Leftrightarrow$ because they occur much more commonly, and it would be unwieldy to use their equivalent expressions in terms of $\sim$ and $\wedge$.

A logical statement is a *tautology* if it is always true, regardless of the truth values of its component statements. One example is $P \vee \sim P$. On the other hand, a logical statement is a *contradiction* if it is always false; one example is $P \wedge \sim P$.

Exercise 2.5 Show that another example of a tautology is $P \Rightarrow (Q \Rightarrow P)$.

Two statements are *logically equivalent*, denoted $\equiv$, if they produce matching truth values. For example, $P \Rightarrow Q$ is logically equivalent to $\sim P \vee Q$ as the following table demonstrates:

P	Q	$P \Rightarrow Q$	$\sim P$	$\sim P \vee Q$
T	T	T	F	T
T	F	F	F	F
F	T	T	T	T
F	F	T	T	T

Another way to think of this is that statements R and S are logically equivalent when $R \Leftrightarrow S$ is a tautology.

Exercise 2.6 Which pairs of statements in Exercises 2.3 and 2.4 are logically equivalent?

Exercise 2.7 Prove that $P \Rightarrow Q$ is logically equivalent to $\sim Q \Rightarrow \sim P$.

Exercise 2.8 Prove that $P \Leftrightarrow Q$ is logically equivalent to $(P \Rightarrow Q) \wedge (Q \Rightarrow P)$.

REMARK 2.3. We will use the logical equivalence

$$P \Leftrightarrow Q \equiv (P \Rightarrow Q) \wedge (Q \Rightarrow P)$$

several times throughout this book. For example, we use it in the next section when we present a proof that the rational numbers can be characterized as real numbers with repeating decimals (Theorem 2.5). In the proof we show that rational numbers have repeating decimal expansions ($P \Rightarrow Q$), and we show separately that a number with a repeating decimal expansion is rational ($Q \Rightarrow P$). This equivalence is also used in the proof of the Pythagorean Theorem in Project 11.1. The theorem claims that a triangle

$\mathcal{T}$ is a right triangle *if and only if* $a^2 + b^2 = c^2$. To prove it we show that right triangles have this arithmetic property ($P \Rightarrow Q$), and that a triangle with this arithmetic property is a right triangle ($Q \Rightarrow P$).

GOING BEYOND THIS BOOK. There are a great number of logic puzzles that involve knights and knaves, which were popularized by Raymond Smullyan in *What is the Name of this Book?* [**Smu11**]. This classic text contains dozens of charming logic puzzles, from the elementary to some that are quite subtle.

2.2. Valid Arguments and Direct Proofs

A mathematical proof provides an argument that begins with the assumed hypotheses and then derives the conclusion using only valid inferences. In this section we present a very small portion of formal logic that describes the structure of mathematical proofs.

A valid argument consists of an assumed set of statements, followed by the application of various rules of inference, to arrive at a set of concluding statements. Two of the most important rules of logic are *modus ponens* and *modus tollens*. The basic structure of *modus ponens* is

$$\begin{array}{c} P \\ P \Rightarrow Q \\ \hline Q \end{array}$$

It simply states that if P is true, and if P implies Q is true, then Q must be true as well.

Modus tollens is similar:

$$\begin{array}{c} \sim Q \\ P \Rightarrow Q \\ \hline \sim P \end{array}$$

Here we claim that if Q is false, and if P implies Q is true, then it must be the case that P is false as well.

A valid argument in this formal setting is just a sequence of valid inferences that lead from the assumptions to the conclusions.

GOING BEYOND THIS BOOK. The field of formal logic is a rich subject that sits in the liminal area between mathematics and philosophy. While it is worthy of a semester's worth of study, this is not a course in formal logic, so we'll end our discussion of formal logic after saying that interested readers can research topics like modal logic, Turing machines, and undecidability. A classic text on the subject is *Computability and Logic* by G. Boolos, J. Burgess, and R. Jeffrey [**BBJ07**].

Direct proofs are the easiest form of mathematical proofs to read. In such arguments you begin with the hypotheses and proceed to directly draw conclusions from them. In very broad strokes, rules of inference like *modus ponens* and *modus tollens* provide the logic behind proof techniques. In a direct proof we use *modus ponens*: we assume P, and then somehow show $P \Rightarrow Q$, and conclude Q. You have already used direct proofs of this form for most of the true statements in Exercises 1.10, 1.11, and 1.12. For example, in part (c) of Exercise 1.10, you can think of P as the statement "$a|b$ and $b|c$" and Q as the statement "$a|c$."

The following claim can be established by a direct proof. Here P is the statement that n is a positive integer, and Q is the displayed equation.

THEOREM 2.4. *Let n be a natural number. Then*

$$1+2+\cdots+n=\frac{n(n+1)}{2}.$$

DIRECT PROOF. The expressions $1+2+\cdots+n$ and $n+(n-1)+\cdots+1$ are both the sum of the first n natural numbers, and we can add them as shown below:

$$\begin{array}{cccccccccccc} & 1 & + & 2 & + & 3 & + & \cdots & + & n-1 & + & n \\ + & n & + & n-1 & + & n-2 & + & \cdots & + & 2 & + & 1 \\ \hline & n+1 & + & n+1 & + & n+1 & + & \cdots & + & n+1 & + & n+1 \end{array}$$

Since there are n columns, each column adding to $n+1$, the total is $n(n+1)$. Thus

$$2\cdot(1+2+\cdots+n)=n(n+1),$$

which implies that

$$1+2+\cdots+n=\frac{n(n+1)}{2}. \qquad \square$$

Exercise 2.9 The 5×4 rectangle in Figure 1 might suggest a geometric proof of Theorem 2.4 that is a variant of the algebraic proof given above. What is the proof?

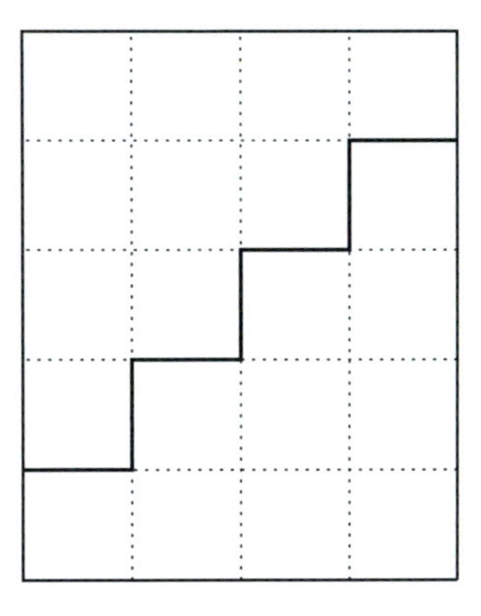

Figure 1. A picture showing that $2(1+2+3+4)=4\cdot 5$.

You can often prove biconditional statements using two direct proofs. For example, there is the well-known characterization of the rational numbers as those real numbers whose decimal expansions eventually repeat. We can use long division to find the decimal expression for $18/7$:

$$\frac{18}{7}=2.\overline{571428}=2.57142857142857\ldots$$

We include the possibility that a decimal expression is finite, such as

$$\frac{27}{25}=1.08=1.08\overline{0}=1.080000\ldots,$$

since this is just a special case where the repeating pattern is all zeros.

Exercise 2.10 Here are two questions that provide some practice in moving between ratios of integers and decimal expansions.

(a) Use long division to find a repeating decimal expression for $1/11$.
(b) Express the real number

$$56.1\overline{67} = 56.1676767\cdots$$

as a ratio of two integers.

The claim that this condition *characterizes* the rational numbers means that a real number is rational if and only if its decimal expansion is repeating. So if P is the proposition "x is a rational number" and Q is the proposition "the decimal expansion of x repeats," then to verify the claim we need to prove the biconditional $P \Leftrightarrow Q$, which by Exercise 2.8 is logically equivalent to $(P \Rightarrow Q) \wedge (Q \Rightarrow P)$.

THEOREM 2.5. *A real number x is a rational number if and only if it has a decimal expansion that eventually repeats.*

PROOF. We have two claims to verify, and we prove each by direct proofs.

Proof of $P \Rightarrow Q$: Here we assume that x is rational, so $x = a/b$ for some integers a and b, with $b > 0$. We can find a decimal expansion for x using long division. At each step in the long division process you get a remainder, which by the Division Algorithm (Theorem 1.12) is an integer between 0 and $b-1$, inclusive. Since there are only finitely many remainders, at some step after the decimal point the remainder that appears is the same as a previous remainder. And if remainder r_j is the same as remainder r_i, then we know remainder r_{j+1} is the same as remainder r_{i+1}, which implies that r_{j+2} is the same as remainder r_{i+2}, etc. This means that the decimal expansion is repeating.

Proof of $Q \Rightarrow P$: Now we may assume that we have a repeating decimal expansion for x:

$$x = n.d_1 d_2 \cdots d_j \overline{d_{j+1} \cdots d_{j+k}},$$

where j and k are integers, with $j \geq 0$ and $k \geq 1$. The repeating pattern begins at the $(j+1)$th digit, and the repeating sequence contains k digits. Multiplying by well-chosen powers of 10 allows us to move the decimal point:

$$10^j \cdot x = nd_1 d_2 \cdots d_j.\overline{d_{j+1} \cdots d_{j+k}}$$

and

$$10^{j+k} \cdot x = nd_1 d_2 \cdots d_j d_{j+1} \cdots d_{j+k}.\overline{d_{j+1} \cdots d_{j+k}}\,.$$

It follows that

$$10^{j+k} \cdot x - 10^j \cdot x = (nd_1 d_2 \cdots d_j d_{j+1} \cdots d_{j+k}) - (nd_1 d_2 \cdots d_j)\,.$$

From this we get an expression of x as the ratio of two integers:

$$x = \frac{(nd_1 d_2 \cdots d_j d_{j+1} \cdots d_{j+k}) - (nd_1 d_2 \cdots d_j)}{10^{j+k} - 10^j}\,.$$

□

In this presentation of a proof of Theorem 2.5 we have highlighted the underlying logic present in the mathematical argument. Namely, a biconditional ($\Leftrightarrow$) can be proven by establishing two conditionals ($\Rightarrow$ and $\Leftarrow$).

The key point of this section is that a mathematical proof should be an instance of a valid argument, where you proceed from a set of assumptions and use valid inferences to attain a set of conclusions. In practice, mathematical proofs are often not explicitly presented in such a fashion, if for no other reason than this produces tedious prose.

2.3. Proofs by Contradiction

Like direct proofs, proofs by contradiction want to establish that a proposition Q follows from a proposition P. The structure of such arguments can seem a bit odd, as the first step is to assume the opposite, $\sim Q$. One then shows that $\sim Q$ implies $\sim P$, which contradicts the fact that P is known to be true. Thus the assumption that $\sim Q$ is true must be wrong, and therefore Q is true.

A classic example of a proof by contradiction is the argument we presented showing that $\sqrt{2}$ is irrational. This begins by assuming the opposite of what is claimed, namely that $\sqrt{2}$ is rational, leading to a contradiction. Immediately after that proof, as part of Exercise 1.20, you should have proven that $1 + \sqrt{2}$ is irrational. Your argument was most likely a proof by contradiction, which might be quite similar to the proof we give of the following more general claim.

PROPOSITION 2.6. *The sum of a rational number and an irrational number is irrational.*

PROOF. Let q be a rational number and let x be an irrational number. Assume to the contrary that $q + x$ is rational. Then there are integers m and n such that $q + x = m/n$; and because q is rational, there are integers a and b such that $q = a/b$, where n and b are positive. Thus

$$\frac{a}{b} + x = \frac{m}{n} \quad \Rightarrow \quad x = \frac{m}{n} - \frac{a}{b} = \frac{mb - an}{nb}.$$

Expressing x as a ratio of two integers – the integer $mb - an$ and the positive integer nb – contradicts our assumption that x is irrational. Thus the hypothesis that $q + x$ is rational must be false, hence $q + x$ is irrational. □

Exercise 2.11 Use a proof by contradiction to show that the product of a natural number and an irrational number is irrational. Can "natural number" be replaced with "integer"?

The proof of the following result is a combination of a proof by contradiction and the consideration of cases.

PROPOSITION 2.7. *Any real-valued solution to $x^3 + x = 1$ is irrational.*

PROOF. Assume to the contrary that there is a rational solution to $x^3 + x = 1$. Express this solution as m/n, written in lowest terms. Then because $(m/n)^3 + (m/n) = 1$ we have

$$m^3 + m \cdot n^2 = n^3 \qquad (2.1)$$

after we multiply both sides of the equation by n^3. But this is impossible, as we show by considering the following cases.

If both m and n are odd, the left-hand side of Equation 2.1 is even while the right-hand side is odd.

If m is odd and n is even, the left-hand side of Equation 2.1 is odd while the right-hand side is even.

If m is even and n is odd, the left-hand side of Equation 2.1 is even while the right-hand side is odd.

Finally, having both m and n even contradicts the assumption that the fraction m/n is in lowest terms. □

Exercise 2.12 Does Proposition 2.7 still hold if you replace the equation $x^3 + x = 1$ with $x^3 + x^2 = 2$?

For the remainder of this section we connect Propostion 2.7 and proofs by contradiction with a sort of question that sometimes arises in a calculus course. If you haven't studied the topics mentioned here then just skim the discussion; it is not referred to elsewhere in this book.

Here is a standard calculus exercise:

(a) Show that $x^3 + x = 1$ has a real solution.
(b) Show that there is only one solution to this equation.
(c) Use Newton's Method to estimate the solution.

Your solution to this problem might go along the following lines.

(a) Note that if $f(x) = x^3 + x - 1$, then solving $x^3 + x = 1$ is equivalent to finding an x where $f(x) = 0$. Since $f(0) = -1$ and $f(1) = 1$, the Intermediate Value Theorem implies $f(x) = 0$ for some x between 0 and 1 because f is continuous on the interval $[0, 1]$.

(b) *Assume to the contrary* that $f(x) = 0$ for two distinct real numbers, $x = a$ and $x = b$. Then $f(a) = f(b)$, so by Rolle's Theorem there must be a real number c in the interval (a, b) where $f'(c) = 0$. But $f'(x) = 3x^2 + 1$, which is always greater than zero. So no such c can exist, and therefore the root of f must be unique.

(c) You might wish to start with $x_0 = 0$ and then find $x_1 = 0 - f(0)/f'(0) = 1$ and $x_2 = 1 - f(1)/f'(1) = 3/4$. At this point you might use a calculator to compute another step or two in the process.

Doing mathematics not only involves gaining comfort with the skills needed for constructing and writing formal proofs; it also involves nurturing your mathematical curiosity. For example, if you worked through the calculus exercise above you might wonder: starting at $x_0 = 0$, will Newton's Method ever produce the exact solution to this equation? In thinking about this for a moment, you should realize that every approximation produced by applying Newton's Method, starting with any initial value that is a rational number, will result in a rational number (or be undefined if some $f'(x_i)$ is zero). Thus you will know that Newton's Method will never give the exact solution once you have established that the solution must be irrational (Proposition 2.7).

A noteworthy result following immediately from other work is often called a "corollary."

COROLLARY 2.8. *Newton's Method, beginning with any initial value that is a rational number, will never produce the exact solution to* $x^3 + x = 1$.

2.4. Converse and Contrapositive

Proofs are often concerned with establishing the truth of an implication $P \Rightarrow Q$. It is important to distinguish the original implication from its *converse*, $Q \Rightarrow P$. A conditional can hold without its converse holding. For example, "puppies are cute" might be rephrased as

$$\text{If } \underbrace{\text{the animal is a puppy,}}_{P} \text{ then } \underbrace{\text{the animal is cute}}_{Q}.$$

This is obviously a true claim. Yet the converse, $Q \Rightarrow P$, which says

$$\text{If } \underbrace{\text{the animal is cute,}}_{Q} \text{ then } \underbrace{\text{the animal is a puppy,}}_{P}$$

is false. For example, pygmy marmosets are cute animals that are not puppies.

The *contrapositive* of $P \Rightarrow Q$ is $\sim Q \Rightarrow \sim P$. A conditional and its contrapositive are logically equivalent, as you showed in Exercise 2.7.

EXAMPLE 2.9. Consider the following claim about an integer n: "If n is divisible by 4, then n^2 is divisible by 4." This is a true statement, as is its contrapositive: "If n^2 is not divisible by 4, then n is not divisible by 4." On the other hand, the converse is false: "If n^2 is divisible by 4, then n is divisible by 4." For example, 6^2 is divisible by 4, yet 6 is not.

If you replace the instances of "4" with "2," the original statement, its converse, and its contrapositive are all true. For proofs, recall Exercise 1.11.

Constructing the contrapositive requires negating two statements. Exercises at the end of this section give you some practice negating statements involving $\Rightarrow$, $\wedge$, and $\vee$, a skill that mathematicians use on a daily basis, as well as some important logical equivalences involving them.

If you worked on Exercise 1.42, which asks you to prove that $\sqrt[3]{2}$ is irrational, then you probably proved or at least used the following lemma.

LEMMA 2.10. *Let $n \in \mathbb{N}$. Then n^3 is even if and only if n is even.*

The proof is not difficult, and it allows us to use a contrapositive.

PROOF. If n is even, then $n = 2m$ for some $m \in \mathbb{N}$. Hence $n^3 = 8m^3$, and so n^3 is even.

To prove the other direction, assume to the contrary that n is not even. So $n = 2m+1$ for some integer m. Then $n^3 = 8m^3 + 12m^2 + 6m + 1$. Since $8m^3 + 12m^2 + 6m$ is even, n^3 is not even. □

Notice that if P stands for "n is even" and Q stands for "n^3 is even," then Lemma 2.10 is asking us to prove $P \Leftrightarrow Q$. We have done so by showing

$$\underbrace{P \Rightarrow Q}_{\text{1st paragraph}} \quad \wedge \quad \underbrace{\sim P \Rightarrow \sim Q}_{\text{2nd paragraph}}.$$

In other words, we used the contrapositive to establish $Q \Rightarrow P$.

Exercise 2.13 For each of the implications $P \Rightarrow Q$ below, state the contrapositive and the converse, and then determine the truth status of the original statement, its contrapositive, and its converse. (Assume that n is an integer and a, b, and x are arbitrary real numbers.)

(a) If n^2 is even, then n is even.
(b) If x^2 is rational, then x is rational.
(c) If $|a+b| < 1$, then $|a| + |b| < 1$.
(d) If the professor is 5 minutes late, then class is cancelled.

Exercise 2.14 Use a truth table to show that the negation of $P \Rightarrow Q$ is logically equivalent to $P \wedge \sim Q$.

Exercise 2.15 Use truth tables to show

(a) $\sim(P \wedge Q)$ is logically equivalent to $\sim P \vee \sim Q$.
(b) $\sim(P \vee Q)$ is logically equivalent to $\sim P \wedge \sim Q$.

The negation of "Derek is rich and good-looking" is then "Derek is not rich or not good-looking." Similarly, the negation of "John is grumpy or sleepy" would be "John is neither grumpy nor sleepy" (or, more formally, "John is not grumpy and not sleepy").

2.5. Quantifiers

The *universal quantifier* is expressed in English as "For all ..." or "For every ...," and it means that a particular property holds for every element in a given set of possibilities.[1] For example,

$$\text{For all integers } n,\ n \leq n^2.$$

A common notation for the universal quantifier is $\forall$, so symbolically you can rewrite the statement just given as $\forall n \in \mathbb{Z}, n \leq n^2$.

Exercise 2.16 Prove that the statement above is in fact true.

Exercise 2.17 To show that the underlying set matters, determine whether the similar statement is true or false:

$$\text{For all rational numbers } q,\ q \leq q^2.$$

The *existential quantifier* is often expressed in English as "There exists ...," meaning that there is an object of the given type that has a particular property. For example,

$$\text{There exists a real number } x \text{ such that } x^3 + x + 1 = 0.$$

A common notation for the existential quantifier is $\exists$, so symbolically you can rewrite the statement just given as $\exists x \in \mathbb{R}, x^3 + x + 1 = 0$. Sometimes an existential claim also includes a claim of uniqueness:

1 We will study the elementary theory of sets in the next chapter; here, we merely need the notion of an element being in a set, as already introduced in Section 1.6.

There is one and only one real number x such that $x^3 + x + 1 = 0$.

The common notation for this is $\exists!$.

Exercise 2.18 Which of the following are true?

(a) $\exists n \in \mathbb{Z}, |n| = 5$
(b) $\exists! n \in \mathbb{Z}, |n| = 5$
(c) $\exists! n \in \mathbb{N}, |n| = 5$
(d) $\forall n \in \mathbb{N}, |n| = 5$

Exercise 2.19 If you studied calculus, review the solutions to parts (a) and (b) of the calculus exercise at the end of Section 2.3, and note the different types of arguments used.

As we mentioned in the previous section, it is important to be able to correctly negate mathematical statements. Universal quantifiers claim that something is always true, so to refute such a claim you just need a single counterexample. If A is a set and $P(a)$ is a statement involving a, then symbolically we write

$$\sim(\forall a \in A, P(a)) \text{ iff } \exists a \in A, \sim P(a).$$

For example, the claim "The square of any integer is positive" is saying

$$\forall z \in \mathbb{Z}, z^2 > 0.$$

The negation is then

$$\exists z \in \mathbb{Z}, z^2 \not> 0,$$

which is to say "$\exists z \in \mathbb{Z}, z^2 \leq 0$." The number $z = 0$ shows that the original claim is false, and its negation is true.

An existentially quantified statement only claims that there is an instance where the statement holds. Negating it requires that you prove no such instance can occur. Thus refuting an existentially quantified statement requires you to establish a universally quantified statement:

$$\sim(\exists a \in A, P(a)) \text{ iff } \forall a \in A, \sim P(a).$$

For example, the claim "there is a real number x whose square is -1" is saying "$\exists x \in \mathbb{R}, x^2 = -1$." The negation is then saying "$\forall x \in \mathbb{R}, x^2 \neq -1$."

Exercise 2.20 Convert the following English sentences into symbolic form. It will be helpful to let H be the set of all humans and T be the set of all (worldly) things.

(a) Every human is mortal (or ... We're all gonna die!).
(b) Somebody scored higher than me on this test.
(c) Nothing is better than apple pie!

Order matters when there are multiple quantifiers in a single claim. For example

$$\forall y \in \mathbb{R}\ \exists x \in \mathbb{R}, y < x$$

Figure 2. We have much to learn from Dilbert.

is a true statement, while

$$\exists x \in \mathbb{R}\ \forall y \in \mathbb{R}, y < x$$

is false.

Exercise 2.21 Write English sentences for the two claims above.

Exercise 2.22 Your roommate, Pat, is having trouble understanding the Dilbert cartoon in Figure 2. Explain it to Pat.

Exercise 2.23 Determine whether each of the following statements is true or false, and explain your reasoning.

(a) $\exists x \in \mathbb{Q}, x^2 + x = -2/9$
(b) $\forall x \in \mathbb{Q}, x^2 + x \in \mathbb{Q}$
(c) $\forall n \in \mathbb{Z}\ \exists x \in \mathbb{R}, x^3 = n$
(d) $\exists x \in \mathbb{R}\ \forall n \in \mathbb{Z}, x^3 = n$

Exercise 2.24 Translate the following English sentences into symbolic form. Let T be the set of (worldly) things, let G be the set of things that glister, and let D be the set of all doors.

(a) Given any two distinct real numbers, there is a rational number between them.
(b) All that glisters is not gold. (From *Merchant of Venice*.)
(c) Nothing can stop the Sandman! (Title of *The Amazing Spiderman* #4)
(d) All doors will not open. (New Jersey Transit)

2.6. Induction

Proofs by mathematical induction are used in a specific setting that appears frequently in mathematics:

(1) There is a statement, $P(n)$, that depends on an integer n.

(2) $P(n)$ is true for a smallest base case $n = n_0$, with additional base cases possibly necessary.
(3) The truth of $P(n+1)$ can be derived from the truth of the prior statements

$$P(n_0), P(n_0+1), \ldots, P(n).$$

This method of proof was previewed in Section 1.3, where we used an induction argument to connect the Fibonacci numbers to the golden ratio. When you see a claim that includes a phrase like "for all $n \in \mathbb{N}$" or "for all non-negative integers n," induction might be a fine way to try to prove it.

2.6.1. Examples of Proofs by Induction. As a first example of the power of induction, let's dive straight into a proof of a result about factorization that we quoted when we were proving Theorem 1.11.

THEOREM 2.11. *Every natural number $n \geq 2$ is either prime or it can be written as a product of primes.*

INDUCTIVE PROOF. This is a proof by mathematical induction that $P(n)$, the statement "n is either prime or the product of primes," is true for all integers $n \geq 2$.

Base Case: $P(2)$ is true because 2 is prime.

Inductive Step: Assume that each of the statements $P(2), P(3), \ldots, P(n)$ is true. Using this assumption we can deduce that $P(n+1)$ is true as well, since $n+1$ is in one of two cases:

(a) If $n+1$ is prime, $P(n+1)$ is true.
(b) Otherwise, $n+1$ is not prime, so $n+1 = ab$ for natural numbers a and b that are not equal to 1 or $n+1$, so we know that $2 \leq a \leq n$ and $2 \leq b \leq n$. By our assumption, regardless of the particular values of a and b, we know that $P(a)$ and $P(b)$ are true, so each of a and b is either prime or the product of primes. Multiply these expressions together and $n+1 = ab$ is expressed as the product of primes.

The Conclusion: We now know that $P(n)$ is true for all $n \geq 2$. □

We hope you appreciate and are perhaps rather astonished by the efficiency of the argument. Once you have (1) an appropriate statement, (2) the truth of a base case or cases, and (3) an argument showing that the truth of any statement follows from the truth of previous statements, you are guaranteed the truth of all statements in the sequence. As we mentioned in Section 1.3, mathematical induction does not provide a direct proof that $P(n)$ is true for all $n \geq n_0$; instead, it shows that for any $n \geq n_0$, you can prove the statement $P(n)$ if called upon to do so. The term "induction" is derived from "induce," to infer the general from the specific. For more information on how induction fits into the logical foundations of mathematics, see *Set Theory: A First Course* by D.W. Cunningham [**Cun16**].

As a second example, let's investigate the following statement, $P(n)$:

"$7^n - 1$ is an integer multiple of 6."

$P(0)$ is the statement "$7^0 - 1$ is an integer multiple of 6," which is true since $7^0 - 1 = 0 = 0 \cdot 6$. $P(1)$ is the statement "$7^1 - 1$ is an integer multiple of 6," which is true since $7^1 - 1 = 6 = 1 \cdot 6$.

Exercise 2.25 Verify that $P(2)$ and $P(3)$ are true as well.

Given the positive results you have obtained so far, you might now believe that $P(n)$ is true for all integers $n \geq 0$. There are several legitimate ways to prove this fact; here, we provide an inductive proof.

THEOREM 2.12. *For all non-negative integers n, $7^n - 1$ is a multiple of 6.*

INDUCTIVE PROOF. This is a proof by mathematical induction that $P(n)$, the statement "$7^n - 1$ is a multiple of 6," is true for all integers n greater than or equal to $n_0 = 0$.

Base Case: $P(0)$ is true because $7^0 - 1 = 0 = 0 \cdot 6$, which shows that $7^0 - 1$ is a multiple of 6.

Inductive Step: Assuming that $P(k)$ is true for all values of k with $0 \leq k \leq n$, let's show that $P(n+1)$ is true. To do this we rewrite $7^{n+1} - 1$ in a format that allows us to use previous cases:

$$7^{n+1} - 1 = 7 \cdot 7^n - 1 = (6+1)7^n - 1 = \underbrace{6 \cdot 7^n}_{A} + \underbrace{7^n - 1}_{B}.$$

Expression A is obviously a multiple of 6, and expression B is a multiple of 6 because we are assuming that $P(k)$ is true when $k = n$. Since Proposition 1.7 tells us that the sum of two multiples of 6 is a multiple of 6, we deduce that $7^{n+1} - 1$ is a multiple of 6. Thus, $P(n+1)$ is true.

The Conclusion: By mathematical induction, $P(n)$ is true for all integers $n \geq 0$. □

Exercise 2.26 Find the moment in the proof where the inductive assumption is used.

As the two examples above illustrate, most proofs by mathematical induction can be presented in a common format; a template is offered in Figure 3 that you might find helpful. Filling in the details for the inductive step is the most challenging task in completing proofs by induction because it's here that you must say why $P(n+1)$ is true in a way that depends on some of the statements $P(n_0), P(n_0+1), \ldots, P(n)$ being true.

The two examples also highlight some of the differences you can encounter in the details. In particular, in the proof of Theorem 2.12 the truth of $P(n+1)$ depends only on the truth of the previous statement $P(n)$ and not on any of the statements $P(n_0)$, $P(n_0+1), \ldots, P(n-1)$. Such arguments are called *simple* induction. A line of falling dominoes is often used as an analogy to explain the concept of simple induction; see Figure 4. Sometimes an inductive step of the type you see in Theorem 2.11 and that we present in Figure 3 is called *strong* induction in order to distinguish it from simple induction.

Exercise 2.27 Minimally modify the template in Figure 3 to give it the structure of simple induction. Then verify that you can use the modified version to prove that $10^n - 1$ is an integer multiple of 9.

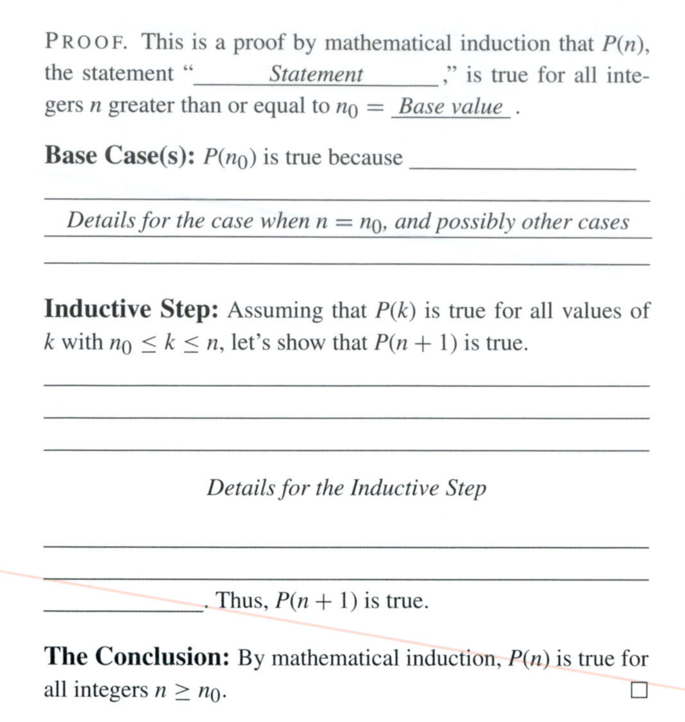

PROOF. This is a proof by mathematical induction that $P(n)$, the statement "_____*Statement*_____," is true for all integers n greater than or equal to $n_0 =$ __*Base value*__.

Base Case(s): $P(n_0)$ is true because __________

Details for the case when $n = n_0$, and possibly other cases

Inductive Step: Assuming that $P(k)$ is true for all values of k with $n_0 \leq k \leq n$, let's show that $P(n+1)$ is true.

Details for the Inductive Step

__________. Thus, $P(n+1)$ is true.

The Conclusion: By mathematical induction, $P(n)$ is true for all integers $n \geq n_0$. □

Figure 3. After you have worked out the details, it is often possible to present an induction proof by filling in this template.

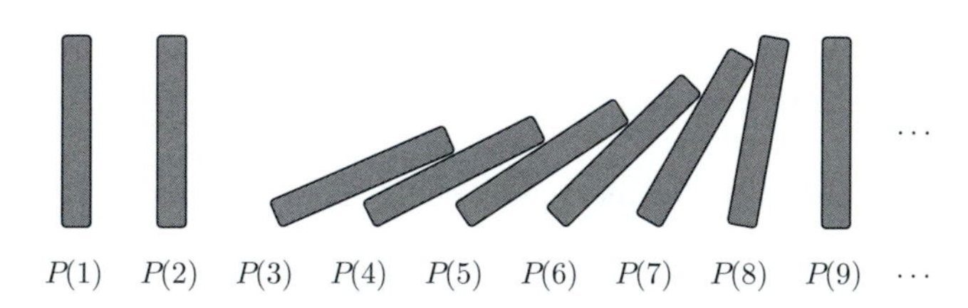

Figure 4. All of the dominoes labelled 3 or higher will fall because both of the following are true: someone knocks over domino 3 (the base case), and domino $n+1$ falls because it is close enough to domino n when that domino falls (the inductive step). This situation is analogous to simple induction proving that a statement $P(n)$ is true for all integers $n \geq 3$.

Exercise 2.28 Express each of the following natural numbers as a product of primes.

(a) 1000
(b) 1001
(c) 1002
(d) 1003

Then comment on the fact that the proof of Theorem 2.11 uses strong induction.

Induction arguments are often used when counting objects of a given type. For example, when you flip a coin n times in a row, you get a finite number of results;

HHTTH and THTHT are two of the $32 = 2^5$ different possibilities when $n = 5$. The number of possible coin-flip sequences for any value of n can be determined using the following important general counting principle.[2]

PROPOSITION 2.13. *Let $n \in \mathbb{N}$, and let s_i be the number of options you have for filling in the ith term in a sequence of length n. Then the total number of sequences of length n is the product $s_1 s_2 \cdots s_n$.*

PROOF. We first remark that if any $s_i = 0$, then the result holds because the product equals 0 and, indeed, there are no sequences as there is no available option for the ith term. So we can assume for the rest of the proof that each $s_i > 0$. We now use induction on n, the length of the sequence.

Base Case: If $n = 1$, it is clear that the number of sequences is s_1, as that is the number of options you have for the first (and only!) term in the sequence.

Inductive Step: We assume that the statement is true for sequences of length $n \geq 1$. We can enumerate the number of sequences of length $n + 1$ using a rectangular array. Such a sequence will have the form (F, L), where F is an allowed sequence for the first n positions, and L is one of the s_{n+1} options available for the last position. By our inductive hypothesis, there are $s_1 s_2 \cdots s_n$ sequences for the first n positions; for notational convenience, let $s = s_1 s_2 \cdots s_n$. We can present all of the sequences of length $n + 1$ in the rectangular array

$$\begin{array}{cccc} F_1L_1 & F_1L_2 & \ldots & F_1L_{s_{n+1}} \\ F_2L_1 & F_2L_2 & \ldots & F_2L_{s_{n+1}} \\ \vdots & \vdots & \ddots & \vdots \\ F_sL_1 & F_sL_2 & \ldots & F_sL_{s_{n+1}} \end{array}$$

which contains exactly

$$s \times s_{n+1} = (s_1 s_2 \cdots s_n) \times s_{n+1}$$

sequences of length $n + 1$.

The Conclusion: By mathematical induction, the total number of sequences of length n is $s_1 s_2 \cdots s_n$. □

COROLLARY 2.14. *There are exactly 2^n heads–tails sequences of length n.*

PROOF. This is an application of Proposition 2.13, where the number of options for each stage in the sequence is 2. □

Similar reasoning tells you that 2^n rows are required for a truth table that involves n initial statements $P_1, P_2, \ldots, P_n$. We will apply the counting principle of Proposition 2.13 several times in later chapters.

2 In this and further proofs by induction, we allow ourselves the flexibility of language not to follow the template in Figure 3 verbatim; but we could, as you should check.

2.6.2. Identifying When and How to Apply Induction. Newcomers to mathematical induction sometimes struggle to know when to use it to prove a result, and whether the induction is simple or strong, and how many base cases to employ. Some guidance is available, as suggested in the following discussion, but a complete answer always depends on the details of your particular situation.

The proof of Theorem 2.11 requires strong induction because you have no control over the factorization of $n + 1$ as ab, so you need to assume the truth of $P(a)$ and $P(b)$ over a wide range of possibilities. On the other hand, the Fibonacci recursion $f_{n+1} = f_n + f_{n-1}$ suggests that induction steps involving the $(n + 1)$th Fibonacci number might only need to reference the two immediately prior cases. For example, without referring to Proposition 1.4, here is a proof that the Fibonacci numbers grow in size at a rate that is at least exponential.

THEOREM 2.15. *For all $n \in \mathbb{N}$, the Fibonacci number f_n is bounded below by $(3/2)^{n-2}$.*

PROOF. This is a proof by mathematical induction that the inequality $f_n \geq (3/2)^{n-2}$ holds for all natural numbers n.

Base Cases: The statements involving f_1 and f_2 can be checked by hand:

$$f_1 = 1 \geq \left(\frac{3}{2}\right)^{1-2} = \frac{2}{3},$$

$$f_2 = 1 \geq \left(\frac{3}{2}\right)^{2-2} = 1.$$

Inductive Step: Assume that $f_k \geq (3/2)^{k-2}$ is true for all values of k with $1 \leq k \leq n$, for some arbitrary integer $n \geq 2$. Then consider the next Fibonacci number, f_{n+1}:

$$\begin{aligned} f_{n+1} &= f_n + f_{n-1} \\ &\geq \left(\frac{3}{2}\right)^{n-2} + \left(\frac{3}{2}\right)^{(n-1)-2} \\ &= \left(\frac{3}{2}\right)^{n-1}\left[\left(\frac{3}{2}\right)^{-1} + \left(\frac{3}{2}\right)^{-2}\right] = \left(\frac{3}{2}\right)^{n-1}\left[\frac{10}{9}\right] \\ &> \left(\frac{3}{2}\right)^{n-1} = \left(\frac{3}{2}\right)^{(n+1)-2}. \end{aligned}$$

In this derivation, we first use the definition of the Fibonacci numbers to express f_{n+1} as $f_n + f_{n-1}$. The inductive hypothesis then allows us to use $f_n \geq (3/2)^{n-2}$ and $f_{n-1} \geq (3/2)^{(n-1)-2}$ in passing from the first line to the second. □

Notice that two base cases were given in the proof above. In general, the required number of base cases depends on details in the inductive step. In the inductive step above, to pass from the first line to the second when $n + 1 = 3$, we need to know that the inequalities hold for f_2 and f_1.

Exercise 2.29 In the proof of Theorem 2.15, would it be acceptable to offer five base cases, checking by hand that $f_i \geq (3/2)^{i-2}$ for $i = 1, 2, 3, 4$, and 5?

Exercise 2.30 In the inductive proof of Proposition 1.4, the statement $P(n)$ is an equation,

$$f_n = \frac{\phi^n - (1-\phi)^n}{\sqrt{5}}.$$

Is simple or strong induction used?

Another setting where induction is useful occurs when an equation or inequality involving a small number of terms is generalized to n terms. The proof can often be presented as simple induction, with the easiest form of the statement as the single base case. For example, you may recall from your prior mathematics experience that if a and b are positive real numbers, then we can write

$$\log_a b = y \text{ when } b = a^y.$$

One of the most important rules for log is the product rule: if a, b, and c are positive real numbers, then

$$\log_a(bc) = \log_a b + \log_a c.$$

Setting $b = c$ gives

$$\log_a b^2 = 2 \cdot \log_a b,$$

and we can use the product rule to prove a more general result.

PROPOSITION 2.16. *If a and b are positive real numbers and $n \in \mathbb{N}$, then*

$$\log_a b^n = n \cdot \log_a b.$$

PROOF. We use induction on n, where $P(n)$ is the displayed equation.

Base Case: Both $\log_a b^n$ and $n \cdot \log_a b$ equal $\log_a b$ when $n = 1$, so $P(1)$ is true.

Inductive Step: Assume that $P(n)$ is true, and consider $P(n+1)$. We can write $\log_a b^{n+1} = \log_a(b^n \cdot b)$, which by the product rule for log gives

$$\log_a b^{n+1} = \log_a b^n + \log_a b.$$

By our inductive assumption, the first term on the right-hand side of this equation equals $n \cdot \log_a b$, so we get

$$\log_a b^{n+1} = n \cdot \log_a b + \log_a b = (n+1) \log_a b,$$

as desired. □

For another example of this type of generalization, look ahead to the statement of Lemma 4.10. The lemma can be proved using simple induction; the base case is handled earlier, in Lemma 4.9.

Exercise 2.31 As caveats that the details of individual situations matter, consider exercises 1.45, 1.46, 1.47, and 1.48 from Chapter 1. Which of these might have nice proofs by mathematical induction? For those that do, would you use simple or strong induction, and how many base cases would you need?

Exercise 2.32 Use mathematical induction to prove that *Bernoulli's Inequality*,

$$(1+x)^n \geq 1+nx,$$

holds for all $x \geq -1$ and all $n \in \mathbb{N}$. Is your proof simple or strong, and how many base cases are needed?

REMARK 2.17. Project 11.2 discusses the two-player game of Chomp, whose analysis leads to an interesting induction argument. It also provides a good example of 'doing' mathematics as outlined in Section 2.8. We recommend flipping to the project now, just to do a bit of exploration of the ideas there and to have a first look at the inductive proof outlined in Exercise 11.12. The project right after it, Project 11.3 on the arithmetic–geometric mean inequality, also involves the use of induction, this time in a less standard format. It is worth your time looking at the outline of the argument developed there, just to see another way in which a proof by induction can be structured.

2.7. Ubiquitous Terminology

Mathematics is written in a concise format that takes a bit of getting used to. In fact, students often struggle because they believe they should just be able to "read" mathematics as opposed to "work through" a mathematics text; but a crucial part of your job as a reader of mathematics is to work to fill in details that were either skipped or condensed as the author wrote up her ideas. More on the process of "doing" mathematics is discussed in the next section. The aim of this section is simply to help you understand how mathematics is usually presented.

A *mathematical statement* is a sentence that is either true or false; it is clear and unambiguous. "Every Fibonacci number is even" is a mathematical statement while "Sure looks like rain" is not.

A *hypothesis* is a statement that is presumed to be true, and a *conclusion* is a statement that can (we hope!) be derived from one or more hypotheses. Here's an example:

If x is irrational, then x^2 is irrational.

Written symbolically, this says: $\forall x \notin \mathbb{Q}, x^2 \notin \mathbb{Q}$. The hypothesis is that you start with an irrational number, and the conclusion is that the square of that number is also irrational. Another example is

Every third Fibonacci number is even.

The hypothesis is "$n = f_{3k}$ for some $k \in \mathbb{N}$" and the conclusion is "$n = 2m$ for some $m \in \mathbb{N}$."

A *counterexample* is a specific example that shows that the hypotheses of a mathematical statement can be satisfied without having the conclusion be satisfied. In the preceding paragraph we wrote, "If x is irrational, then x^2 is irrational." This statement is true in many cases; it is true for $x = \pi$, as one example. However, $\sqrt{2}$ is irrational, but $(\sqrt{2})^2 = 2$ is not. And the presentation of this single counterexample establishes that the statement is false.

A *proof* of a mathematical statement is a valid chain of inferences that explains how you get from the hypotheses to the conclusions. In theoretical mathematics, proof is the standard and in fact only means of establishing truth. All the evidence you collect, short of a mathematical proof, only shows the plausibility of a statement, not its veracity.

In any mathematics book you will find a number of words that usually appear in a special font or style at the start of a paragraph. Often they signal the beginning of a mathematical statement.

A **definition** is a statement that precisely describes the meaning of a word, phrase, or notation. Often the term being defined is presented in italics (or is underlined when you are working at a chalkboard).

A **conjecture** is a mathematical statement that has neither been proven nor disproven. Usually there is some sort of evidence, experimental or intuitive, that strongly supports the idea that the conjecture is true.

A **theorem** is a mathematical statement that has been proven. Directly related to this term are

- **proposition**: a theorem that is not so grand;
- **lemma**: a theorem that is either a significant step in the proof of a theorem or proposition, or a theorem that is used in the proofs of many theorems or propositions;
- **corollary**: a theorem that follows immediately, either by specialization or by a short argument, from a preceding lemma, proposition, or theorem.

Typically a chapter or article in mathematics will begin with a brief introduction, followed by one or more definitions. Then there will perhaps be an example or two, which motivate the statement of a theorem. This is followed by a proof of the theorem, and then possibly by some corollaries that might be of interest in their own right.

Exercise 2.33 Revisit examples of definitions, theorems, propositions, lemmas, and corollaries that have already occurred in the book to this point.

Exercise 2.34 Exercise 1.5 asked you to "guess" a certain connection between two mathematical objects, which you can then use to fill in the blanks for Exercise 2.61. Do you believe that you generated enough evidence in Exercise 1.5 to be able to call your guess a conjecture?

2.8. The Process of Doing Mathematics

Having briefly described how mathematics is presented, we ought to also describe how mathematics is actually done. Doing mathematics is different from reading mathematics. Doing mathematics is a more challenging and more rewarding process; it is not as mechanical as you might think based on your previous experiences with arithmetic, geometry, and algebra (and calculus, if you've studied that). You often start with an idea that something should be true, or perhaps with a problem that has been assigned, or even just with a new definition that you would like to better understand. Then the fun begins.

Something – perhaps experimenting with a definition or simply being assigned a particular problem in your textbook – leads you to a conjecture. You then accumulate evidence for the conjecture, possibly by examining particular examples or by seeing that the conjecture is true when various additional hypotheses are added.

Quite often your evidence will force you to clarify or revise your conjecture, which will then force you to revisit the evidence you've already accumulated, as well as mull over new directions to explore.

If you are persistent, you might glimpse a proof of your conjecture. This is the moment when you say "That ought to be true because ..." At this point, it is time to start carefully working on the details. This is very important, because the details are often difficult to work out, and they also will frequently force you to clarify and revise the outline of a proof you had previously thought you had seen. And, in many cases, the details will actually point out to you reasons why you need to clarify and revise the conjecture that you thought you had glimpsed how to prove.

2.8.1. A Circle of Coins. Alice and Bob play a game where they alternate moves, and Alice starts. There is a circle of 12 coins, and on each move a player can remove either one coin or two coins that were originally adjacent on the circle. For example, on Alice's first move she might remove the coin at position 3, as shown in Figure 5. Then Bob might remove the consecutive pair of coins at positions 8 and 9; but Bob may not remove the two coins at positions 2 and 4, because they were not originally adjacent.

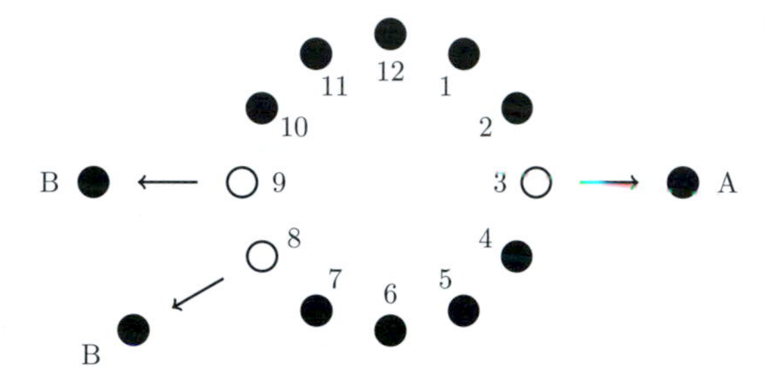

Figure 5. Alice and Bob's first two moves.

The player who removes the final coin wins. Does either player have a winning strategy in this game?

The first step in answering this question is found in the following exercise.

Exercise 2.35 Grab a friend and play the coin game many times, keeping a record of who started and who won. If you begin to develop strategies, keep notes on the strategies and how effective they are. Don't read any further until you are exhausted from play!

Based on your results, you might conjecture that Bob has an advantage, being able to respond to Alice's moves. But how can you describe Bob's advantage? Can you determine and describe a way that Bob always wins, regardless of Alice's moves?

In chess there is a strategy that beginning players often attempt, of simply mirroring the moves of their opponent for the first couple of moves. In the 12 coins game there is a similar approach that involves taking coins on the opposite side of the circle. Notice what would happen in the example above if Bob takes only the coin at position 9: two separate and identical arcs of five coins would remain. This puts Bob in a great position to respond to any of Alice's future moves: Bob can play "symmetrically" by taking the coin(s) from the position(s) halfway around the circle. For example, if Alice next takes the two coins at positions 6 and 7, Bob can take the coins at positions 12 and 1. If Alice then takes the coin at position 11, Bob can take the coin at position 5. Here's the key point: by playing symmetrically, Bob will always have a move to make, so he can never lose! All Bob has to do on his first move is prepare this symmetry. Can you see how this is done?

Exercise 2.36 Play the game many times again with the second player always attempting the symmetric strategy. Once you are convinced that the strategy works, try to write down an explanation.

At this point, we have a conjecture, supported by many examples from Exercises 2.35 and 2.36, and we believe we have written down an argument that demonstrates our conjecture. It is now time to try to write up a proof. As one possible approach, we define what we mean by symmetric and separate configurations of coins, and then we express the key idea as a lemma.

DEFINITION 2.18. Let S and T be two disjoint sets of coins from the original 12 coins in a circle. Then S and T are *symmetric* configurations if whenever a coin is in S then the coin halfway around the circle is in T, and similarly whenever a coin is in T then the coin that is halfway around the circle is in S.

The configurations S and T are *separate* if there are positions m and n that have had their coins removed, and S lies entirely on one side of the line through positions m and n, and T lies entirely on the other side.

LEMMA 2.19. *If the current game position consists of two separate, symmetric configurations of coins, the second player has a winning strategy.*

PROOF. Call the two configurations of coins S and T. Because they are separate, each player can remove coins from only one of the configurations. If the first player plays in S, then because S and T are symmetric, the second player is able to make the corresponding play in T. Similarly, if the first player plays in T, the second player can make the corresponding play in S. The resulting game position after the two players have played consists of two separate and symmetric configurations of coins. Thus we are once again in the situation where, whatever play the first player makes, the second player can make a symmetric move.

The number of remaining coins decreases with each play, and so it must eventually reach 0. Because the configurations are symmetric, the second player always has a move to make, so the second player wins. □

Finally we are able to write up a result that employs the lemma just proven.

PROPOSITION 2.20. *The second player has a winning strategy in the 12 coins in a circle game.*

PROOF. After Alice's first move there are 10 or 11 coins remaining and they form a single arc. If the arc contains 11 coins, Bob can remove the single coin in the middle of the arc, leaving two separate and symmetric arcs of coins. If the arc contains 10 coins, Bob can remove the pair of coins in the middle of the arc, again leaving two separate and symmetric arcs of coins. Thus Bob can ensure that, after the first two moves of the game, the remaining coins form two separate and symmetric configurations. By Lemma 2.19, Bob has a winning strategy as the second player. □

It feels great to have a proof in hand, but always be on the lookout for improvements, including the possibility of generalizing the result and proof to cover other situations as well. For example, it is natural to wonder if there is anything special about the number 12. If there are 11 or 20 coins to start, will that change the result? Is Definition 2.18 useful when starting with a different number of coins, and does Lemma 2.19 still hold? Certainly if there are only two coins the result does change, as Alice can then win on the first move. After you've had a chance to think about how the result might depend on the initial number of coins, visit Exercise 2.74 at the end of this chapter.

REMARK 2.21. When asked who wins the n coins in a circle game, the experienced mathematician might respond with a short confident statement like, "The second player, of course – she can just play symmetrically!" And while this captures the most important single idea for analyzing the game, you will find that she has to qualify these remarks when you mention the cases of $n = 1$ and $n = 2$, or the differences that might arise depending on whether n is even or odd. The point is that the details do matter, and it's important to present them as part of an argument that both highlights crucial ideas and covers all cases.

2.8.2. Egyptian Fractions. In this example, we leave more of the discovery work for you.

A small number of mathematical papyri have survived from ancient Egypt. One of the interesting aspects of these artifacts is the treatment of fractions. The Rhind Papyrus dates to the second intermediate period and it contains a table expressing the fractions $2/n$ for all odd n from 3 to 101 in terms of unit fractions. In contemporary notation, the table begins:

$$
\begin{aligned}
\frac{2}{3} &= \frac{1}{2} + \frac{1}{6} & \frac{2}{5} &= \frac{1}{3} + \frac{1}{15} \\
\frac{2}{7} &= \frac{1}{4} + \frac{1}{28} & \frac{2}{9} &= \frac{1}{6} + \frac{1}{18} \\
\frac{2}{11} &= \frac{1}{6} + \frac{1}{66} & \frac{2}{13} &= \frac{1}{8} + \frac{1}{52} + \frac{1}{104} \\
\frac{2}{15} &= \frac{1}{10} + \frac{1}{30} & \frac{2}{17} &= \frac{1}{12} + \frac{1}{51} + \frac{1}{68} \\
\frac{2}{19} &= \frac{1}{12} + \frac{1}{76} + \frac{1}{114} & \frac{2}{21} &= \frac{1}{14} + \frac{1}{42}
\end{aligned}
$$

Additional decompositions into unit fractions occur in other parts of the document, including exercises that ask how to divide a given number of loaves of bread among ten people. The answer in the case of six loaves is that each person gets $1/2 + 1/10$.[3]

These examples motivate the following definition. Let a/b be a rational number strictly between 0 and 1. An *Egyptian fraction* representation expresses a/b as a finite

3 Here we see one reason why unit fractions may have been considered important. The fact that $6/10 = 1/2 + 1/10$ tells us that to get an equal subdivision we should slice five of the six loaves in half, and the remaining loaf into ten pieces, which will allow us to give equal shares to each of the ten people.

sum of distinct unit fractions. The first question to ask is: *Do all rational numbers have Egyptian fraction expressions?* And the natural first step to take is to consider examples.

Exercise 2.37 Find Egyptian fraction decompositions for $2/5$, $3/7$, and $8/11$.

One way that you might have approached the exercise above is to use the largest unit fractions possible. For example, since $1/3 < 3/7 < 1/2$, the largest unit fraction we could try to use in an Egyptian fraction representation of $3/7$ is $1/3$. This gets us started with

$$\frac{3}{7} = \frac{1}{3} + \frac{2}{21},$$

but we then need to find an expression for $2/21$. A representation for $2/21$ is given in the Rhind Papyrus, and applying that we get

$$\frac{3}{7} = \frac{1}{3} + \frac{1}{14} + \frac{1}{42}.$$

We could also have continued with our strategy of using the largest unit fraction smaller than $2/21$, which is $1/11$. With this method we get

$$\frac{3}{7} = \frac{1}{3} + \frac{1}{11} + \frac{1}{231}.$$

The method of using the largest possible unit fraction is referred to as a "greedy algorithm," and it was used by Fibonacci to prove that every rational number between 0 and 1 has an Egyptian fraction representation.

Exercise 2.38 Apply the greedy algorithm to discover Egyptian fraction expressions for four rational numbers of your choice that are of the form a/b with $a, b \in \mathbb{N}$ and $b > a > 1$. Be very organized in your work, and try to discover a reason why this process will always result in an Egyptian fraction expression. Then try to prove it.

Exercise 2.39 Our work presented above shows that $3/7$ has two distinct Egyptian fraction decompositions. Use the fact that $1 = 1/2 + 1/3 + 1/6$ to prove that every rational number strictly between 0 and 1 admits multiple Egyptian fraction decompositions.

Exercise 2.40 We have indicated that every rational number between 0 and 1 has an Egyptian fraction representation and that these expressions are not unique. What other questions might you ask about such representations? Come up with at least two questions, and don't try to answer them.

GOING BEYOND THIS BOOK. The mathematics of ancient Egypt is both fascinating and frustrating, with a handful of artifacts indicating the existence of interesting perspectives, but only very few of them exist. A great introduction to this part of the history of mathematics is Imhausen's survey focused primarily on Egyptian fractions [**Imh06**]. If you enjoy reading that article, then you should also read Imhausen's book *Mathematics in Ancient Egypt* [**Imh16**].

2.9. Writing Up Your Mathematics

> The point of rigour is *not* to destroy all intuition; instead, it should be used to destroy *bad* intuition while clarifying and elevating *good* intuition. It is only with a combination of both rigorous formalism and good intuition that one can tackle complex mathematical problems.
>
> —Terence Tao

The two themes of this chapter – the structure of formal proof and the creativity needed to produce mathematics – combine quite nicely when you are trying to present mathematics. It is often the case that the process of explaining ideas and writing proofs is itself a catalyst for continued creativity. This is mentioned in the previous section, but it merits some additional attention.

2.9.1. Returning to Egyptian Fractions. In the previous section we indicated that Fibonacci's greedy algorithm will always produce an Egyptian fraction expression for any rational number between 0 and 1. In Exercise 2.38 we asked you to experiment with some specific rational numbers. We hope that you found a pattern something like the following:

$$\begin{aligned}\frac{\mathbf{12}}{13} &= \frac{1}{2} + \frac{\mathbf{11}}{26}\\ &= \frac{1}{2} + \frac{1}{3} + \frac{\mathbf{7}}{78}\\ &= \frac{1}{2} + \frac{1}{3} + \frac{1}{12} + \frac{\mathbf{1}}{156}.\end{aligned}$$

We have tried to indicate via boldface the insight you might have gained in your experiments. Namely, it appears that the numerators of the "remainder" fractions are decreasing. And if that is indeed true, it explains why this process will always terminate in an Egyptian fraction expression.

So why should the numerators of the remainders be decreasing? First, we know what the remainder is:

$$\frac{a}{b} - \frac{1}{g} = \frac{ag - b}{bg}.$$

We also know that the algorithm calls for us to use the largest possible unit fraction. If $1/g$ is the largest unit fraction less than or equal to a/b, then $1/(g-1) > a/b$. This tells us that $b > ag - a$, which after rearranging terms gives us $a > ag - b$.

The discussion above – from "We hope that you found ..." to "... rearranging terms gives us $a > ag - b$" – is a distilled version of all the scratch work that leads to the following proof of Fibonacci's result.

LEMMA 2.22. *Let a/b be a rational number between* 0 *and* 1*, and let $g \in \mathbb{N}$ be the smallest natural number such that $1/g \le a/b$. If*

$$\frac{a}{b} = \frac{1}{g} + \frac{c}{d},$$

where c/d is in lowest terms, then c is a non-negative integer where $c < a$.

PROOF. The fact that c/d is non-negative follows from $a/b - 1/g \geq 0$, so c is non-negative.

Since

$$\frac{a}{b} - \frac{1}{g} = \frac{ag - b}{bg}$$

and c/d expresses $(ag - b)/bg$ in lowest terms, we know that $c \leq ag - b$, so it suffices to show that $a > ag - b$.

Because a/b is less than 1 we know that $g \geq 2$. We also know that because $1/g$ is the largest unit fraction less than or equal to a/b, then $1/(g - 1) > a/b$. Cross-multiplying then shows that $b > ag - a$, which implies that $a > ag - b$. Thus $c < a$. □

THEOREM 2.23. *Every rational number between 0 and 1 has an Egyptian fraction representation. Further, the number of unit fractions produced by Fibonacci's greedy algorithm in the expression for a/b is at most a.*

PROOF. Fibonacci's greedy algorithm begins with the largest unit fraction less than or equal to a/b. We apply induction based on the numerator a. We are unconcerned with the value of b – it is an arbitrary natural number.

The base case is when $a = 1$, in which case $a/b = 1/b$ is already in Egyptian fraction form.

The inductive hypothesis is that if a/b is a rational number between 0 and 1 with numerator $1 \leq a \leq n$, then Fibonacci's greedy algorithm produces an Egyptian fraction expression for a/b with at most a terms. Consider the next case, where $a = n + 1$ and again b is arbitrary, with a/b between 0 and 1. Then by Lemma 2.22,

$$\frac{a}{b} = \frac{n+1}{b} = \frac{1}{g} + \frac{c}{d},$$

with $c \leq n$. If $c = 0$, then $a/b = 1/g$, so a/b has an Egyptian fraction expression with just one term (and so at most a terms). If $c > 0$, by the induction hypothesis the greedy algorithm applied to c/d produces an Egyptian fraction expression with at most n terms. None of these n terms can be $1/g$, or else $a/b \geq 1/g + 1/g \geq 1/(g-1)$ and thus g was not smallest denominator in Lemma 2.22. This then gives an Egyptian fraction expression for a/b with at most $n + 1$ terms. □

Notice that in working out how to present our proof we improved our intended result. We initially aimed only to show that Fibonacci's greedy algorithm works. We ended up showing that the numerator bounds the number of unit fractions that will occur in Fibonacci's Egyptian expression for a/b. If in response to Exercise 2.40 you asked about the maximum number of terms needed for a given rational number, you now have a partial answer. This fact also highlights that the table in the Rhind Papyrus is not presenting the shortest expressions for $2/n$, since many entries contain more than two terms, while the greedy algorithm will always yield a representation as the sum of two unit fractions.

Exercise 2.41 There are many things that appear in our proof of Fibonacci's result that did not appear in the distilled version of our scratch work, and may not have appeared in your work on Exercise 2.38. Please think about the following.

(a) Why did we use a "lemma" in presenting this proof? Was that a good strategy in terms of clarifying our argument?

(b) In the proof of the lemma we mention "Because a/b is less than 1 we know that $g \geq 2$." This was not noted in our scratch work. Why is it helpful to mention it?
(c) Identify at least one other place where we included information or steps that did not appear in our scratch work.

2.9.2. A Pattern in the Fibonacci Sequence. In Section 2.7 you encountered the statement "Every third Fibonacci number is even." This might have (we hope!) prompted you to do a little investigating: $f_3 = 2$, $f_6 = 8$, $f_9 = 34, \ldots$ And having done some initial investigation you may have even said "I bet that's true!" If you were asked to actually prove the claim, you would probably notice that every third Fibonacci number is preceded by two odd Fibonacci numbers. This should prompt you to revise your conjecture: "The Fibonacci sequence continually repeats the 'odd-odd-even' pattern." And you might already see an approach to proving the result, based on two facts you have known for quite some time: "odd plus odd is even" and "even plus odd is odd."

At this point you are ready to start writing down your ideas in the standard theorem–proof format. However, your first attempt at stating the theorem might lack some necessary precision, looking something like:

THEOREM? *The Fibonacci sequence goes odd–odd–even, forever.*

In order to get the details correct, notice that for any $n \in \mathbb{N}$, f_{3n} is even. And while it makes the odd-odd-even pattern less transparent, an accurate and unambiguous statement of your result is:

THEOREM 2.24. *Let f_i be the ith term of the Fibonacci sequence. Then, for each $n \in \mathbb{N}$, f_{3n} is even, and all of the other terms in the Fibonacci sequence are odd.*

The phrase "for each $n \in \mathbb{N}$" provides a hint that the proof of this result will involve mathematical induction. And initial attempts to write up such a proof will almost certainly lead to the recognition that each $n \in \mathbb{N}$ can be written in the form $3m, 3m+1$, or $3m+2$, and that you will need to divide your proof into cases. The results of these explorations might result in a proof that looks something like the following.

PROOF. By Theorem 1.12 (the Division Algorithm), each $n \in \mathbb{N}$ can be written in exactly one of the following three forms: $n = 3m$, $n = 3m+1$, or $n = 3m+2$, where m is an integer. Thus Theorem 2.24 is claiming that the Fibonacci numbers of the form f_{3m} are even, while those of the form f_{3m+1} and f_{3m+2} are odd. This claim is true for $f_1 = 1$, $f_2 = 1$, and $f_3 = 2$, and these facts then form the base cases for a proof by induction.

We assume that the claim holds for the Fibonacci numbers up to f_{3m} for some $m \in \mathbb{N}$. We need to show that the claim holds up to $f_{3(m+1)} = f_{3m+3}$. To do this we consider the three Fibonacci numbers that follow f_{3m}.

If the index is of the form $3m+1$, then, by the definition of the Fibonacci sequence, $f_{3m+1} = f_{3m} + f_{3m-1}$. By the inductive hypothesis, f_{3m} is even and $f_{3m-1} = f_{3(m-1)+2}$ is odd. Thus f_{3m+1} is odd.

If the index is of the form $3m+2$, then, by the definition of the Fibonacci sequence, $f_{3m+2} = f_{3m+1} + f_{3m}$. By the inductive hypothesis, f_{3m} is even, and by the previous case, f_{3m+1} is odd. Thus f_{3m+2} is odd.

If the index is of the form $3m+3$, then, by the definition of the Fibonacci sequence, $f_{3m+3} = f_{3m+2} + f_{3m+1}$. By the previous cases, f_{3m+1} and f_{3m+2} are both odd. Thus f_{3m+3} is even.

The result therefore holds for all $n \in \mathbb{N}$ by induction. □

2.9.3. Writing Down versus Writing Up. The previous two examples are our attempts to illustrate, much more efficiently than what happens in real life, the process by which you move from insights to mathematical arguments. The process of exploring a topic and gaining an insight is followed by the need to write down your ideas. However, as we hope your work on the examples showed you, *writing down* your mathematics is not the same as the process of *writing up* your mathematics. After writing down your initial thoughts, it is almost always the case that you realize that you have missed a detail or explained something poorly, forcing you to revise or completely rework what you thought you had finished. Only after a round or two of revision will you be in a place where you can begin to write up your ideas in a manner that others will benefit from reading.

Here is one final example that illustrates the need for revision as you move from insights to arguments.

EXAMPLE 2.25. A collection of lines in the plane is in *general position* if no two lines are parallel and no three lines meet in a common point. After some doodling you might notice the following two facts:

(a) 1 line in general position cuts the plane into 2 pieces.
(b) 2 lines in general position cut the plane into 4 pieces.

Given this limited amount of information, if you are bold you might quickly guess that the pattern continues. That is, 3 lines in general position will cut the plane into 6 pieces (if you think the pattern is "multiples of 2"). Or you might think that 3 lines in general position will cut the plane into 8 pieces (if you think the pattern is "powers of 2"). As you begin writing down your ideas, you will almost certainly notice that 3 lines in general position actually cut the plane into 7 pieces, letting you know that two cases are not enough to unearth the general pattern.

The correct response to this minor setback is to gather more information to see if you can find a more accurate conjecture. Introduce some helpful notation: let $p(n)$ be the number of regions in the plane formed by n lines in general position. Then additional experimentation will verify the following table of values:

n	$p(n)$
1	2
2	4
3	7
4	11
5	16
6	22

Exercise 2.42 If you insert a row at the top of the table for $n = 0$, what value should you write for $p(0)$?

After examining the results in this table for as long as necessary, you might notice that the difference between consecutive values of $p(n)$ increases by 1 as you go down the table, and this might eventually lead you to conjecture the formula

$$p(n) = p(n-1) + n\,.$$

The recursive nature of this formula – the fact that it references the value of $p(n-1)$ in describing $p(n)$ – suggests looking for proof by induction. This argument would likely show that adding the $(n+1)$th line will add $n+1$ new regions. Having gotten this far in the process, we turn the rest over to you! In Exercise 2.76 we ask you to first write down, and then write up, a proof of this result.

Most of the examples in Sections 2.8 and 2.9 have led you to theorems that can be proven by induction, and several exercises at the end of the chapter give you additional opportunities to practice this important proof technique. For not necessarily inductive explorations, consider Exercises 2.72, 2.75, 2.77, 2.80, and 2.82.

GOING BEYOND THIS BOOK. If you enjoyed Sections 2.8 and 2.9, consider reading Tao's *Solving Mathematical Problems* [**Tao06**]. It discusses how to approach problem solving in mathematics, illustrating various strategies through interesting examples.

2.10. End-of-Chapter Exercises

Exercises you can work on after Section 2.1

2.43 Construct the truth table for each of the following expressions.

(a) $P \vee (\sim P \wedge Q)$
(b) $(P \vee Q) \Leftrightarrow (\sim P \wedge Q)$
(c) $(P \Rightarrow Q) \Rightarrow (P \Rightarrow R)$

2.44 Resolve these two knights and knaves puzzles. For the second one, you might need to think outside of the truth table.

(a) You meet Alice, Bob, and Cecelia in the Land of Knights and Knaves. Alice tells you "All of us are knaves" and Bob says "Exactly one of us is a knight." What are Alice, Bob, and Cecelia?
(b) You meet Alice, Bob, and Cecelia in the Land of Knights and Knaves. You ask Alice "How many of you are knights?" Regrettably you cannot quite hear her reply. You ask Bob, "What did Alice say?" He replies, "Alice said there is one knight among us," which prompts Cecelia to exclaim, "Bob is lying!" What are Alice, Bob, and Cecelia?

2.45 Show that the statement $P \veebar Q$, where $\veebar$ is the symbol for exclusive-or, is logically equivalent to $\sim(P \Leftrightarrow Q)$.

2.46 Show that $P \Rightarrow (Q \vee R)$ is logically equivalent to the statement $(P \wedge \sim Q) \Rightarrow R$ and also to $(P \Rightarrow Q) \vee (P \Rightarrow R)$.

2.47 Find an expression using only the connectives $\sim$ and $\wedge$ that is logically equivalent to $P \Rightarrow Q$. Then do the same for $P \Leftrightarrow Q$. Do you think you could do the same for any logical expression, including those with more than two variables?

2.48 Determine which of the following statements are logically equivalent to each other.

(a) $\sim P \vee Q$
(b) $\sim(P \Rightarrow Q)$
(c) $P \Rightarrow Q$
(d) $P \Rightarrow \sim Q$
(e) $P \wedge \sim Q$
(f) $\sim Q \Rightarrow \sim P$
(g) $\sim Q \vee \sim P$

Exercises you can work on after Sections 2.2–2.4

2.49 Find the negation of each of the following statements. Then determine which statement is true: the original or its negation?

(a) 7 is prime and 9 is prime.
(b) $\pi < 355/113$ or $\pi = 355/113$.
(c) $-3 \leq -2$ and $|-3| \leq |-2|$.
(d) $\sqrt{2}$ is rational or 0 is negative. (Be careful with the opposite of "negative.")

2.50 Consider the statement: *If you get admitted to Lafayette College, then you won't enroll anywhere else.*

(a) What is the contrapositive of this statement?
(b) What is the converse of this statement?
(c) What is the negation of this statement?

2.51 The sums discussed in Theorem 2.4 are called *triangular numbers*, with the notation

$$T_n = 1 + 2 + 3 + \cdots + n\,.$$

Prove that for every integer greater than 1, $T_{n-1} + T_n = n^2$.

2.52 Use a proof by contradiction to show that there do not exist integers a and b such that

$$6a - 15b = 50\,.$$

2.53 Prove that if x is a real number where $x^2 \geq x$, then either $x \geq 1$ or $x \leq 0$.

2.54 Prove that if $x \in \mathbb{R}$, with $x > 0$ and x irrational, then $\sqrt{x}$ is also irrational.

Exercises you can work on after Section 2.5

2.55 Write the following as English sentences. You might not be able to determine whether these statements are true or false until you've worked though later parts of the book, but feel free to make guesses anyway.

(a) $\exists x \in \mathbb{Z}, x^3 + 3 = 0$
(b) $\exists x \in \mathbb{R}, x^3 + 3 = 0$
(c) $\forall x \in \mathbb{R}\ \exists y \in \mathbb{Q}, |x - y| < 1/1000$
(d) $\forall x \in \mathbb{R}\ \forall n \in \mathbb{N}\ \exists y \in \mathbb{Q}, |x - y| < 1/n$

2.56 Write the symbolic version of each statement below, and determine if it is true or false.

(a) There is a natural number that is less than its square.
(b) There is a real number whose cube is -1.
(c) The square of any real number is non-negative.

2.57 Negate each of the following statements, and then determine whether the negated statement is true or false.

(a) For all $r \in \mathbb{R}$, there exists $x \in \mathbb{Z}$ such that $r < x$.
(b) There exists an $n \in \mathbb{N}$ such that for every $m \in \mathbb{N}$, $n|m$.
(c) There exists an $n \in \mathbb{N}$ such that for every $m \in \mathbb{N}$, $m|n$.
(d) For all $x \in \mathbb{Q}$, there exists $y \in \mathbb{Z}$ such that $|x - y| < 1/2$.
(e) For every $q \in \mathbb{Q}$, there exists $z \in \mathbb{Z}$ such that $|3z - q| < 2$.

2.58 Determine whether each of the following statements is true or false.

(a) $\sim(\forall n \in \mathbb{Z}, n^2 > n)$
(b) $\sim(\exists q \in \mathbb{Q}, \sqrt{q} \in \mathbb{Q})$

(c) $\sim(\exists! x \in \mathbb{R}, x^2 = 1)$
(d) $\sim(\forall n \in \mathbb{N}\ \forall z \in \mathbb{Z}, \frac{z}{n} \in \mathbb{Q})$

2.59 We have given some examples that indicate that order matters when it comes to quantified statements. The following collection of statements pair alternate orderings of the quantifiers, and you should determine which of these statements are true and which are false.

(a) $\forall x \in \mathbb{R}\ \exists y \in \mathbb{R}, x + y = 0$
(b) $\exists y \in \mathbb{R}\ \forall x \in \mathbb{R}, x + y = 0$
(c) $\forall x \in \mathbb{R}\ \exists n \in \mathbb{N}, nx \in \mathbb{Z}$
(d) $\exists n \in \mathbb{N}\ \forall x \in \mathbb{R}, nx \in \mathbb{Z}$
(e) $\forall x \in \mathbb{R}\ \exists n \in \mathbb{Z}, nx = 0$
(f) $\exists n \in \mathbb{Z}\ \forall x \in \mathbb{R}, nx = 0$

2.60 Statements with three quantifiers can be particularly complicated. The following expressions are all similar, but each has a distinctive choice of order and quantification of the three variables. Which are true statements? Which are false?

(a) $\forall x \in \mathbb{R}\ \forall y \in \mathbb{R}\ \forall z \in \mathbb{R}, x + y < z$
(b) $\forall x \in \mathbb{R}\ \forall y \in \mathbb{R}\ \exists z \in \mathbb{R}, x + y < z$
(c) $\exists x \in \mathbb{R}\ \forall y \in \mathbb{R}\ \exists z \in \mathbb{R}, x + y < z$
(d) $\exists z \in \mathbb{R}\ \forall y \in \mathbb{R}\ \forall x \in \mathbb{R}, x + y < z$
(e) $\forall z \in \mathbb{R}\ \forall x \in \mathbb{R}\ \exists y \in \mathbb{R}, x + y < z$

Exercises you can work on after Section 2.6

Inductive arguments should be used for all of the exercises in this section, except where otherwise noted. Also, revisit Proposition 1.4 and a handful of exercises in Chapter 1 (1.24, 1.45, 1.46, 1.48, 1.49, and 1.51), and prove them now with inductive arguments if you didn't do this earlier.

2.61 Using the results of Exercises 1.5 and 2.34, complete this sentence:

"The number of ways to write n as an ordered sum of 1s and 2s is ______."

Then prove that your statement is true for all positive integers n.

2.62 Prove that if $n \geq 3$, the sum of the interior angles of a convex n-gon is

$$(n - 2) \times 180^\circ.$$

2.63 Prove that for all integers $n \geq 3$

$$\left(1 + \frac{1}{n}\right)^n < n.$$

2.64 Prove that for all non-negative integers n

$$\sum_{i=1}^{2^n} \frac{1}{i} \geq 1 + \frac{n}{2}.$$

2.65 Prove that for all $n \in \mathbb{N}$

$$1 + 2 + \cdots + n = \frac{n(n+1)}{2}.$$

Compare your inductive proof with the direct proof of Theorem 2.4.

2.66 Prove that for all $n \in \mathbb{N}$

$$1 + 3 + 5 + \cdots + (2n - 1) = n^2.$$

Can you also find a direct proof, either similar to the one given for Theorem 2.4, or based on a geometric figure like the one from Exercise 2.9?

2.67 Prove that for all $n \in \mathbb{N}$

$$1^2 + 2^2 + 3^2 + \cdots + n^2 = \frac{n(n+1)(2n+1)}{6}.$$

2.68 A *Tower of Hanoi* consists of n disks stacked in size order on a rod, with two empty rods nearby; see Figure 6 for an example with $n = 5$. Solving the puzzle requires you to move the entire stack to another rod, obeying the following rules:

(a) You can only move one disk at a time from one rod to another;
(b) You may only move a top disk; and
(c) No disk may be placed on a smaller disk.

Prove the statement

"The n-disk Tower of Hanoi puzzle can be solved in $2^n - 1$ moves."

Figure 6. A Tower of Hanoi problem with five disks.

2.69 Sixty-four balls are put into a number of separate piles. At each step you choose two piles, A and B, which contain a and b balls respectively. If $a \leq b$ you can remove a balls from pile B and put them into pile A. Prove that it is always possible to get all the balls into a single pile.

2.70 For which integers n is $6 \cdot 7^n - 2 \cdot 3^n$ a multiple of 4? Prove your claim. Can you also find a direct proof?

Exercises you can work on after Sections 2.7–2.9

2.71 Conjecture the values of $n \in \mathbb{N}$ for which $2n + 1 < 2^n$, and then prove your claim.

2.72 The sequence $(1, 1, 1, 2, 5)$ has a remarkable property: its sum and its product are the same:

$$1 + 1 + 1 + 2 + 5 = 1 \cdot 1 \cdot 1 \cdot 2 \cdot 5 = 10\,.$$

(a) Find another increasing sequence of five natural numbers whose product equals its sum.
(b) Find a third increasing sequence of five natural numbers whose product equals its sum.
(c) Make a conjecture about how many increasing sequences of five natural numbers have this property.
(d) Prove your conjecture.

2.73 Complete the following statement, and then prove your claim.

$$\frac{1}{1 \cdot 2} + \frac{1}{2 \cdot 3} + \frac{1}{3 \cdot 4} + \cdots + \frac{1}{n \cdot (n+1)} = \underline{\qquad\qquad\qquad}$$

$$\text{for all } n \in \mathbb{N} \text{ with } n \geq \underline{\quad}.$$

2.74 The n coins in a circle game is the same as the 12 coins game from Section 2.8.1, except that the number of coins initially in the circle is any natural number n. For what values of n does Alice have a winning strategy? For what values does Bob have a winning strategy? Prove your conjectures. Don't feel compelled to use the definitions and proofs given for the 12 coins game.

2.75 A virus is planted on some set of squares in an $n \times n$ chessboard. Any square that shares two or more edges with infected squares becomes infected, and this process repeats until no more squares can be infected. In Figure 7 we show two examples.

(a) Continue the process of infecting squares using the two examples shown in Figure 7. Does the infection eventually cover the entire board?
(b) Construct an initial infection of a 7×7 chessboard with only six infection sites. Does the infection eventually cover the entire board?
(c) Continue experimenting with initial infections. In particular, show by example that it is possible for an infection on a 7×7 board to start with more than seven initial infection sites but not result in the total infection of the board.
(d) Prove that you must infect at least 7 squares on a 7×7 board in order for the infection to spread to the entire board. As a hint for one possible approach, you might think about the perimeter of the infected region.
(e) Generalize your result to $n \times n$ chessboards.

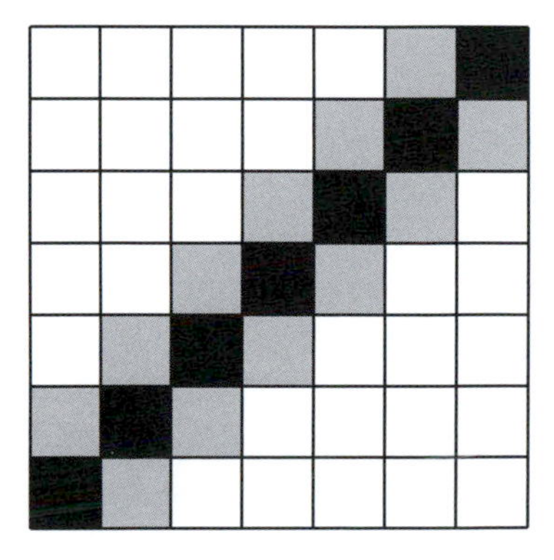

Figure 7. Two infected 7×7 chessboards. The black squares indicate the original location of the virus. The gray squares show where the contagion spreads after the first step.

2.76 Finish the process begun in Example 2.25, where we outlined how experimentation might lead to the conjecture that n lines in general position divide the plane into $p(n)$ pieces, with $p(1) = 2$ and $p(n) = p(n-1)+n$ for $n \geq 2$. First write down the details of the induction argument to prove to yourself that you have the correct approach; then write up your work in theorem–proof style that is suitable for others to read.

If you are dissatisfied that you have only a recursive formula for $p(n)$, continue to Exercise 2.87.

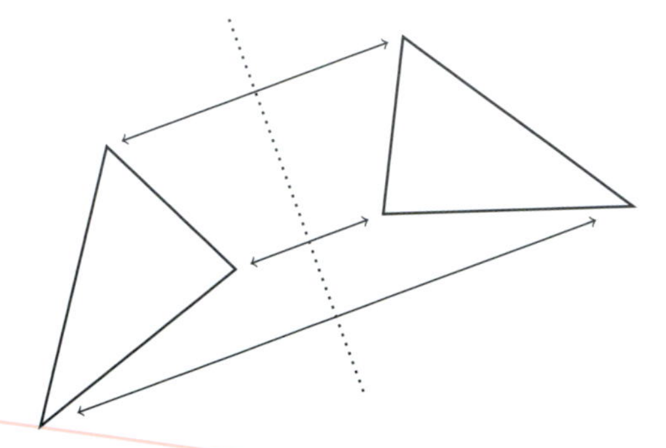

Figure 8. Reflection across the dotted line takes one triangle to the other.

2.77 A *reflection across a line* in the plane will move a triangle to a congruent copy, as shown in Figure 8.

(a) A single reflection exchanges the two triangles in Figure 8. Find two congruent triangles in the plane where a single reflection cannot exchange them. Can you move one of your triangles exactly onto the other one using a sequence of two reflections? three reflections? four reflections?
(b) Given two congruent triangles, what is the maximum number of reflections needed to move one triangle exactly onto the other?
(c) Prove your answer to the previous question is correct.
(d) Double-check your proof by applying it to the congruent triangles shown in Figure 9.[4]

2.78 The *Lucas numbers* are defined by setting $\ell_0 = 2$, $\ell_1 = 1$, and

$$\ell_{n+1} = \ell_n + \ell_{n-1}$$

for all $n \in \mathbb{N}$.

(a) Compute the next ten Lucas numbers.
(b) Prove that for all $n \in \mathbb{N}$, $\ell_n = 2f_{n-1} + f_n$, where f_n is the nth Fibonacci number.
(c) Is there a similar formula if instead of setting $\ell_0 = 2$ and $\ell_1 = 1$ we had defined $\ell_0 = 3$ and $\ell_1 = 2$?

4 It is not uncommon when writing geometric arguments to allow the figures you have in mind to overly influence the presentation of the argument. So it is a good practice to double-check the presentation of the argument against an alternate figure or two.

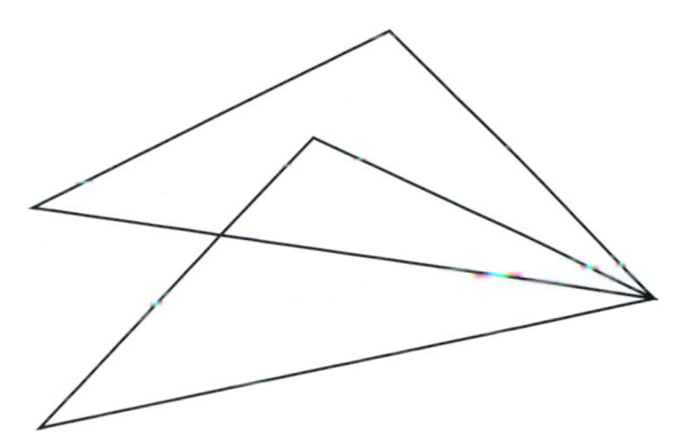

Figure 9. Two congruent triangles in the plane.

(d) Define a *generalized Fibonacci sequence* to be an integer sequence that satisfies

$$\gamma_{n+1} = \gamma_n + \gamma_{n-1},$$

where the initial terms are $\gamma_0 = a$ and $\gamma_1 = b$ for some $a, b \in \mathbb{N}$. Conjecture a relationship between this sequence and the Fibonacci sequence and then prove your conjecture.

(e) Do we need to restrict ourselves to $a, b \in \mathbb{N}$? Would your result hold if $a, b \in \mathbb{Z}$? What about $a, b \in \mathbb{R}$?

2.79 Here's a popular two-player game to play with a stack of $n \geq 1$ chips: alternating turns, each player can remove either 1, 2, or 3 chips from the stack on each turn, with the player removing the last chip losing. For example, if $n = 6$, the first player could remove 2 chips to leave 4, and then the second player could remove 1 chip to leave 3, so that the first player can remove 2 to leave 1, forcing the second player to lose.

Analyze this game for any $n \geq 1$, assuming both players use optimal strategy. Then write up your conclusions in the form of a theorem with proof.

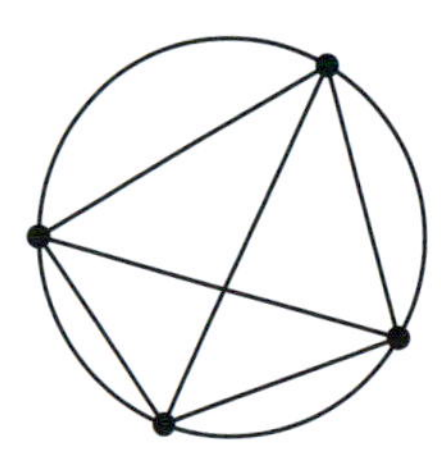

Figure 10. Three arrangements of points on a circle.

2.80 Given n distinct points on a circle, you can draw all of the line segments connecting those points, thereby dividing the disk bounded by the circle into a number of regions. The situation for $n = 2, 3$, and 4 is shown in Figure 10. As you can see, 2 points result in 2 regions; 3 points result in 4 regions; and 4 points result in 8 regions.

Given this data, it would be natural to guess that the following statement might be true: "The line segments connecting n points on the circle will produce 2^{n-1} regions in the disk." As you will see, however, this is another cautionary tale giving warning that surprises can occur beyond the first few cases.

(a) Verify that this guess also holds for 5 points on the circle.
(b) Verify that this guess is false, by examining the case of 6 points on the circle.
(c) Explain why the original phrasing of the statement needs to be improved, to include the idea of the *maximum* number of possible regions formed.
(d) Let $m(n)$ denote the maximum number of regions generated by n points. Then your previous work on this problem should have shown:

n	$m(n)$
2	2
3	4
4	8
5	16
6	31

Verify that all of these values match the fourth-degree polynomial

$$m(n) = \frac{1}{24}\left(n^4 - 6n^3 + 23n^2 - 18n + 24\right).$$

(e) The formula above is correct, and can be proven by induction. There are, however, better approaches to this problem, known as Moser's Circle Problem. We expect those approaches are beyond what you might generate on your own in a reasonable amount of time, and so we aren't asking you to prove the formula. Rather, you should read more about this and related topics in Conway and Guy's discussion of the problem in [**CG96**]. You might also read Richard Guy's "The strong law of small numbers," which provides several cautionary tales about spurious patterns in mathematics [**Guy88**].

More Exercises!

2.81 Exercise 1.31 asks you to explore which Fibonacci numbers are divisible by 3. You should have formed the conjecture that $3|f_k$ if and only if $k = 4n$ for some $n \in \mathbb{N}$.

(a) Prove that for any $k > 4, f_k = 3f_{k-3} + 2f_{k-4}$.
(b) Use this formula and an induction argument to prove that $3|f_{4n}$ for every $n \in \mathbb{N}$.
(c) Use this formula and a proof by contradiction to show that if $3|f_k$, then $k = 4n$ for some $n \in \mathbb{N}$.

Figure 11. A triomino oriented horizontally.

2.82 A *monomino* is a 1×1 square; a *domino* is a 1×2 rectangle; and a *triomino* is a 1×3 rectangle (as shown in Figure 11).

(a) Show that you can tile a 4×4 board with one monomino and five triominoes; the pieces may be oriented in any direction. (A tiling requires the board to be covered without any pieces overlapping.)

(b) Show that you can tile a 5×5 board with one monomino and eight triominoes.
(c) For what positive integers n can you tile an $n \times n$ board using exactly one monomino and some number of triominoes?

2.83 An *L-shaped tile* is like a domino, except that it is made of three 1×1 squares in the shape of an 'L' (as in Figure 12).

(a) Prove that a $2^n \times 2^n$ square with a 1×1 square corner missing can be tiled by L-shaped tiles; the tiles may be oriented in any direction.
(b) Is it possible to use L-shaped tiles to tile a $2^n \times 2^n$ board that is missing a single square in any position?

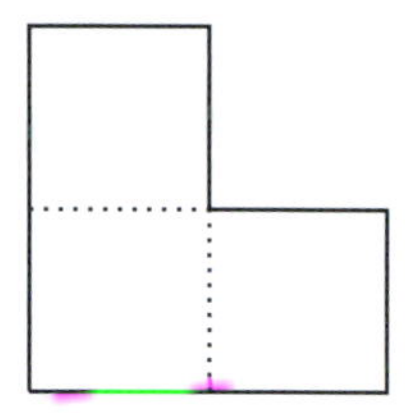

Figure 12. An L-shaped tile.

2.84 You were introduced to Egyptian fraction expressions in subsection 2.8.2. If q is a rational number strictly between 0 and 1, then an *Engel expansion* for q is an Egyptian fraction expression of the form

$$q = \frac{1}{a_1} + \frac{1}{a_1 a_2} + \cdots + \frac{1}{a_1 a_2 \cdots a_k},$$

where the a_i are natural numbers and $a_1 \leq a_2 \leq \cdots \leq a_k$. For example,

$$\frac{5}{7} = \frac{1}{2} + \frac{1}{2 \cdot 3} + \frac{1}{2 \cdot 3 \cdot 4} + \frac{1}{2 \cdot 3 \cdot 4 \cdot 7}.$$

(a) Explain why the assumption that $0 < q < 1$ tells us $a_1 \geq 2$, that is, $a_1 \neq 1$.
(b) Find an Engel expansion for $17/20$.
(c) Devise a greedy algorithm that will produce the Engel expansion for any rational number between 0 and 1.
(d) Prove that every rational number between 0 and 1 has a finite Engel expansion. That is, prove that the method you just devised in part (c) actually works. This might be harder than proving that Fibonacci's greedy algorithm works.
(e) We know that Egyptian fraction decompositions are not unique. Are the more specialized Engel expansions unique?

2.85 A function f is *additive* if for all real numbers x and y

$$f(x+y) = f(x) + f(y).$$

This is called *Cauchy's functional equation* in honor of his work on such functions in the 1800s.

In this exercise you will prove that every additive function has the property that

$$f(q \cdot x) = q \cdot f(x) \tag{2.2}$$

for each $q \in \mathbb{Q}$ and each $x \in \mathbb{R}$. The argument is done via cases, focusing on the expression of q as $q = m/n$ for $m \in \mathbb{Z}$ and $n \in \mathbb{N}$.

(a) Use the fact that $f(0+0) = f(0) + f(0)$ to show that Equation 2.2 holds for $q = 0$.
(b) Use an induction argument to establish Equation 2.2 for $q = m$, with $m \in \mathbb{N}$.
(c) Establish Equation 2.2 for $q = 1/n$, $n \in \mathbb{N}$, by examining

$$f\left(n \cdot \left(\frac{x}{n}\right)\right).$$

(d) Establish Equation 2.2 for all rational numbers $q = m/n$, where $q > 0$.
(e) Establish Equation 2.2 for all rational numbers $q = m/n$, where $q < 0$, by examining $f(q \cdot x + |q| \cdot x)$.
(f) Suppose that f is an additive function, and $f(2/3) = 4/5$. What do you know about the value of $f(4/5)$, or $f(\sqrt{2})$?

2.86 Let $n \in \mathbb{N}$. Suppose that $2n$ points in the plane are given, with n of them colored orange and n of them colored blue. Also assume that the points are in general position, meaning that no three points are collinear. A line in the plane is called an *equitable line*, or *e-line* for short, if it passes through one orange and one blue point, and for each side of the line, the number of orange points equals the number of blue points on that side.

(a) Prove that there is an e-line when $n = 2$ and when $n = 3$.
(b) Prove that there is an e-line for any $n \in \mathbb{N}$.
(c) Prove that there are at least two e-lines for any $n > 1$.

2.87 Prove that

$$p(n) = \frac{n^2 + n + 2}{2}$$

is a non-recursive solution to the lines in the plane problem of Example 2.25 and Exercise 2.76.

2.88 What is wrong with the following proof that all people are the same height?

PROOF? We will prove by mathematical induction that in any set of n people, all of the people have the same height. For the base case, just note that it is certainly true that in any set of 1 people, all of the people have the same height!

As the inductive step, assume that for $1 \leq k \leq n$, in any set of k people, all of the people have the same height. We want to show now that in any set of $n + 1$ people, all of the people have the same height. Well, take one person, X, out of that set: by the inductive step, the remaining n people all have the same height. Then, replace that person and take a different person, Y, out of the set: again, the remaining n people must all have the same height. Now, think about the people who are not X and Y: by the two previous sentences they have the same height, and they must also have the same height as X and Y. Thus, all $n + 1$ people have the same height. □

2.89 A *Baby Tetris brick* is one of the three shapes built out of 1×1 squares shown in Figure 13. A Baby Tetris *winning configuration* is a tiling of a $2 \times n$ rectangle by Baby Tetris bricks. Figure 14 shows three winning configurations for a 2×5 rectangle.

Let T_n be the number of distinct winning configurations for a $2 \times n$ rectangle. Our goal is to find a nice formula for T_n.

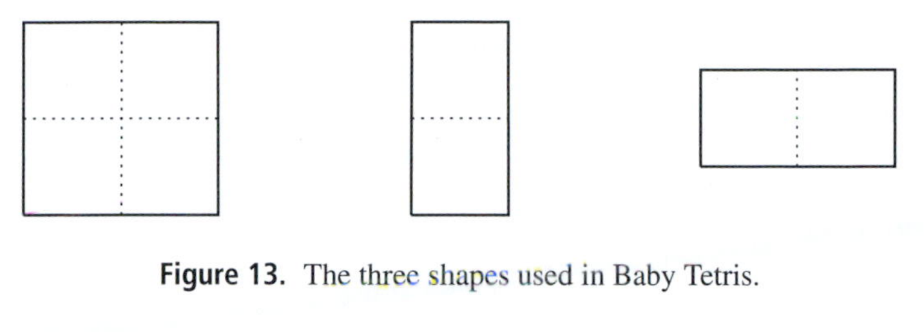

Figure 13. The three shapes used in Baby Tetris.

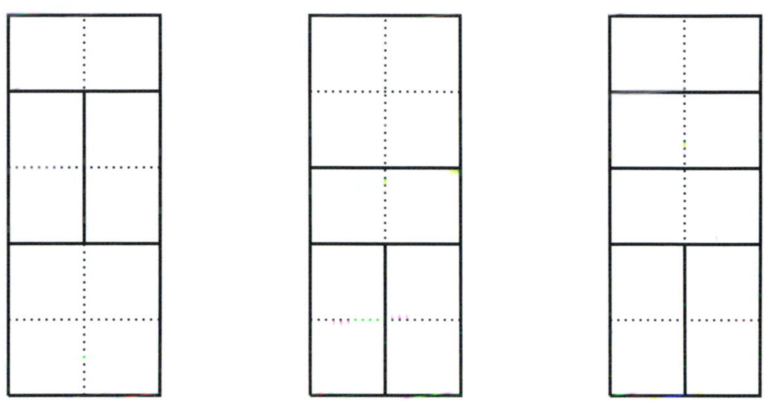

Figure 14. Three winning configurations in Baby Tetris.

(a) Determine the values of T_1, T_2, and T_3.
(b) Prove that $T_n = T_{n-1} + 2T_{n-2}$.
(c) Use this formula to compute T_4, T_5, and T_6.
(d) Conjecture a formula for T_i that does not reference prior terms in the series.[5] As a hint, first try to conjecture a formula for $3T_i$.
(e) Prove your conjecture.

2.90 William Thurston was a truly exceptional mathematician whose work in low-dimensional topology has had profound impact. In his article "On proof and progress in mathematics" [**Thu94**], Thurston argues that doing mathematics is a great deal richer than a "one-dimensional scale (speculation versus rigor)." Read that article, consider your own experiences, and then describe a specific aspect of mathematics that is not well described by the speculation versus rigor scale.

2.91 Timothy Gowers won the Fields Medal for his contributions to functional analysis. In addition to his research contributions, Gowers is an excellent expositor, as one can see by reading his *Mathematics: A Very Short Introduction* [**Gow02**]. Read Gowers' essay "Is mathematics discovered or invented" in [**Gow11**]. Which verb do you feel most accurately describes mathematical research? Are you *discovering* patterns and insights, or *creating* them?

5 This should be similar to, but more elementary than, the formula we proved that expressed the Fibonacci numbers in terms of the golden ratio ϕ in Section 1.3.

If You Have Studied Calculus

2.92 In a single-variable calculus course you learn how to locate extreme values.

(a) State the definition of an *absolute maximum* using quantifiers.
(b) State the definition of a *relative minimum* using quantifiers.

In case it helps, in Figure 15 we have drawn a function that has a relative minimum (when $x = 1$) and an absolute maximum (when $x = 3$).

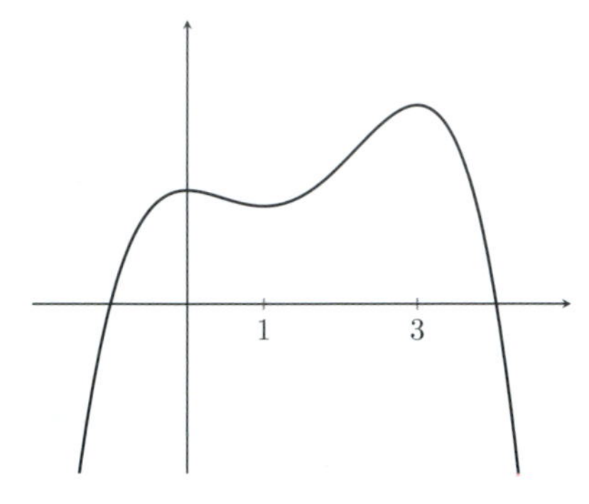

Figure 15. A function with a relative maximum, a relative minimum, and an absolute maximum.

2.93 Use standard integral computation techniques combined with an induction argument to prove that

$$\int_0^\infty x^n e^{-x}\, dx = n!$$

for all $n \in \mathbb{N}$. Does the result also hold for $n = 0$?

2.94 The *harmonic mean* of the n positive numbers $x_1, x_2, \ldots, x_n$ is

$$H = \frac{n}{\frac{1}{x_1} + \frac{1}{x_2} + \cdots + \frac{1}{x_n}}.$$

If you are interested in baseball statistics, Bill James' *power-speed number* is the harmonic mean of a player's total number of home runs and stolen bases.

Just to give yourself a bit of comfort with this mean, consider the following problem:

(a) Trisha drives r_1 miles per hour on her commute to work in the morning, and r_2 miles per hour on her return commute home. Show that her average speed is the harmonic mean of r_1 and r_2.

The following sequence of problems should lead you to a proof that the harmonic mean is always bounded from above by the arithmetic mean.

(b) Consider the case of two positive real numbers x_1 and x_2. Show that proving

$$\frac{2}{\frac{1}{x_1} + \frac{1}{x_2}} \leq \frac{x_1 + x_2}{2}$$

for all positive real numbers x_1 and x_2 reduces to proving

$$\frac{x_1}{x_2} + \frac{x_2}{x_1} \geq 2 .$$

(c) Prove that it suffices to show

$$x + \frac{1}{x} \geq 2$$

for all $x > 0$.

(d) Use calculus to find the minimum value of $f(x) = x + 1/x$ for positive values of x, thereby establishing that the harmonic mean is less than or equal to the arithmetic mean for any two positive real numbers.

(e) Use an induction argument to prove that the harmonic mean of n positive real numbers is less than or equal to the arithmetic mean of those numbers for all $n \in \mathbb{N}$.

(f) Gather together all of your work for this problem, and write up a clear statement of the harmonic-arithmetic mean inequality and a proof of this result.

3 Sets

When you read advanced mathematics books you will see the word "set" show up with great frequency. Since the late 1800s, set theory has been a fundamental approach to describing mathematical objects. For example, if you study abstract algebra you will encounter a statement that begins

A group is a *set* with a binary operation ...

Similarly, in real analysis you study the *set*, $\mathbb{R}$, of real numbers and properties of various sub*sets* of it.

3.1. Set Builder Notation

A *set* is nothing more than a collection of objects, and the objects are referred to as its *elements*. When a set is quite small, it is usually described by listing its elements inside of squiggly braces. For example, $\{1, 2, 3\}$ is a set whose elements are the first three natural numbers. One very small yet very important set is the *empty set*. This is a set that has no elements, and it is commonly denoted by $\emptyset$ or $\{\}$.

Sets are often described using *set builder notation*. For example, the set of positive "perfect squares," which we give the temporary name S, can be described as

$$S = \{n \in \mathbb{N} \mid n = m^2 \text{ for some } m \in \mathbb{N}\}.$$

The initial statement, $n \in \mathbb{N}$, describes what sort of objects are being considered for membership in S, while the second half describes all of the conditions that are placed on membership. The vertical bar that separates the two halves can be read as "such that," so that as a sentence we have "S is the set of all natural numbers n such that n is equal to m^2 for some natural number m."

Exercise 3.1 Complete the statement: $\mathbb{N} = \{z \in \mathbb{Z} \mid \_\_\_\_\_\_ \}$.

We will let $\mathbb{P}$ denote the set of all *prime numbers*:

$$\mathbb{P} = \{2, 3, 5, 7, 11, 13, 17, \ldots\}.$$

(Unlike $\mathbb{N}$ and $\mathbb{Z}$, there is no standard notation for the primes.) This set can be defined using set builder notation:

$$\mathbb{P} = \{p \in \mathbb{N} \mid p \geq 2, \text{ and if } p = mn \text{ for } m, n \in \mathbb{N}, \text{ then } m = 1 \text{ or } n = 1\}.$$

Here is an example of a "family" of sets defined using set builder notation. For any $n \in \mathbb{N}$, we can define

$$\mathbb{Q}_n = \left\{q \in \mathbb{Q} \mid q = \frac{m}{n^k} \text{ for some } k, m \in \mathbb{Z}\right\}.$$

In other words, $\mathbb{Q}_n$ consists of those rational numbers that can be expressed with a denominator that is a power of n. For example $1/2 \in \mathbb{Q}_2$; on the other hand, $1/2 \notin \mathbb{Q}_3$ for the following reason. Suppose that

$$\frac{1}{2} = \frac{m}{3^k}$$

for some integers m and k. If $k \leq 0$ you would have $1/2$ equaling an integer, so k must be positive. But then cross-multiplying would yield $3^k = 2m$, and hence 3^k would be even for some $k > 0$, a contradiction.

Exercise 3.2 Explain why modifying the definition of $\mathbb{Q}_n$ to become

$$\mathbb{Q}_n = \left\{q \in \mathbb{Q} \mid q = \frac{m}{n^k} \text{ for some } k, m \in \mathbb{Z} \text{ with } k \geq 0\right\}$$

does not change its contents.

Exercise 3.3 Prove that $4/9 \in \mathbb{Q}_3$, $4/9 \notin \mathbb{Q}_2$, and $5/12 \in \mathbb{Q}_6$. We will return to the sets $\mathbb{Q}_n$ at several points, so it is worth spending some time now to understand what elements they contain.

Exercise 3.4 You can explain the existential quantifier using set theory. If A is a set and $\mathrm{P}(a)$ is a statement about $a \in A$, explain why

$$\exists a \in A, \mathrm{P}(a) \quad \text{and} \quad \{a \in A \mid \mathrm{P}(a)\} \neq \emptyset$$

are either both true or both false. Then do the same for the universal quantifier

$$\forall a \in A, \mathrm{P}(a) \quad \text{and} \quad \{a \in A \mid \mathrm{P}(a)\} = A.$$

3.2. Sizes and Subsets

If S is a set with finitely many different elements, then $|S|$ denotes the number of different elements in S. For example,

$$\left|\left\{-\pi, 0, \frac{1+\sqrt{5}}{2}\right\}\right| = |\{\text{Huey, Dewey, Louie}\}| = 3,$$

and

$$|\{n \in \mathbb{Z} \mid n = m^2 \text{ for some } m \in \{-1, 0, 1, 2\}\}| = 3$$

as well. If S is not finite then we simply say that S is an infinite set.

We say that A is a *subset* of B, and we write $A \subseteq B$, if every $x \in A$ is also an element of B. For example,

$$\{1, 2\} \subseteq \{1, 2, 3\}, \quad \mathbb{N} \subseteq \mathbb{Z}, \quad \text{and} \quad \mathbb{Z} \subseteq \mathbb{Z}.$$

If $A \subseteq B$, and it is known that there is a $b \in B$ with $b \notin A$, then we write $A \subset B$ and say that A is a *proper* subset of B. Thus, $\{1, 2\} \subset \{1, 2, 3\}$ and $\mathbb{N} \subset \mathbb{Z}$. As with $\notin$, we use the symbols $\nsubseteq$ and $\not\subset$, so that we may write $\mathbb{Z} \not\subset \mathbb{Z}$.[1]

1 There is an alternate notational convention for subsets. In this notation, $A \subset B$ allows for the possibility that A and B are equal, and the notation $A \subsetneq B$ indicates that A is a proper subset of B.

A *Hasse diagram* is sometimes used to display subset relationships in a finite collection of sets, with $A \subset B$ if you can follow an ascending path of edges from A to B. For example, Figure 1 shows the subset relationships among all eight possible subsets of $\{a, b, c\}$.

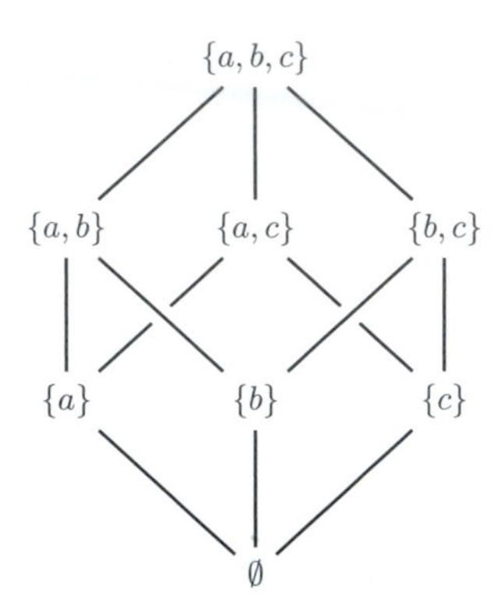

Figure 1. A Hasse diagram in the shape of a projected cube, showing all eight subsets of $\{a, b, c\}$.

Exercise 3.5 Show that if $n \in \mathbb{N}$, then $\mathbb{Z} \subseteq \mathbb{Q}_n$. Then, explain why $\mathbb{Z} \subset \mathbb{Q}_n$ for $n \geq 2$.

EXAMPLE 3.1. It is important to understand that $\in$ and $\subseteq$ correspond to different concepts. Let $A = \{1, 2, 3, \{1, 4\}\}$ be the 4-element set whose members consist of three integers and a set of two integers. Then the following are true statements:

(a) $2 \in A$
(b) $\{1, 3\} \subset A$
(c) $\{1, 4\} \in A$
(d) $|A| = 4$

Exercise 3.6 Referring to Example 3.1, explain why the following statements are false:

(a) $4 \in A$
(b) $\{1, 4\} \subseteq A$
(c) $\{1, 2\} \in A$
(d) $A \subset \mathbb{N}$

Two sets are *equal* if they contain exactly the same elements. That is, $A = B$ if both of the following are true: $a \in A$ implies $a \in B$, and $b \in B$ implies $b \in A$. Alternatively, using subset notation we say that $A = B$ if both $A \subseteq B$ and $B \subseteq A$.

In the proof of the proposition below, we show that two sets are equal in a standard way. Notice that the word "let" begins both halves of the argument, as you need to start each half with any possible element of the given set.

PROPOSITION 3.2. $\mathbb{Z} = \mathbb{Q}_1$.

PROOF. We will first show that $\mathbb{Z} \subseteq \mathbb{Q}_1$, and then that $\mathbb{Q}_1 \subseteq \mathbb{Z}$.

To show that $\mathbb{Z} \subseteq \mathbb{Q}_1$, let $z \in \mathbb{Z}$. If we set $m = z$ and $k = 1$, then we know that $z = z/1 = m/1^k \in \mathbb{Q}_1$ as well.

To show that $\mathbb{Q}_1 \subseteq \mathbb{Z}$, let $q \in \mathbb{Q}_1$. By the definition of $\mathbb{Q}_1$, there exist integers m and k such that $q = m/1^k$. Since $1^k = 1$ for any k, we know that $q = m/1 = m \in \mathbb{Z}$. □

Exercise 3.7 Determine all of the subset and proper subset relationships among the famous sets $\mathbb{N}$, $\mathbb{Q}$, $\mathbb{R}$, and $\mathbb{Z}$.

Exercise 3.8 Determine all sets S such that $|S| = 2$ and $S \subseteq \{1, 2, 3, 4\}$.

For $n \in \mathbb{N}$, we define

$$n\mathbb{Z} = \{m \in \mathbb{Z} \mid n|m\}.$$

In this definition, the two vertical bars have different meanings that should be clear from context: "$n\mathbb{Z}$ equals the set of all integers m such that n divides m." For example, $7\mathbb{Z} = \{\ldots, -14, -7, 0, 7, 14, \ldots\}$.

Exercise 3.9 Prove that $6\mathbb{Z} \subset 3\mathbb{Z}$.

3.3. Union, Intersection, Difference, and Complement

Given two sets A and B, you can combine them into one set:

$$A \cup B = \{x \mid x \in A \text{ or } x \in B\}.$$

The resulting set is called the *union* of A and B. Similarly, you can restrict to common elements:

$$A \cap B = \{x \mid x \in A \text{ and } x \in B\}.$$

The resulting set is the *intersection* of A and B.

These operations are commonly visualized using *Venn diagrams*. A 2-set Venn diagram is a way of illustrating how two sets may interact with each other, as in Figure 2. The entire shaded region represents $A \cup B$ while the smaller, darker region is $A \cap B$. The rectangle acts as a "universal" set containing both A and B as subsets; we will say more about universal sets later in this section.

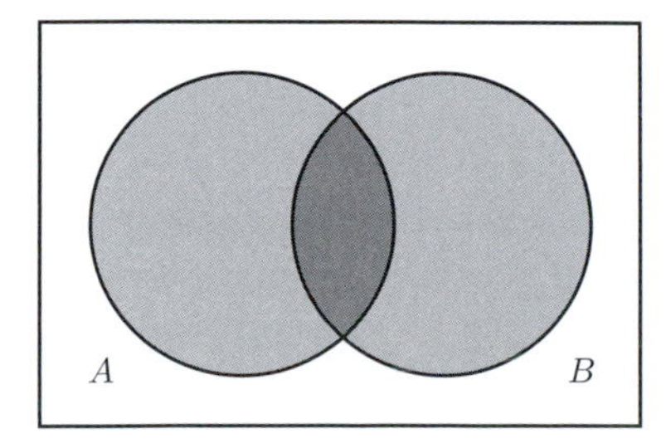

Figure 2. A 2-set Venn diagram.

You can also have a set that is the *difference* of two sets:

$$A - B = \{a \in A \mid a \notin B\}.$$

Notice that in this definition of set difference we do not assume that $B \subseteq A$. Another common way to write $A - B$ is $A \setminus B$.

REMARK 3.3. We emphasize that this type of "subtraction" has absolutely nothing to do with the difference of real numbers. For example,

$$\{\text{moose}, 3, 7\} - \{3, -4, \pi\} = \{\text{moose}, 7\}$$

doesn't involve the calculation of $7 - (-4)$, and certainly not moose $- \pi$!

The *symmetric difference* of two sets is

$$A \triangle B = \{x \mid x \in A - B \text{ or } x \in B - A\}.$$

Put another way, $A \triangle B = (A - B) \cup (B - A)$. This set is illustrated in Figure 2 by the two somewhat crescent-shaped regions that are lightly shaded.

EXAMPLE 3.4. If $A = \{1, 3, 5\}$ and $B = \{-1, 0, 1, 2\}$, then

(a) $A \cap B = \{1\}$
(b) $A \cup B = \{-1, 0, 1, 2, 3, 5\}$
(c) $A \setminus B = \{3, 5\}$
(d) $B - A = \{-1, 0, 2\}$
(e) $A \triangle B = \{-1, 0, 2, 3, 5\}$

Exercise 3.10 If $C = \{1, 2, 3, 4, 5\}$ and $D = \mathbb{P}$, the set of primes, determine the following sets.

(a) $C \cap D$
(b) $C \cup D$
(c) $C \setminus D$
(d) $D - C$
(e) $C \triangle D$

Exercise 3.11 Two sets A and B are said to be *disjoint* if they have no common elements; that is, $A \cap B = \emptyset$. Explain whether each of the following statements is logically equivalent to $A \cap B = \emptyset$.

(a) $A \setminus B = A$ and $B \setminus A = B$
(b) $A \triangle B = A \cup B$

We often want to talk about the things that are not in a set A. However, there is a small problem. For example, you might want to point out that

$$\frac{1}{2} \notin \mathbb{Z},$$

but it would be rather silly to point out that

$$\text{The Himalayas} \notin \mathbb{Z}.$$

In a less silly vain, suppose someone says "Pick a number not equal to 1." Are you allowed to pick -3? the golden ratio ϕ? or maybe the imaginary number $\mathbf{i} = \sqrt{-1}$? The term "number" hasn't described a specific domain of discourse, making the request vague.

When talking about elements that are not in a given set, it is always best to make sure that a *universal set*, the set of elements under consideration, has been made explicit. If U is the universal set, and if $A \subseteq U$, then the *complement* of A consists of all of the elements in U that aren't in A:

$$A^c = U - A = \{u \in U \mid u \notin A\}.$$

For example, if we let $\mathbb{R}$ be the universal set, then $\mathbb{Q}^c = \mathbb{R} - \mathbb{Q}$ is the set of irrational numbers.

Exercise 3.12 Prove that $\mathbb{Q}^c = \mathbb{R} - \mathbb{Q} \neq \emptyset$.

REMARK 3.5. Because the universal set is not always clear in a given situation, some mathematicians avoid referring to complements of sets, preferring to always write $U - A$ instead of A^c.

Exercise 3.13 In this exercise you'll examine the identity:

$$A - B = A - (A \cap B)\,.$$

(a) Explain why $A - B = A - (A \cap B)$ by appealing to a Venn diagram.
(b) Now prove that $A - B = A - (A \cap B)$ without referring to a Venn diagram. Instead argue that any element of $A - B$ is an element of $A - (A \cap B)$, and vice versa.

3.4. Many Laws and a Few Proofs

There are many elementary facts about set operations that are used quite frequently. The following theorem lists a number of these essentially arithmetic identities. Proofs of two of these identities are presented below; finding proofs for the rest is Exercise 3.51.

THEOREM 3.6. *Let A, B and C be sets. Then the following formulas hold.*

Idempotent Laws

(1) $A \cup A = A$
(2) $A \cap A = A$

Associative Laws

(3) $(A \cup B) \cup C = A \cup (B \cup C)$
(4) $(A \cap B) \cap C = A \cap (B \cap C)$

Commutative Laws

(5) $A \cup B = B \cup A$
(6) $A \cap B = B \cap A$

Distributive Laws

(7) $A \cup (B \cap C) = (A \cup B) \cap (A \cup C)$
(8) $A \cap (B \cup C) = (A \cap B) \cup (A \cap C)$

PROOF OF (3). We first show that $(A \cup B) \cup C \subseteq A \cup (B \cup C)$. To do this, let $x \in (A \cup B) \cup C$. By the definition of the union of two sets, this implies that $x \in A \cup B$ or $x \in C$, which (again by the definition of union) means that $x \in A$ or $x \in B$ or $x \in C$. This in turn means that $x \in A$ or $x \in B \cup C$, which implies that $x \in A \cup (B \cup C)$. Thus, we have shown that $(A \cup B) \cup C \subseteq A \cup (B \cup C)$.

The other half of the proof is Exercise 3.14. □

Exercise 3.14 Prove that $A \cup (B \cup C) \subseteq (A \cup B) \cup C$, which completes the proof of (3).

PROOF OF (8). We first show that $A \cap (B \cup C) \subseteq (A \cap B) \cup (A \cap C)$. Let $x \in A \cap (B \cup C)$. This implies, by the definition of intersection, that $x \in A$ and $x \in B \cup C$. In particular, $x \in B$ or $x \in C$. If $x \in B$, then $x \in A \cap B$, so $x \in (A \cap B) \cup (A \cap C)$. If $x \in C$, then $x \in A \cap C$, and thus $x \in (A \cap B) \cup (A \cap C)$. The conclusion is that $x \in (A \cap B) \cup (A \cap C)$, proving that $A \cap (B \cup C) \subseteq (A \cap B) \cup (A \cap C)$.

The other half of the proof is Exercise 3.15. □

Exercise 3.15 Prove that $(A \cap B) \cup (A \cap C) \subseteq A \cap (B \cup C)$, which completes the proof of (8).

The associative laws tell us that the manner in which we build the unions and intersections of three sets, two-at-a-time, doesn't matter. Section 3.5 gives definitions of the union and intersection of an arbitrary number of sets, in a way that is consistent with the associative laws we have just proven and that leads to more general versions of the distributive and other laws.

The following theorem is stated in terms of complements; a complement-free version can be found in Exercise 3.53.

THEOREM 3.7. *Let A and B be subsets of a universal set U. Then*

(1) $\emptyset^c = U$
(2) $(A^c)^c = A$
(3) $A \subseteq B$ *if and only if* $B^c \subseteq A^c$
(4) DEMORGAN'S LAWS
  (a) $(A \cap B)^c = A^c \cup B^c$
  (b) $(A \cup B)^c = A^c \cap B^c$

PROOF OF (3). To prove the "only if" half, we use the assumption that $A \subseteq B$ to prove that $B^c \subseteq A^c$. Since we want to show that $B^c \subseteq A^c$, we let $x \in B^c$; our goal is to conclude that $x \in A^c$.

Since

$$x \in B^c = \{y \in U \mid y \notin B\},$$

we know that $x \notin B$. We are assuming that $A \subseteq B$, so we know that $x \notin A$ as well. Since $x \notin A$, x is in the set $\{y \in U \mid y \notin A\}$, which is exactly the statement that $x \in A^c$.

As you might have expected, the proof of the "if" half is Exercise 3.16. □

Exercise 3.16 Finish the proof of (3) in Theorem 3.7.

Figure 3 shows a Venn diagram for three sets A, B, and C. When dealing with two or three sets at a time, Venn diagrams are very useful for helping you build instinct and then verify claims; these are important parts of the process of mathematical exploration and explanation. However, Venn diagrams have their limitations, as they might not be helpful for situations involving more than three sets (see Section 3.5 and Exercises 3.57, 3.65, and 3.88). The proofs given so far in this section are the type of general "element" arguments already seen in the proof of Proposition 3.2 and in part (b) of Exercise 3.13. These types of arguments are ubiquitous in many mathematical settings and require practice.

Exercise 3.17 Use a 3-set Venn diagram to investigate the claim that symmetric difference is associative: for any sets A, B, and C,

$$(A \triangle B) \triangle C = A \triangle (B \triangle C).$$

Exercise 3.18 Let A and B be finite sets. Venn diagrams may be useful for investigating the following:

(a) Prove $|A \cup B| = |A| + |B| - |A \cap B|$.

(b) If C is another finite set, is there an analogous formula for $|A \cup B \cup C|$? If so, prove that your formula is correct.

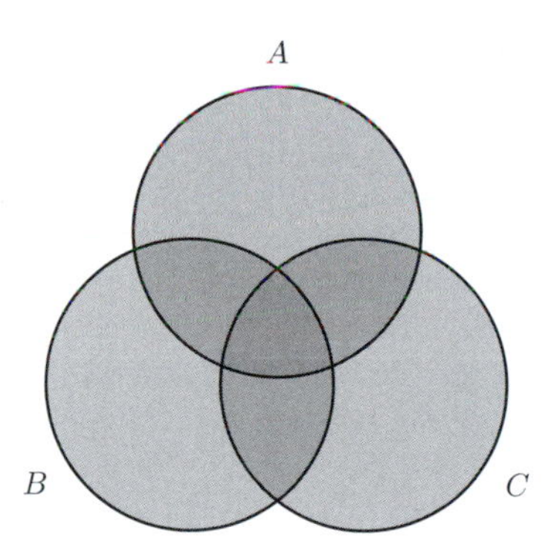

Figure 3. The standard 3-set Venn diagram.

3.5. Indexing

Section 3.3 discussed the intersection and union of two sets, and Section 3.4 extended these to three sets. What if we want to discuss intersections and unions of more than three sets? In particular, how could we discuss intersections and unions of an infinite number of sets? As an example to motivate these general questions, we look first at a specific infinite collection of nested open intervals on the real number line $\mathbb{R}$.

Recall that $(a, b) = \{ x \in \mathbb{R} \mid a < x < b \}$. For each $n \in \mathbb{N}$ define an associated open interval

$$S_n = \left(\frac{n-1}{n}, 1 \right).$$

For instance, $S_1 = (0, 1)$, $S_2 = (1/2, 1)$, and $S_3 = (2/3, 1)$.

Exercise 3.19 Indicate the first five of these sets on a number line to see why we call them "nested."

What is the union of all of these intervals? Generalizing the definition of union in Section 3.3, it should consist of all real numbers that are in at least one of the sets $S_1, S_2, S_3, \ldots$ Notationally, we write

$$\bigcup_{n\in\mathbb{N}} S_n,$$

which can be read as "the union of the S_ns, taken over all natural numbers n."

PROPOSITION 3.8.

$$\bigcup_{n\in\mathbb{N}} S_n = (0, 1).$$

PROOF. We prove that the two sides of the equation are equal by showing that each is a subset of the other.

Let $x \in \bigcup_{n\in\mathbb{N}} S_n$. This means that $x \in S_i$ for some $i \in \mathbb{N}$. Since $S_i \subseteq (0, 1)$ for each $i \in \mathbb{N}$, we know that $x \in (0, 1)$.

Alternatively, let $x \in (0, 1)$. Since $S_1 = (0, 1)$, we know that $x \in S_1$, and therefore $x \in S_1 \subseteq \bigcup_{n\in\mathbb{N}} S_n$. □

What is the intersection of these intervals? It should consist of all of the real numbers that are in every one of the sets $S_1, S_2, S_3, \ldots$

PROPOSITION 3.9.

$$\bigcap_{n\in\mathbb{N}} S_n = \emptyset.$$

PROOF. In trying to prove that a set is empty, it is a common tactic to begin by assuming to the contrary that the set is not empty. So to establish the equation we begin by assuming to the contrary that there is some $x \in \bigcap_{n\in\mathbb{N}} S_n$. Then this x must satisfy

$$(n - 1)/n < x < 1$$

for all $n \in \mathbb{N}$. In particular, if we consider all n of the form $n = 10^k$ for natural numbers k, then we must have $0.9 < x < 1$, and $0.99 < x < 1$, and $0.999 < x < 1$, and so on. Can any such x exist? The answer is "no"; but a rigorous proof must be postponed until we encounter the Archimedean Property in Section 8.2. □

Now let's define the general concept of indexing. An *index set* is simply a set I that helps describe a collection of mathematical objects. In the case of sets, the individual sets are written as S_i, where $i \in I$. In the example above, the index set $I = \mathbb{N}$, because there is one set for each natural number. Other options for index sets are $I = \{a, b\}$ when there are only two sets involved, and $I = \mathbb{R}$ when there's a set corresponding to each real number.

We can now define unions and intersections of any indexed collection of sets:

$$\bigcup_{i\in I} S_i = \{x \mid x \in S_i \text{ for some } i \in I\},$$

$$\bigcap_{i\in I} S_i = \{x \mid x \in S_i \text{ for all } i \in I\}.$$

Exercise 3.20 Convince yourself that, in the case when I has two elements, these new definitions of union and intersection express the same concept as the definitions given in Section 3.3.

For example, let $I = \{1, 2, 4\}$, and define S_i to be the set of points in the plane $\mathbb{R}^2$ that lie on the graph of $f(x) = \sin(ix)$ for $0 \leq x \leq 2\pi$. Then

$$\bigcup_{i \in I} S_i$$

consists of all of the points on the curves shown in Figure 4, while the intersection contains only three of them:

$$\bigcap_{i \in I} S_i = \{(0, 0),\ (\pi, 0)\ (2\pi, 0)\}.$$

Exercise 3.21 Let $I = \{1, 2, 4\}$, and define C_i to be the set of points in $\mathbb{R}^2$ that lie on the graph of $f(x) = \cos(ix)$ for $0 \leq x \leq 2\pi$. Draw a graph of the indexed union of the sets C_i, and determine their intersection.

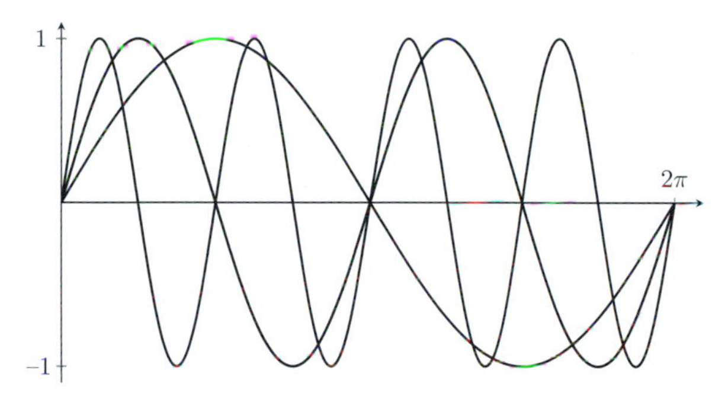

Figure 4. The curves $\sin(x)$, $\sin(2x)$, and $\sin(4x)$.

REMARK 3.10. In Section 3.4 we presented a number of results about intersections and unions in the context of two or three sets. You are asked to prove versions of these results in the context of indexed sets in Exercise 3.57 at the end of this chapter. Even if you decide not to prove the results, it would be worth your time to glance at the statements in that exercise.

3.6. Cartesian Product

In this section, we define the Cartesian product of two sets, a concept that will be essential when we discuss functions and relations.

DEFINITION 3.11. The *Cartesian product* of two sets A and B, denoted $A \times B$, is the set

$$A \times B = \{(a, b) \mid a \in A \text{ and } b \in B\}.$$

The elements of $A \times B$ are ordered pairs, with the first element in the pair from A and the second element from B. For example, if $A = \{1, 2\}$ and $B = \{1, 2, 3\}$, then $A \times B$ has six elements:

$$A \times B = \{(1, 1), (1, 2), (1, 3), (2, 1), (2, 2), (2, 3)\}.$$

Ordered pairs are different than 2-element sets since the order of the listed elements matters: for example, $\{1,2\} = \{2,1\}$ as sets, but $(1,2) \neq (2,1)$.[2]

You are no doubt familiar with the use of ordered pairs of real numbers for graphing points in the "Cartesian" plane $\mathbb{R}^2 = \mathbb{R} \times \mathbb{R}$, where $A = \mathbb{R}$ and $B = \mathbb{R}$ form the horizontal and vertical axes, respectively. For example, the graph of $f(x) = x^2 + 1$ is the subset of $A \times B$ defined by the set of ordered pairs

$$\{(x, x^2 + 1) \mid x \in \mathbb{R}\}.$$

However, many of the sets A and B that you may encounter won't allow for a similar 2-dimensional graphical representation.

Exercise 3.22 If $A = \emptyset$ and B is any set, what is $A \times B$?

Exercise 3.23 What is $|A \times B|$ in terms of $|A|$ and $|B|$? Does your answer hold if one of A and B is infinite, or empty?

REMARK 3.12. It would be nice to be able to define the concept of an ordered pair in terms of sets, since sets are our building blocks; and in fact we can do this by defining (x, y) to be the set $\{\{x\}, \{x, y\}\}$. However, most of the time when we're dealing with the Cartesian product of two sets, we write the elements in the standard ordered pair notation (x, y).

Exercise 3.24 Prove that $(1,2) \neq (2,1)$ using the technical definition given above that $(x, y) = \{\{x\}, \{x, y\}\}$.

3.7. Power

Given a set A, the collection of all subsets of A is called the *power set* of A. We will write this set as $\mathcal{P}(A)$; another common notation is 2^A, for reasons made apparent in Theorem 3.14.

EXAMPLE 3.13.

(a) If $A = \{1,2\}$, then $\mathcal{P}(A) = \{\emptyset, \{1\}, \{2\}, \{1,2\}\}$.
(b) The power set of the empty set is a set whose only element is the empty set: $\mathcal{P}(\emptyset) = \{\emptyset\}$, which can also be written $\{\{\}\}$. This set is not equal to the empty set itself because $\emptyset \notin \emptyset$ but $\emptyset \in \mathcal{P}(\emptyset)$.
(c) When dealing with power sets, be careful with $\in$ and $\subseteq$. Check that $\mathbb{N} \subseteq \mathbb{Z}$, while $\mathbb{N} \in \mathcal{P}(\mathbb{Z})$.

Exercise 3.25 The empty set $\emptyset$ is the only set A with $|A| = 0$, and we have seen that $|\mathcal{P}(\emptyset)| = 1$.

(a) Show that if A is any set with $|A| = 1$, then $|\mathcal{P}(A)| = 2$.
(b) Show that if A is any set with $|A| = 2$, then $|\mathcal{P}(A)| = 4$.
(c) Determine the size of $\mathcal{P}(A)$ if $|A| = 3$.

2 The ordered pair notation is not to be confused with open interval notation $(a, b) = \{x \in \mathbb{R} \mid a < x < b\}$. Context should make the distinction clear.

Exercise 3.26 What is $\mathcal{P}(\mathcal{P}(\emptyset))$? How about $\mathcal{P}(\mathcal{P}(\mathcal{P}(\emptyset)))$?

We hope that by now you've conjectured a formula for $|\mathcal{P}(A)|$ in terms of $|A|$ when A is a finite set. We provide two proofs of the theorem below, the first using induction and the second using the counting principle stated in Proposition 2.13.

THEOREM 3.14. *Let A be a finite set with n elements. Then $\mathcal{P}(A)$ is a finite set with 2^n elements.*

INDUCTIVE PROOF. The base case was established in Exercise 3.25: if $|A| = 0$, then we must have $A = \emptyset$, so

$$|\mathcal{P}(A)| = |\{\emptyset\}| = 1 = 2^0.$$

Thus the theorem is true for $n = 0$.

For the inductive step, we assume that the theorem statement is true for all sets of size at most n. Let A be any set with $|A| = n + 1$, and let α be any element of A. Then A can be expressed as

$$A = A' \cup \{\alpha\},$$

where $A' = A - \{\alpha\}$ is a set with n elements.

Every $B \in \mathcal{P}(A)$ is in exactly one of the following two cases:

(a) $\alpha \notin B$, in which case $B \subseteq A'$,
(b) $\alpha \in B$, in which case $B = B' \cup \{\alpha\}$ for some $B' \subseteq A'$.

The subsets B in case (a) are exactly the elements of $\mathcal{P}(A')$, and so by induction there are 2^n such subsets. The subsets B in case (b) also correspond to the elements of $\mathcal{P}(A')$: every such $B \subseteq A$ corresponds to a unique $B' \subseteq A'$, and vice versa. Thus, we conclude that

$$|\mathcal{P}(A)| = |\mathcal{P}(A')| + |\mathcal{P}(A')| = 2^n + 2^n = 2^{n+1}.$$

□

DIRECT PROOF. Since A is finite we can list its elements in some particular order:

$$a_1, \text{ then } a_2, \text{ then } a_3, \ldots, \text{ then } a_n.$$

Any subset of A can be uniquely described by a sequence of yes/no answers to the n questions, "Is a_i in A?" As examples: the empty set corresponds to a sequence of length n where each answer is "no"; the set A corresponds to each answer being "yes"; and if $n \geq 4$, the subset $\{a_1, a_2, a_4\} \subseteq A$ corresponds to "yes, yes, no, yes, no, no, no, ..., no."

By Proposition 2.13, the total number of such yes/no sequences is 2^n, so $\mathcal{P}(A)$ has 2^n elements. □

Exercise 3.27 Let $A = \{a, b, c, d\}$ and $\alpha = d$. Referring to the inductive proof above, verify that there are eight elements in each of the two cases, and determine all of the sets B' in case (b).

Figure 5 is a visual aid that might help you better understand the inductive step in the first proof of Theorem 3.14. It is an enhanced Hasse diagram showing $\mathcal{P}(\{a, b, c, d\})$ as the disjoint union of two bold "cubes," each the size of $\mathcal{P}(\{a, b, c\})$. With $A = \{a, b, c, d\}$,

$A' = \{a, b, c\}$, and $\alpha = d$, it is clear that the size of $\mathcal{P}(A)$ is twice the size of $\mathcal{P}(A')$: a thin edge connects a subset B that contains d with its corresponding subset $B' \subseteq A'$.

For even moderately sized n, the power set of an n-element set S contains an enormous number of different subsets. Applying Theorem 3.14 to $A = \mathcal{P}(S)$, we see that $\mathcal{P}(A) = \mathcal{P}(\mathcal{P}(S))$ has $(2^n)^n = 2^{(n^2)}$ subsets, which is greater than 30 million when $n = 5$. Among these many subsets are those of a special type that we will consider again in Chapters 5 and 6:

DEFINITION 3.15. Given a set S and an index set I, a collection $\mathcal{B} = \{S_i\}$ of non-empty subsets of S is a *partition of S* if the following two conditions hold:

(a) $S_i \cap S_j = \emptyset$ when $i \neq j$.

(b) $\bigcup_{i \in I} S_i = S$.

The subsets S_i are called *blocks*.

EXAMPLE 3.16. The blocks of a partition of a set S "cover" the set without overlap. If $S = \{a, b, c, d\}$, then the following subsets of $\mathcal{P}(S)$ are two of many possible partitions of S:

$$\{\{a\}, \{b, c\}, \{d\}\} \quad \text{and} \quad \{\{a, b, d\}, \{c\}\}.$$

The first partition has three blocks, of sizes 1, 2 and 1; the second has two blocks.

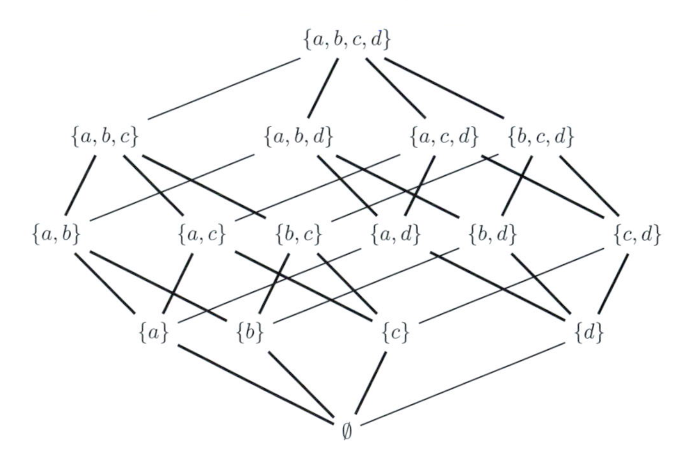

Figure 5. The power set of $\{a, b, c, d\}$, displayed as a "hypercube."

Exercise 3.28 The set $S = \{a, b, c\}$ has five partitions. Find them.

Exercise 3.29 This problem explores some partitions of $\mathbb{N}$.

(a) Show that $\mathbb{N}$ can be partitioned into two blocks, using "even" and "odd."
(b) Find a partition of $\mathbb{N}$ into infinitely many blocks.
(c) Find a partition of $\mathbb{N}$ into three blocks, each of infinite size.
(d) Find a partition of $\mathbb{N}$ into infinitely many blocks, each of size 2.

3.8. Counting Subsets

In this chapter, we have already presented arguments that help us determine the sizes of certain finite sets: $|A \cup B|$ in Exercise 3.18, $|A \times B|$ in Exercise 3.23, and $|\mathcal{P}(A)|$ in Section 3.7. The goal of this section is to determine the sizes of two subsets of $\mathcal{P}(A)$: the set of all k-element subsets of A, and the set of partitions of A into k blocks. For example, Exercise 3.8 and the middle level of Figure 5 show that the number of 2-element subsets of a 4-element set is 6. We begin by first developing several basic enumeration techniques with wide applicability beyond sets.

3.8.1. Sampling. Let S be a set of n elements. We might wish to select, or *sample*, k elements from S, one element at a time, keeping track of the order of our choices. We then have two natural possibilities: after an element is selected, it is either still available for future selection ("sampling with replacement"), or it is removed from future consideration ("sampling without replacement"). For example, if $S = \{a, b, c\}$, then the sequence (a, b, b, a) is one ordered sample of size $k = 4$, but only if we are sampling with replacement.

Exercise 3.30 Use the counting principle of Proposition 2.13 to prove that the number of ways to sample k times with replacement from an n-element set S is equal to n^k.

Let $(n)_k$ be the number of ways to sample k times from an n-element set S without replacement. Recall that if m is any positive integer, *m factorial* is defined as

$$m! = m \cdot (m-1) \cdots 2 \cdot 1,$$

and we set $0! = 1$.

PROPOSITION 3.17. *If $0 \le k \le n$, then*

$$(n)_k = n(n-1)\cdots(n-k+1) = \frac{n!}{(n-k)!}.$$

PROOF. If $k > 0$, there are n ways to select a first element from S. Once it is chosen, there are $n-1$ ways to select the second element from S. With the first two elements chosen, there are $n-2$ ways to select the third element from S. This process continues until the kth element is chosen from among $n-(k-1) > 0$ options. By Proposition 2.13, the product of these values is $(n)_k$, giving the first equality above. The second equality follows from rewriting $n(n-1)\cdots(n-(k-1))$ as

$$\frac{n(n-1)\cdots(n-(k-1))(n-k)(n-(k+1))\cdots 2 \cdot 1}{(n-k)(n-(k+1))\cdots 2 \cdot 1}.$$

If we define $(n)_0 = 1$ for all non-negative integers n, then the equalities hold for $k = 0$ as well: the expression $n(n-1)\cdots(n-(k-1))$ is an empty product and so equals 1, and there's only one way to pick nothing. □

Exercise 3.31 Compute $(4)_2$, and explain what it counts in terms of the set $\{a, b, c, d\}$.

A *permutation* of S is the special case of sampling without replacement when $k = n$. In this case, we are ordering all of the elements of S, and we get the formula

$$(n)_n = n(n-1)\cdots 2 \cdot 1 = n!$$

Exercise 3.32 There are many ascending paths of four edges from $\emptyset$ to $\{a, b, c, d\}$ in the Hasse diagram of Figure 5; for example,

$$\emptyset \to \{b\} \to \{b, d\} \to \{b, c, d\} \to \{a, b, c, d\}.$$

How many different such paths are there?

3.8.2. Combinations. What if instead of an ordered sample, we wish to choose a subset of size k from the elements of S? In this context, the subset is often called a *combination*, or a *k-combination*. A combination is by definition unordered; but we can use ordered sampling without replacement to prove the following formula for $\binom{n}{k}$, which counts k-combinations.

THEOREM 3.18. *The number of k-combinations of an n-element set is*

$$\binom{n}{k} = \frac{(n)_k}{k!} = \frac{n!}{k!(n-k)!}.$$

PROOF. Every ordered sample of size k can be produced in two steps: first choose the corresponding k-element subset of S, and then permute those k elements. Moreover, every selection of a k-element subset of S and a permutation of its elements determines a unique ordered sample of size k. Since there are $k!$ ways to permute any k elements, we must have

$$(n)_k = \binom{n}{k} \cdot k!,$$

which immediately leads to the desired formula for $\binom{n}{k}$. □

The symbol $\binom{n}{k}$ is referred to as "n choose k" because it counts the number of ways of choosing k elements from among n possible options.

Exercise 3.33 Explain the relationship between $\binom{4}{2}$, $(4)_2$, and $2!$ in terms of the set $\{a, b, c, d\}$, referring back to Exercise 3.31.

Exercise 3.34 Prove that the formula

$$\binom{n}{0} + \binom{n}{1} + \cdots + \binom{n}{n-1} + \binom{n}{n} = 2^n$$

holds for all $n \in \mathbb{N}$ by explaining why the left-hand side counts all of the subsets of an n-element set.

Several end-of-chapter exercises give you practice with permutations and combinations. For further discussion and application, you can then proceed directly to Section 9.3 if you wish, where the common terminology "binomial coefficient" for $\binom{n}{k}$ is explained and Pascal's Triangle is presented.

3.8.3. Partitions. Let $\left\{ {n \atop k} \right\}$ represent the number of partitions of an n-element set S into k blocks, with n and $k \in \mathbb{N}$. There is no formula for $\left\{ {n \atop k} \right\}$ akin to the one given in Theorem 3.18 for $\binom{n}{k}$. However, some cases are not too difficult to analyze. For instance,

$\left\{ {n \atop 1} \right\} = 1$, because the only possible partition of S into one block has S as the only block. And $\left\{ {n \atop n} \right\} = 1$ as well, because the only possible partition of S into n blocks has a single element of S in each block.

A more interesting case is $\left\{ {n \atop 2} \right\}$ for $n \geq 2$. Every partition of S into two blocks can be obtained by letting S_1 be a subset of S and then setting $S_2 = S - S_1$. However, if you do this for every subset $S_1 \subseteq S$, two issues arise:

(a) Every partition is generated twice.
(b) In two instances, $\{S_1, S_2\}$ is not a legitimate partition.

Exercise 3.35 Explain the two issues above, and prove that $\left\{ {n \atop 2} \right\} = 2^{n-1} - 1$.

Exercise 3.36 Prove that $\left\{ {n \atop n-1} \right\} = \binom{n}{2}$ for $n \geq 2$.

The numbers $\left\{ {n \atop k} \right\}$ are called Stirling numbers of the second kind. A partial table of values is given in Figure 6, and a recursive formula is presented in Exercise 3.84. The total number of partitions of an n-element set S is called the nth Bell number, B_n; thus,

$$B_n = \left\{ {n \atop 1} \right\} + \left\{ {n \atop 2} \right\} + \cdots + \left\{ {n \atop n} \right\}.$$

The first six Bell numbers are $1, 2, 5, 15, 52$, and 203. You showed that $B_3 = 5$ in Exercise 3.28.

1	1					
2	1	1				
3	1	3	1			
4	1	7	6	1		
5	1	15	25	10	1	
6	1	31	90	65	15	1
n / k	1	2	3	4	5	6

Figure 6. Stirling numbers of the second kind for $k, n \leq 6$.

3.9. A Curious Set

In this section we provide a cautionary tale about set theory.

Let's attempt to define a set whose elements are themselves sets:

$$\text{RUSSELL} = \{A \mid A \notin A\}.$$

An appropriate response at first glance is "Hunh?" But then you can start finding elements of RUSSELL. For instance, $\mathbb{Z} \notin \mathbb{Z}$ so $\mathbb{Z} \in$ RUSSELL; and $\emptyset \notin \emptyset$, so $\emptyset \in$ RUSSELL. At this point you might think that every set is an element of RUSSELL. However, consider the set:

$$\mathcal{E} = \{\emptyset, \{\emptyset\}, \{\emptyset, \{\emptyset\}\}, \ldots\}.$$

If you have a sufficiently robust notion of what is implied by the ellipses in the definition of $\mathcal{E}$, then it might appear that $\mathcal{E} \in \mathcal{E}$.

Let's try to determine if RUSSELL $\in$ RUSSELL. Well, if we assume that RUSSELL $\in$ RUSSELL, then by the definition of RUSSELL we know that RUSSELL $\notin$ RUSSELL, which contradicts our assumption. On the other hand, if we assume that RUSSELL $\notin$ RUSSELL, then by the definition of RUSSELL we know that RUSSELL $\in$ RUSSELL, which again is a contradiction. Thus it is neither possible for RUSSELL to be in RUSSELL nor for RUSSELL not to be in RUSSELL. (If your head hurts at this point, that's okay. Step away from the book and take a stroll outside until the pain subsides.)

This example was introduced to the mathematics community around the turn of the last century by Bertrand Russell. It pointed out that a naive[3] treatment of set theory, similar to what is occurring in this chapter, may initially appear to be benign but is in fact not logically consistent. This led to a number of axiomatic approaches to set theory, the most famous of which is the Zermelo–Fraenkel set theory.

GOING BEYOND THIS BOOK. If you are interested in the fact that set theory is considerably more complicated than our toolkit approach in this chapter indicates, then you have a number of options. You might investigate topics like "Zermelo Fraenkel" and "Axiom of Choice." You might read a more extensive treatment; we recommend Cunningham's book [**Cun16**] and Hajnal and Hamburger's text [**HH99**]. And if you are interested in the history around Russell's Paradox, we recommend Grattan-Guinness' article [**GG78**], but caution that this paper will be easier to follow after you have learned Cantor's Diagonal Argument (see the proof of Theorem 7.15).

3 There is a branch of set theory, called "naive set theory," which we are not referencing here! By "naive" we mean "lacking experience or understanding."

3.10. End-of-Chapter Exercises

Exercises you can work on after Sections 3.1 and 3.2

3.37 List every element in each of the following sets.

(a) $\{n \in \mathbb{N} \mid 7|n \text{ and } n \leq 30\}$
(b) $\{n \in \mathbb{Z} \mid n^2 \leq |n|\}$
(c) $\{x \in \mathbb{R} \mid x^4 - 4x^2 + 3 = 0\}$

3.38 Determine all of the subset and proper subset relationships among the following sets.

(a) $S_1 = \{n \in \mathbb{N} \mid 2 \nmid n \text{ and } n > 2\}$
(b) $S_2 = \{a^2 - b^2 \mid a, b \in \mathbb{N}\}$
(c) $S_3 = \{a^2 - b^2 \mid a, b \in \mathbb{N},\ a > b, \text{ and exactly one of } a, b \text{ is odd}\}$

3.39 Which, if any, of the following statements are true?

(a) $\mathbb{Q}_4 \subseteq \mathbb{Q}_2$
(b) $\mathbb{Q}_4 \subset \mathbb{Q}_2$
(c) $\mathbb{Q}_{18} \subseteq \mathbb{Q}_{12}$
(d) $\mathbb{Q}_{12} \subseteq \mathbb{Q}_{18}$

3.40 For any natural number n, is there any difference between the set $\mathbb{Q}_n$ and the set

$$\hat{\mathbb{Q}}_n = \left\{q \in \mathbb{Q} \mid q = \frac{m}{n^k} \text{ in lowest terms for some } k, m \in \mathbb{Z}\right\} ?$$

3.41 Find three different elements that are in both $6\mathbb{Z}$ and $8\mathbb{Z}$. Is there a convenient way to describe the set of all elements that are in both $6\mathbb{Z}$ and $8\mathbb{Z}$?

3.42 What conditions on a and b guarantee that $a\mathbb{Z} \subseteq b\mathbb{Z}$? Is your condition logically equivalent to $a\mathbb{Z} \subseteq b\mathbb{Z}$?

3.43 For $n \in \mathbb{N}$, let ${}^n\mathbb{Z} = \{z \in \mathbb{Z} \mid z = n \cdot m \text{ for some } m \in \mathbb{Z}\}$. Is this set equal to $n\mathbb{Z}$?

3.44 For $k \in \mathbb{N}$ define

$$T_k = \{z \in \mathbb{Z} \mid |z - \sqrt{2}| < k\}.$$

Find a formula expressing $|T_k|$ in terms of the value of k.

Exercises you can work on after Section 3.3

3.45 Let $A = \{-5, -2, 0, 4, 7\}$ and define $[n] = \{1, 2, \ldots, n\} \subset \mathbb{N}$. Determine each of the following.

(a) $A \cap \mathbb{N}$
(b) $A \cap [5]$
(c) $(A - \mathbb{N}) \cup [2]$
(d) $(A \triangle [5]) \cap [6]$
(e) $([5] - [6]) \cup ([4] - [3])$

3.46 Describe the following sets.

(a) $2\mathbb{Z} \cap 3\mathbb{Z}$
(b) $2\mathbb{Z} \cup 3\mathbb{Z}$

3.47 One way to develop an instinct for logical connectives is through their parallel constructions in set theory. For example, when thinking of P $\wedge$ Q you can imagine a Venn diagram with sets P and Q containing those situations where P is true and those situations where Q is true. Thus P $\wedge$ Q would correspond to the intersection of P and Q: $\wedge$ is analogous to $\cap$. Similarly, $\vee$ is analogous to $\cup$.

(a) Explain why $\Rightarrow$ is analogous to $\subseteq$.
(b) What is $\Leftrightarrow$ analogous to?
(c) What is exclusive-or analogous to?

3.48 Considering the three sets

$$A = \{ m \in \mathbb{Z} \mid \sqrt{m+2} \in \mathbb{Z} \}, \quad B = \{-5, -2, -1, 0, 1, 2, 3, 6\},$$

$$C = \{ n \in \mathbb{N} \mid n \text{ is a prime less than } 26 \}$$

as subsets of the universal set $\mathbb{Z}$, determine the following.

(a) $B \cap C$
(b) $A \cap C$
(c) $(\mathbb{N} - A)^c \cap B$

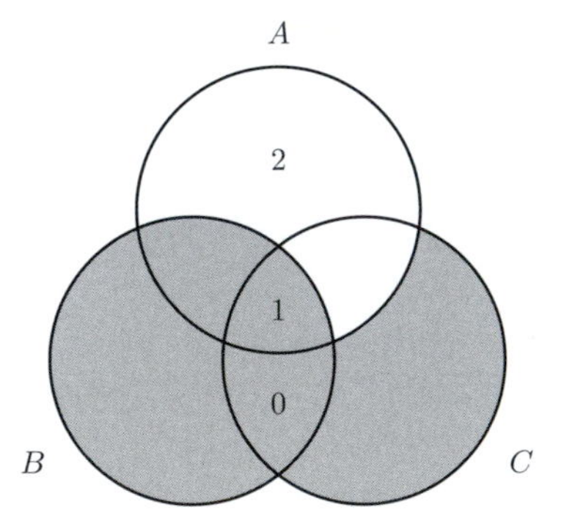

Figure 7. A 3-set Venn diagram with highlighted regions for Exercise 3.49.

3.49 Figure 7 shows a 3-set Venn diagram for the sets A, B, and C. Using only the processes of intersection, union, and complement, find expressions that specify certain regions. For example, the connected region labelled 0 is

$$A^c \cap B \cap C.$$

Find expressions for:

(a) the region labelled 1,
(b) the region labelled 2,
(c) the union of the shaded regions.

3.50 Let U be an infinite set with $S_1 \subseteq U$ and $S_2 \subseteq U$. How many different possible subsets of U can be constructed using the sets S_1 and S_2 and the symbols (,), $\cup$, $\cap$, and ^{c}? Each set and symbol can be used a finite number of times.

Exercises you can work on after Section 3.4

3.51 Prove the remaining six items in Theorem 3.6.

3.52 Prove the remaining three items in Theorem 3.7.

3.53 Prove the following generalization of Theorem 3.7.

Let A, B, and C be sets. Then the following hold.

(1) $C - \emptyset = C$
(2) $C - (C - A) = C \cap A$
(3) $A \cap C \subseteq B \cap C$ if and only if $C - B \subseteq C - A$
(4) DeMorgan's Laws
  (a) $C - (A \cap B) = (C - A) \cup (C - B)$
  (b) $C - (A \cup B) = (C - A) \cap (C - B)$

3.54 Is the following claim correct:

$$A \cap (B \triangle C) = (A \cap B) \triangle (A \cap C)\,?$$

If so, prove it; and, if not, provide a counterexample.

3.55 Let A and B be two sets. Prove that the following are equivalent.

(a) $A \subseteq B$
(b) $A \cup B = B$
(c) $A \cap B = A$
(d) $B^c \subseteq A^c$

The claim that "the following are equivalent" means that you need to show that each statement is logically equivalent to every other statement in the list above. This can be done in any number of ways, one of which is to establish the following cycle of implications:

$$(d) \Rightarrow (a) \Rightarrow (b) \Rightarrow (c) \Rightarrow (d).$$

Exercises you can work on after Sections 3.5 and 3.6

3.56 For each natural number n, let S_n be an infinite set.

(a) What can you conclude about the size of $\bigcup_{n \in \mathbb{N}} S_n$, and why?
(b) What can you conclude about the size of $\bigcap_{n \in \mathbb{N}} S_n$, and why?

3.57 Verify the following generalizations of results from Section 3.4. Here I is an arbitrary index set for the collection of sets $\{S_i \mid i \in I\}$.

Distributive Laws

(1) $A \cup \left(\bigcap_{i \in I} S_i\right) = \bigcap_{i \in I} (A \cup S_i)$

(2) $A \cap \left(\bigcup_{i \in I} S_i\right) = \bigcup_{i \in I} (A \cap S_i)$

DeMorgan's Laws

(3) $\left(\bigcup_{i \in I} S_i\right)^c = \bigcap_{i \in I} (S_i^c)$

(4) $\left(\bigcap_{i \in I} S_i\right)^c = \bigcup_{i \in I} (S_i^c)$

3.58 Let D_i be the disk of radius 1 centered at the point $(i, 0)$ in $\mathbb{R}^2$:

$$D_i = \{(x, y) \in \mathbb{R}^2 \mid (x - i)^2 + y^2 \leq 1\}.$$

(a) Plot the points in $\bigcup_{i \in \{1,2,3\}} D_i$.
(b) Plot the points in $\bigcap_{i \in \{1,2,3\}} D_i$.
(c) Plot the points in $\bigcup_{i \in \mathbb{R}} D_i$.

3.59 For each $r \in \mathbb{R}$ let $\ell_r = \{(x, rx) \in \mathbb{R}^2 \mid x \in \mathbb{R}\}$. Each ℓ_r is a line in the plane that goes through the origin.

(a) Plot the points in the sets ℓ_{-1}, ℓ_0, ℓ_1, and ℓ_2.
(b) Show that $\bigcap_{i \in \{-1,0,1,2\}} \ell_i = \{(0, 0)\}$.
(c) What is $\bigcup_{i \in \{-1,0,1,2\}} \ell_i$?
(d) What is $\bigcup_{i \in \mathbb{R}} \ell_i$?

3.60 (This exercise assumes you have studied the equations of lines and planes in $\mathbb{R}^3$.) For each $k \in \mathbb{R}$ let T_k be the set of all points $(x, y, z) \in \mathbb{R}^3$ satisfying the equation $x + y + kz = 0$. Describe the sets $\bigcap_{k \in \mathbb{R}} T_k$ and $\bigcup_{k \in \mathbb{R}} T_k$.

3.61 For each $i \in \mathbb{Z}$ let A_i be the half-open interval $[i, \infty)$ on the real line.

(a) What is $\bigcap_{i \in \{-3,2,7\}} A_i$?
(b) What is $\bigcap_{i \in \{-3,2,7\}} A_i^c$?
We emphasize that this means $A_{-3}^c \cap A_2^c \cap A_7^c$, not $(A_{-3} \cap A_2 \cap A_7)^c$. Part (4) of Exercise 3.57 reinforces this.
(c) What is $\bigcap_{i \in \mathbb{Z}} A_i$?
(d) What is $\bigcup_{i \in \mathbb{Z}} A_i$?

3.62 Either explicitly describe a collection of sets S_i, $i \in \mathbb{N}$, so that the following conditions are all true simultaneously, or prove that this is impossible.

(a) For each $i \in \mathbb{N}$, S_i is an infinite subset of $\mathbb{Z}$.
(b) $S_{i+1} \subset S_i$ for each $i \in \mathbb{N}$.
(c) $\bigcap_{i \in \mathbb{N}} S_i$ is an infinite set.

3.63 Recall that $n\mathbb{Z} = \{m \in \mathbb{Z} \mid n|m\}$.

(a) Show that if $I = \{n_1, \ldots, n_k\}$ is any finite subset of $\mathbb{N}$, then $\bigcap_{i \in I} i\mathbb{Z}$ is an infinite set.
(b) Show that $\bigcap_{k \in \mathbb{N}} k\mathbb{Z} = \{0\}$.

3.64 Draw each of the following Cartesian products as a subset of $\mathbb{R}^2$, where (a, b) and $[a, b]$ are open and closed intervals of real numbers.

(a) $\{1, 2\} \times \{-3, -1, 1, 3\}$
(b) $\{1, 2\} \times [2, 5]$
(c) $[-2, 3] \times (1, 4)$
(d) $[-2, 3] \times \mathbb{Z}$
(e) $\mathbb{N} \times \mathbb{Z}$
(f) $[-2, 3] \times \mathbb{Q}$

Admittedly, for each of the last few products you can't easily depict the entire set, but you should demonstrate that you know exactly which points in $\mathbb{R}^2$ are elements of the product.

3.65 For sets A, B, C, and D, prove the following.

(a) $A \times (B \cup C) = (A \times B) \cup (A \times C)$
(b) $(A \times B) \cap (C \times D) = (A \cap C) \times (B \cap D)$
(c) $(A \times B) \cup (C \times D) \subseteq (A \cup C) \times (B \cup D)$

Further, show by example that for (c) the containment can be proper.

3.66 Let A and B be subsets of a universal set U. Then $A^c \times B^c$ and $(A \times B)^c$ are both subsets of $U \times U$. How are they related to each other?

3.67 Consider the following subset of $\mathbb{N} \times \mathbb{N}$:

$$S = \{(m, n) \in \mathbb{N} \times \mathbb{N} \mid m \leq n \leq 2m\}.$$

The set S can be represented as a subset of the lattice of integer points contained inside the first quadrant in $\mathbb{R}^2$, as in Figure 8. In this exercise you will use this graphical representation to illustrate various claims.

(a) Show how Figure 8 can be used to illustrate why this statement is false:

$$\forall m \in \mathbb{N}\, \forall n \in \mathbb{N}, m \leq n \leq 2m.$$

(b) Show how Figure 8 can be used to illustrate the truth of

$$\forall m \in (\mathbb{N} \setminus \{1\})\, \exists n \in \mathbb{N}, m < n < 2m.$$

(c) Is the following claim true or false? How does Figure 8 illustrate your answer?

$$\exists m \in \mathbb{N}\, \forall n \in \mathbb{N}, m \leq n \leq 2m.$$

(d) Is the following claim true or false? How does Figure 8 illustrate your answer?

$$\exists m \in \mathbb{N}\, \exists n \in \mathbb{N}, m \leq n \leq 2m.$$

Exercises you can work on after Section 3.7

3.68 Which of the following statements about the power set of the real numbers are true, and which are false?

(a) $\mathbb{Z} \in \mathcal{P}(\mathbb{R})$
(b) $\mathbb{Z} \subset \mathcal{P}(\mathbb{R})$
(c) $\mathcal{P}(\mathbb{Z}) \subset \mathcal{P}(\mathbb{R})$

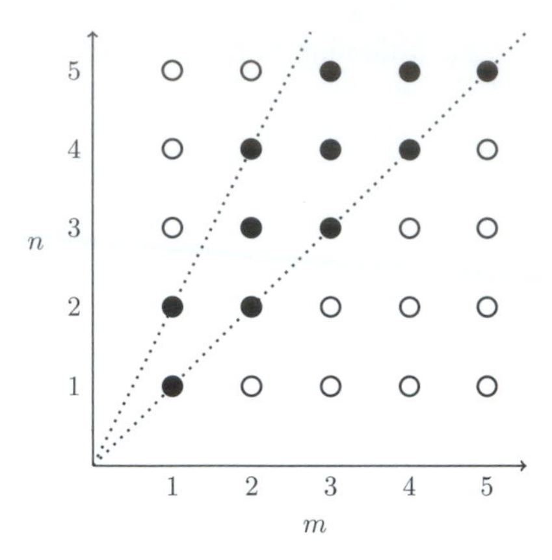

Figure 8. The set of all $(m, n) \in \mathbb{N} \times \mathbb{N}$ satisfying $m \leq n \leq 2m$ is illustrated by the filled circles.

3.69 Let A and B be arbitrary sets, and consider the following pairs of sets. Is one always contained in the other? Are they always equal? Provide proofs or counterexamples.

(a) $\mathcal{P}(A) \cup \mathcal{P}(B)$ and $\mathcal{P}(A \cup B)$
(b) $\mathcal{P}(A) \cap \mathcal{P}(B)$ and $\mathcal{P}(A \cap B)$
(c) $\mathcal{P}(A) - \mathcal{P}(B)$ and $\mathcal{P}(A - B)$

3.70 Is there a set A and an element a such that $a \in A$ and $a \in \mathcal{P}(A)$?

3.71 If $\mathcal{P}(A) = \mathcal{P}(B)$, must $A = B$?

3.72 In this exercise you will consider power sets of certain subsets of the natural numbers. As in Exercise 3.45, for each $n \in \mathbb{N}$ let $[n] = \{1, 2, \dots, n\}$.

(a) Prove that $\mathcal{P}([m]) \subseteq \mathcal{P}([n])$ if and only if $m \leq n$.
(b) Prove that

$$\bigcup_{n \in \mathbb{N}} \mathcal{P}([n]) \subset \mathcal{P}(\mathbb{N}).$$

To show the containment is proper, you must exhibit an explicit member of $\mathcal{P}(\mathbb{N})$ that is not contained in the union of the $\mathcal{P}([n])$s.

3.73 Which of the following are partitions of the set $\{1, 2, 3, 4, 5\}$?

(a) $\{\{1, 2, 3, 4, 5\}\}$
(b) $\{\{1, 5\}, \{2, 3, 4\}\}$
(c) $\{\{1\}, \{2\}, \{3\}, \{4\}, \{5\}\}$

3.74 This problem explores partitions of $\mathbb{Z}$.

(a) Find a partition of $\mathbb{Z}$ into blocks of finite size, where no two blocks have the same size.
(b) Find a partition of $\mathbb{Z}$ into infinitely many blocks, each of infinite size.

3.75 Let $\mathcal{B}$ and $\mathcal{B}'$ be two partitions of a set S. Then $\mathcal{B}'$ is said to be a *refinement* of $\mathcal{B}$ if every block of $\mathcal{B}'$ is a subset of a block of $\mathcal{B}$.

(a) Explain why $\{\{1,5\},\{2,4\},\{3\}\}$ is a refinement of $\{\{1,5\},\{2,3,4\}\}$.
(b) Explain why $\{\{1\},\{2\},\{3\},\{4,5\}\}$ is not a refinement of $\{\{1,5\},\{2,3,4\}\}$.
(c) Find all of the refinements of $\{\{1,5\},\{2,3,4\}\}$.

3.76 Let S be any set and let $\mathcal{B}$ be the partition of S into single elements,

$$\mathcal{B} = \{\{s\} \mid s \in S\}.$$

The notion of a refinement is defined in Exercise 3.75. Prove that $\mathcal{B}$ is a refinement of every partition of S.

Exercises you can work on after Section 3.8

3.77 0010011 is an example of a binary sequence of length 7 containing four 0s and three 1s. How many different binary sequences of length $m+n$ contain m 0s and n 1s?

3.78 Four balls labeled a, b, c, and d are to be thrown into seven baskets numbered $1,2,\ldots,7$. How many ways can this be done if ...

(a) ... no two balls can end up in the same basket?
(b) ... there is no restriction on the number of balls in the same basket?
(c) ... the only restriction is that at least two baskets must be occupied?

3.79 As in Exercise 3.78, suppose that four balls are to be thrown into seven baskets numbered $1,2,\ldots,7$; but now the balls are unlabeled and indistinguishable from each other. "Stars and bars" is one approach to counting the number of different possibilities. Consider all of the different ways that you can create a sequence of six bars and four stars, such as

$$| \, | \star \star | \star | \, | \, | \star .$$

Each $\star$ represents a ball, and each bar separates two adjacent baskets, so that the example just given puts two balls in basket 3, one ball in basket 4, and one ball in basket 7.

How many different assignments of balls to baskets are possible?

3.80 A subset S of $\{1,2,3,\ldots,n\}$ is *diffuse* if it does not contain any two consecutive integers. Prove that the number of k-element diffuse subsets of $\{1,2,3,\ldots,n\}$ is $\binom{n-k+1}{k}$.

If you enjoy this problem, take a look at Exercise 3.87.

3.81 Suppose that A and B are disjoint sets, with $|A| = m$ and $|B| = n$.

(a) What's $|\mathcal{P}(A \cup B)|$?
(b) Express the number of k-combinations of $A \cup B$ as a single binomial coefficient.
(c) In the case when $k \le m$ and $k \le n$, explain why the number of k-combinations of $A \cup B$ can also be expressed as

$$\binom{m}{0}\binom{n}{k} + \binom{m}{1}\binom{n}{k-1} + \cdots + \binom{m}{k}\binom{n}{0}.$$

(d) In fact, the expression in part (c) gives the correct count even when $k > m$ or $k > n$, as long as you have the appropriate definition for $\binom{a}{b}$ when $b > a$. What is the correct approach?

3.82 Each card in a standard deck has one of 13 values (Ace, 2, 3, ..., 10, Jack, Queen, King) and one of four suits ($\clubsuit$, $\diamondsuit$, $\heartsuit$, $\spadesuit$).

(a) How many cards are in a standard deck?
(b) The deck is shuffled well, and five cards are dealt face-up in order on a table. How many different ways can this be done?
(c) In games like poker, the order of the cards in your hand doesn't matter. How many different 5-card hands can be dealt from a standard deck?
(d) One particularly powerful hand in poker is four-of-a-kind, where four of the five cards match in value. How many different four-of-a-kind hands are possible?
(e) Another powerful hand in poker is called a full house: it contains three cards of one value and two cards of a different value. How many different ways can you create a full house?

3.83 How would you define the values of $\left\{ {n \atop k} \right\}$ when $k < 1$ or $k > n$?

3.84 Prove that

$$\left\{ {n+1 \atop k} \right\} = k\left\{ {n \atop k} \right\} + \left\{ {n \atop k-1} \right\}$$

for natural numbers n and k ($n \geq k$). As a hint, focus on which partitions contain element $n + 1$.

3.85 Use the results of Exercises 3.35, 3.36, and 3.84 to compute the next row in the table of Stirling numbers of the second kind (those with entries of "?" in Figure 9).

1	1						
2	1	1					
3	1	3	1				
4	1	7	6	1			
5	1	15	25	10	1		
6	1	31	90	65	15	1	
7	?	?	?	?	?	?	?
n / k	1	2	3	4	5	6	7

Figure 9. Stirling numbers of the second kind for Exercise 3.85.

More Exercises!

3.86 The game of Chomp, explored more generally in Project 11.2, is a 2-player game that can be played on a 5×7 board. On each turn in this game, a player takes a right-angled "bite" from the northeast: a square is removed along with all other squares to the north and east of it. An example is shown in Figure 10, where we have illustrated two moves in the game.

(a) Prove that the possible stages of the game correspond to edge paths from the northwest corner of the board to the southeast corner that are only allowed to move down or right with each step.
(b) Prove that the number of possible stages in 5×7 Chomp is $\binom{12}{7}$.

3.87 Let $\varphi(n)$ be the number of subsets of $\{1, 2, \ldots, n\}$ that do not contain consecutive integers.

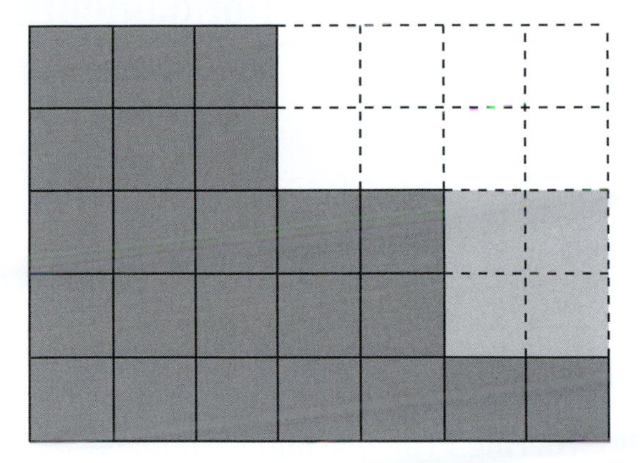

Figure 10. Two plays into the game Chomp. The eight unshaded squares were eaten first; the four lightly shaded squares were eaten second.

(a) Compute $\varphi(n)$ for $n = 1, 2, 3, 4$ and 5. (Do not forget about the empty set.)
(b) Conjecture a connection between $\varphi(n)$ and the Fibonacci numbers, and prove that your conjecture is correct.

3.88 We've seen Venn diagrams for two and three sets at a time. Is it possible to draw a Venn diagram to represent four or more sets? (If you get frustrated, you might consult "The search for simple symmetric Venn diagrams" by Ruskey, Savage, and Wagon [**RSW06**].)

3.89 Call a subset $A \subseteq \mathbb{N}$ *cofinite* if $A^c = \mathbb{N} - A$ is a finite set.

(a) Prove that the union of any two cofinite sets is cofinite.
(b) Prove that the intersection of any two cofinite sets is cofinite.
(c) Let $\{A_i\}$ be an indexed family of cofinite sets A_i. Prove that the union $\bigcup_{i \in I} A_i$ is cofinite.
(d) Let $\{A_i\}$ be an indexed family of cofinite sets A_i. Show by example that the intersection $\bigcap_{i \in I} A_i$ may not be cofinite.
(e) Construct a family of cofinite sets $\{A_i\}$, where $i \in \mathbb{N}$, such that
  (1) $A_{i+1} \subset A_i$,
  (2) $\bigcap_{i \in \mathbb{N}} A_i$ is infinite.
(f) Prove or disprove: Every $A \subset \mathbb{N}$ is the intersection of cofinite subsets of $\mathbb{N}$.

4 The Integers and the Fundamental Theorem of Arithmetic

In this chapter, you will preview a few topics that you otherwise might first see in a course in algebra. The focal point of this chapter is a proof of the Fundamental Theorem of Arithmetic (Theorem 4.11), which says that the natural numbers greater than 1 have unique prime decompositions, up to the ordering of the terms.

4.1. The Well-Ordering Principle and Criminals

We begin with a proof of an important property of the natural numbers, the *Well-Ordering Principle*, which guarantees the existence of minimal elements in sets of natural numbers. The concept of a minimal element, though, makes sense for any set of numbers.

DEFINITION 4.1. A set $S \subseteq \mathbb{R}$ is said to have a *minimal element* if there is an $m \in S$ such that $m \leq s$ for all $s \in S$. That is, m is smaller than all other elements of S.

PROPOSITION 4.2. *If $S \subseteq \mathbb{R}$ has a minimal element, then it has a unique minimal element.*

PROOF. If m and m' are both minimal elements of S, then we know $m \leq m'$ because m must be less than or equal to every element in S. However, it is also true that $m' \leq m$ because m' must be less than or equal to every element in S. Thus $m = m'$. □

Here are a few examples illustrating the idea of minimal elements in sets of numbers.

(a) The minimal element of $\mathbb{N}$ is the number 1.
(b) The integers $\mathbb{Z}$ have no minimal element.
(c) The minimal element of the closed interval $[a, b] = \{x \in \mathbb{R} \mid a \leq x \leq b\}$ is a.
(d) The open interval $(a, b) = \{x \in \mathbb{R} \mid a < x < b\}$ has no minimal element. To prove this, assume to the contrary that there is some $x \in (a, b)$ that is minimal. Then $(a + x)/2$ is still in the open interval (a, b), and $(a + x)/2 < x$. This contradicts the claim that x is the minimal element.
(e) The set of positive rational numbers is

$$\mathbb{Q}^+ = \{q \in \mathbb{Q} \mid q > 0\}.$$

The set $\mathbb{Q}^+$ has no minimal element. For assume to the contrary that there is some q that is a minimal element in $\mathbb{Q}^+$. Then the number $q/2$ is also a rational number and greater than 0, and so $q/2 \in \mathbb{Q}^+$. But $q/2 < q$, contradicting the claim that q is smaller than all other elements of $\mathbb{Q}^+$.

Exercise 4.1 Recall the definition of $\mathbb{Q}_n$ in Chapter 3, and define

$$\mathbb{Q}_n^+ = \{q \in \mathbb{Q}_n \mid q > 0\}.$$

Does $\mathbb{Q}_5^+$ have a minimal element?

THEOREM 4.3 (Well-Ordering Principle for $\mathbb{N}$). *Every non-empty subset of* $\mathbb{N}$ *contains a minimal element.*

PROOF. Let $S \subseteq \mathbb{N}$ be a non-empty subset. Our proof that S contains a minimal element is by induction, with the base case being the situation where S contains the number 1. Since $S \subseteq \mathbb{N}$ and 1 is minimal in $\mathbb{N}$, it must also be the minimal element of S.

Our inductive hypothesis is that every subset $S \subseteq \mathbb{N}$ where

$$S \cap \{1, 2, \ldots, n\} \neq \emptyset$$

contains a minimal element. Assume then that

$$S \cap \{1, 2, \ldots, n, n+1\} \neq \emptyset .$$

If S contains a natural number less than $n + 1$, then by our inductive hypothesis, S contains a minimal element. Otherwise S contains no natural number less than $n + 1$, but it must contain $n + 1$, in which case $n + 1$ is the minimal element of S. □

A powerful proof technique combines the Well-Ordering Principle with a proof by contradiction. As a first example, we can prove that the inequality

$$n < 2^n$$

holds for all $n \in \mathbb{N}$ using this method. The argument begins by assuming to the contrary that there is some n where $n \geq 2^n$. That is, we are assuming the set of counterexamples is a *non-empty* subset of $\mathbb{N}$. Then by the Well-Ordering Principle there is a minimal counterexample, which we denote by m. Because $1 < 2^1$ we know that $m > 1$. This leads to a contradiction, because

$$m \geq 2^m \Rightarrow m - 1 \geq 2^m - 1 \geq 2^m - 2^{m-1} = 2^{m-1} .$$

So $m - 1$ is also a counterexample, contradicting the fact that m was the *smallest* counterexample.

EXAMPLE 4.4. Here is a question inspired by the scoring in American football: what integers can be expressed in the form $3a + 7b$, where a and b are non-negative integers? The best first step is to consider examples, and after checking non-negative integers up to 15, we conjecture that the set of all such integers is

$$\{0, 3, 6, 7, 9, 10, 12, 13, 14, 15, \ldots\} = \{0, 3, 6, 7, 9, 10\} \cup \{n \in \mathbb{N} \mid n \geq 12\} .$$

To prove this, assume to the contrary that there are counterexamples in

$$\{n \in \mathbb{N} \mid n \geq 12\} .$$

By the Well-Ordering Principle there must be a minimal counterexample, m, which we know by our earlier computations must be greater than 15. Since m is the minimal counterexample and $m \geq 15$, it must be the case that $m - 3 = 3a + 7b$ for some non-negative integers a and b, because $12 \leq m - 3 < m$. But then we have $m = 3(a + 1) + 7b$, which contradicts the claim that m is a counterexample.

Exercise 4.2 What postage can you make using only 10 cent stamps and 4 cent stamps?

This proof strategy is sometimes referred to as the *minimal criminal* strategy. This basic structure is to first assume to the contrary that there are counterexamples to a given claim. Then by the Well-Ordering Principle you can discuss the smallest counterexample – the minimal criminal. You then derive a contradiction, thereby proving that the set of counterexamples is empty.

GOING BEYOND THIS BOOK. Minimal criminal arguments often employ steps that are reminiscent of standard proofs by induction, and indeed it is often simply a matter of stylistic preference as to whether to use a minimal criminal argument or induction. We use the minimal criminal strategy in the proof of the Fundamental Theorem of Arithmetic (Theorem 4.11). It also arises in work related to the four-color theorem, an important result in graph theory that is nicely described in Robin Wilson's *Four Colors Suffice* [**Wil14**].

The Well-Ordering Principle is directly related to the principle of induction, each implying the other, and the connections between such topics is developed nicely in [**Cun16**].

4.2. Integer Combinations and Relatively Prime Integers

The "easy" part of the Fundamental Theorem of Arithmetic was proved by induction as Theorem 2.11 in Section 2.6:

> *Every natural number $n \geq 2$ is either prime or it can be written as a product of primes.*

We are now left to show that the factorization of n into the product of primes is unique up to the order of the factors. To facilitate this part of the proof, we will first develop a relationship between two mathematical concepts that are important on their own: integer combinations and relatively prime integers.

DEFINITION 4.5. If a and b are real numbers, an *integer combination* of a and b is an expression of the form $ma + nb$, where $m, n \in \mathbb{Z}$.

For example, $2 \cdot 8 + (-3) \cdot 3$ is an integer combination of 8 and 3 that equals 7. Also, Example 4.4 considers integer combinations of 3 and 7, but with restrictions on the values of m and n.

Exercise 4.3 Find an integer combination of 8 and 3 that equals 1. (It is possible to do this in many different ways.) Then try to find an integer combination of 8 and 6 that equals 1, or explain why this is impossible. What is the essential difference between these two cases?

DEFINITION 4.6. The integers a and b are said to be *relatively prime* if the only positive integer d such that $d|a$ and $d|b$ is $d = 1$.

For example, 48 and 35 are relatively prime, while 5802 and 6111 are not.

Exercise 4.4 Verify that pairs of consecutive Fibonacci numbers f_i and f_{i+1} are relatively prime for $1 \leq i \leq 8$. Why do we need to specify *consecutive* Fibonacci numbers?

Exercise 4.5 Which integers are relatively prime to 0?

It is important to notice a fundamental difference between the definitions of "prime" and "relatively prime": the term "prime" can only be applied to a single integer, while "relatively prime" applies to a pair of integers. For example, 3 is prime, as is 7, while -14 and 21 are not. The set $\{3, 7\}$ contains a relatively prime pair of integers, as does $\{3, -14\}$, while $\{7, -14\}$ and $\{-14, 21\}$ do not.

Here's a result that will be used in the proof of Lemma 4.9 below.

LEMMA 4.7. *If a and b are relatively prime, so are $a - b$ and b.*

PROOF. This proof is by contradiction. Suppose that a and b are relatively prime but $a-b$ and b are not. Then there is some positive integer d greater than 1 such that $d|(a-b)$ and $d|b$. By Exercise 1.17 we know that $d|(a-b)+b$, which means that $d|a$. Since $d|a$ and $d|b$, a and b are not relatively prime, contradicting our original supposition. □

Exercise 4.6 For each of the following statements about integers a, b, and c, either prove the statement is true in general or provide a specific counterexample.

(a) If a and b are relatively prime, so are $a - b$ and $a + b$.
(b) If a and b are relatively prime, so are a and $a + b$.
(c) If a and b are relatively prime, so are a and $ac + b$.
(d) If a and b are relatively prime, so are a and $a + bc$.

You saw in Exercise 4.3 that there's an integer combination of 8 and 3 that equals 1, but no such integer combination of 8 and 6; the reason is that 8 and 3 are relatively prime but 8 and 6, having a common factor of 2, are not. We now prove the general result about this correspondence. The second half of the proof uses mathematical induction quite cleverly: we have two positive integers a and b, but we induct on the single value $n = a + b$.

THEOREM 4.8. *Let a and b be positive integers. There is an integer combination of a and b that equals 1 if and only if a and b are relatively prime.*

PROOF. We first prove the "only if" part, namely

$$\text{an integer combination of } a \text{ and } b \text{ equals } 1 \Rightarrow a \text{ and } b \text{ are relatively prime},$$

by proving the contrapositive: assuming that a and b are not relatively prime, we will show that no integer combination can be equal to 1.

If a and b are not relatively prime, there must exist some positive integer d greater than 1 such that $d|a$ and $d|b$. This means that $a = dx$ and $b = dy$ for some integers x and y, which implies that any integer combination $ma + nb$ of a and b can be written as

$$ma + nb = m(dx) + n(dy) = d(mx + ny).$$

Thus any integer combination of a and b is divisible by d, so no integer combination can be equal to 1.

We now prove the "if" part of the theorem,

a and b are relatively prime $\Rightarrow$ *an integer combination of a and b equals 1*,

using induction on $a + b$.

Suppose a and b are relatively prime positive integers. The smallest possible value of $a + b$ is 2, when $a = b = 1$. In this base case for the induction argument, there is certainly an integer combination of a and b that equals 1:

$$1 \cdot 1 + 0 \cdot 1 = 1.$$

To show that the inductive step also holds, we consider any relatively prime positive integers a and b with $a + b > 2$. Since a and b are assumed to be relatively prime, and since $a + b > 2$ implies that at least one of a and b is greater than 1, they cannot be equal to each other. Therefore we know that either $1 \leq a < b$ or $1 \leq b < a$. Since the definition of relatively prime given above is symmetric with respect to a and b, these two cases are really no different, so without loss of generality we may assume that $1 \leq b < a$.

The key to the inductive step is to apply it to $a' = a - b \geq 1$ and $b' = b \geq 1$, which represents a "smaller" case since $a' + b' < a + b$. Since a' and b' are relatively prime by Lemma 4.7, we may assume that there are integers m and n such that $1 = ma' + nb'$. A calculation now shows how this leads directly to an integer combination of a and b that equals 1, as desired:

$$1 = ma' + nb' = m(a - b) + nb = ma + (n - m)b.$$

□

Exercise 4.7 Is Theorem 4.8 true if at least one of a and b equals 0?

Exercise 4.8 Is Theorem 4.8 true for all integers a and b?

Exercise 4.9 For each of the following triples of numbers, find specific integer combinations of a and b that equal c. For parts (a) and (b), you may find it helpful to first find a combination of a and b that equals 1.

(a) $a = 9, b = 14, c = 3$
(b) $a = -11, b = 19, c = 2$
(c) $a = 21, b = 34, c = 1$

4.3. The Fundamental Theorem of Arithmetic

We now return to the Fundamental Theorem of Arithmetic, with the goal of finally completing a full proof of this well-known fact. We first use the main result of the previous section to prove an important lemma.

LEMMA 4.9. *Let $a, b \in \mathbb{N}$, and let p be a prime number. If $p|ab$, then $p|a$ or $p|b$.*

The structure of this proof will be to show that if $p \nmid a$ then $p|b$, which is equivalent to showing that $p|a$ or $p|b$.[1]

PROOF. Suppose that p is a prime, $p|ab$, and $p \nmid a$. Since $p \nmid a$ and the only positive integers dividing the prime p are 1 and p, we conclude that p and a must be relatively prime. By Theorem 4.8, this means that there are integers m and n such that $1 = mp + na$. Multiplying both sides of this equation by b gives

$$b = b(mp) + b(na) = (bm)p + n(ab).$$

Notice that p divides both terms on the right-hand side of the equation because $p|p$ and $p|ab$, so by Exercise 1.17 we know that p divides their sum, b, as desired. □

To prove the Fundamental Theorem of Arithmetic, it will be best to use a more general version of Lemma 4.9.

LEMMA 4.10. *Let p be a prime number, and let $n \in \mathbb{N}$. If $p|n$ and n can be written as a product of natural numbers*

$$n = a_1 a_2 \cdots a_k,$$

then $p|a_i$ for some i.

Exercise 4.10 Prove Lemma 4.10 by induction on k.

THEOREM 4.11 (Fundamental Theorem of Arithmetic). *There is a unique way, up to the order of the terms, to express a natural number $n \geq 2$ as a product of primes.*

PROOF. Theorem 2.11 states that any integer greater than 1 can be expressed as some product of primes, including the case when the product contains only one prime, so we only need to show the uniqueness of the expression. An equivalent way to state this part of the theorem uses an efficient "standard form" for prime factorizations: any product of primes can be written as

$$p_1^{a_1} p_2^{a_2} \dots p_s^{a_s},$$

where the distinct primes appearing in the product are listed in strictly increasing order $p_1 < p_2 < \cdots < p_s$, and each exponent a_i is a positive integer telling how many times the prime p_i appears in the product. For example, $336 = 2^4\, 3^1\, 7^1$. Using this notion of standard form, the uniqueness part of the theorem states that if n is an integer greater than 1 such that both

$$n = p_1^{a_1} p_2^{a_2} \cdots p_s^{a_s}$$

and

$$n = q_1^{b_1} q_2^{b_2} \cdots q_t^{b_t},$$

where p_i and q_j are sequences of increasing primes, then $s = t$, and $a_i = b_i$ and $p_i = q_i$ for all i.

1 If this is a confusing claim, you might want to revisit the discussion of logic in Section 2.1. In particular, study the example where "logically equivalent" is first defined.

Suppose there is some natural number that has two distinct factorizations. Then by the Well-Ordering Principle (Theorem 4.3), there is a minimal natural number n that has at least two distinct factorizations. Let the standard form of any two such factorizations be the ones just given above.

We know that n is not prime, because then we must have $s = t = 1$, $a_1 = b_1 = 1$, and $p_1 = q_1 = n$. So any standard form expression of n must involve at least two prime factors (which might not be distinct). Since p_1 divides n, it must divide $q_1^{b_1} q_2^{b_2} \cdots q_t^{b_t}$. By Lemma 4.10, p_1 must divide $q_j^{b_j}$ for some j; and so by Lemma 4.10 again, it divides q_j. Since q_j is itself prime, it must be the case that $p_1 = q_j$. Thus n/p_1 can be factored in two distinct ways, the expressions being

$$n/p_1 = p_1^{a_1-1} p_2^{a_2} \cdots p_s^{a_s}$$

and

$$n/p_1 = q_1^{b_1} q_2^{b_2} \cdots q_j^{b_j-1} \cdots q_t^{b_t},$$

where if $a_1 - 1$ or $b_j - 1$ is zero we simply remove this term from our expression. These two expressions are distinct if the original expressions are distinct, because the exponent of the same prime simply decreased by 1. Thus we see that n/p_1 can be factored in distinct ways, contradicting the claim that n is the smallest such integer. Thus there cannot be a non-empty set of counterexamples, and therefore all natural numbers $n \geq 2$ have unique factorizations up to the order of the terms. □

At this point you should pause and reflect on quite an accomplishment: you have worked your way through a proof of a very important number theoretical result, a proof that required induction, direct arguments, indirect arguments, and the support of several lemmas and simpler theorems.

Exercise 4.11 Outline the proof of the Fundamental Theorem of Arithmetic, including the "easy" part, on a single page (and remember it forever).

4.4. LCM and GCD

With the Fundamental Theorem of Arithmetic in hand, we are in a great position to generalize Theorem 4.8 by introducing two important concepts: the "greatest common divisor" and the "least common multiple" of a pair of integers.

DEFINITION 4.12. Let a and $b \in \mathbb{N}$. The *greatest common divisor* of a and b, written GCD(a, b), is the largest positive integer d such that $d|a$ and $d|b$. The *least common multiple* of a and b, written LCM(a, b), is the smallest positive integer m such that $a|m$ and $b|m$.

It is immediate from the definition of GCD that two natural numbers a and b are relatively prime if and only if GCD$(a, b) = 1$.

EXAMPLE 4.13. If $a = 18$ and $b = 30$, the set of all positive common divisors of a and b is $\{1, 2, 3, 6\}$, so GCD$(18, 30) = 6$. The set of all positive common multiples of 18 and 30 is the infinite set $\{90, 180, 270, \ldots\}$, so LCM$(18, 30) = 90$.

Exercise 4.12 Find two natural numbers a and b such that $\text{GCD}(a,b) = 15$ and $\text{LCM}(a,b) = 225$. Is the pair of numbers unique? Could you have $\text{GCD}(a,b) = 15$ and $\text{LCM}(a,b) = 220$ instead?

It is important to make sure that concepts like GCD and LCM are "well defined." Could there be some positive integers a and b for which $\text{GCD}(a,b)$ and $\text{LCM}(a,b)$ don't exist or don't give unique answers? We'll leave the question for LCM to Exercise 4.13 and answer the question for GCD here. Since 1 is a common divisor of both a and b, and since no positive common divisor can be greater than $\min(a,b)$, the minimum of a and b, the set of positive common divisors is a non-empty and finite subset of $\mathbb{N}$. Since a finite, non-empty subset of $\mathbb{N}$ must have a single largest element, the GCD of a and b exists and is unique.

Exercise 4.13 Show that the concept of LCM is well defined.

A simple variant of the standard form of prime factorizations introduced in the proof of the Fundamental Theorem of Arithmetic gives us a quick way to compute GCDs and LCMs. Suppose that a and b are positive integers greater than 1, so that there is a unique prime factorization of each. If we let $p_1, p_2, \ldots, p_n$ list all of the primes that are divisors of a or b, we can write

$$a = p_1^{a_1} p_2^{a_2} \cdots p_n^{a_n} \quad \text{and} \quad b = p_1^{b_1} p_2^{b_2} \cdots p_n^{b_n},$$

where all of the exponents are non-negative but some may be equal to 0. Then

$$\text{GCD}(a,b) = p_1^{\min(a_1,b_1)} p_2^{\min(a_2,b_2)} \cdots p_n^{\min(a_n,b_n)}$$

and

$$\text{LCM}(a,b) = p_1^{\max(a_1,b_1)} p_2^{\max(a_2,b_2)} \cdots p_n^{\max(a_n,b_n)}.$$

For example, the primes that divide 18 and 30 are 2, 3, and 5, so we write $18 = 2^1 3^2 5^0$ and $30 = 2^1 3^1 5^1$ and compute

$$\text{GCD}(18,30) = 2^1 3^1 5^0 = 6 \quad \text{and} \quad \text{LCM}(18,30) = 2^1 3^2 5^1 = 90,$$

matching our earlier answers.

Exercise 4.14 Prove that this method really does compute GCDs and LCMs.

Exercise 4.15 Let $a, b \in \mathbb{N}$. Prove that $ab = \text{GCD}(a,b) \cdot \text{LCM}(a,b)$.

We end this section by examining an important relationship between integer combinations and GCDs. You were asked in an earlier section to show that it was possible to write 1 as an integer combination of 8 and 3, but you could not do the same for 8 and 6. The problem with 8 and 6 is that 2 is a common divisor, and so 2 is a common divisor of $8m$ and $6n$ for any $m, n \in \mathbb{Z}$, and of $8m + 6n$ by Exercise 1.17. The following theorem is the appropriate generalization of Theorem 4.8.

THEOREM 4.14. *Let a and b be positive integers. The smallest positive integer combination of a and b is $\text{GCD}(a,b)$, and every integer combination of a and b is a multiple of $\text{GCD}(a,b)$.*

PROOF. Let $d = \text{GCD}(a, b)$. The fact that every integer combination of a and b is a multiple of d follows from an argument just like that given above for 8 and 6: since $d|a$ and $d|b$, for any integers m and n we find that $d|am$ and $d|bn$, so by Exercise 1.17 we know that $d|(am + bn)$.

To prove that the smallest positive integer combination of a and b is d, we will first show that a/d and b/d are relatively prime integers. They are certainly integers, since $d|a$ and $d|b$. They are also relatively prime, as the following short proof by contradiction shows. If a/d and b/d are not relatively prime, an integer c greater than 1 would divide both a/d and b/d, which would mean that both a/cd and b/cd are integers. But this would mean that cd is a common divisor of a and b that is larger than d, a contradiction.

Since a/d and b/d are relatively prime positive integers, by Theorem 4.8 there are integers m and n such that $m(a/d) + n(b/d) = 1$. Multiplying through by d gives us $ma + nb = d$, as desired. □

Exercise 4.16 Find an integer combination of 21 and 33 that equals 3.

In the spirit of the mathematical process described in Section 2.8, we encourage you to end your study of this section with generalizations of GCD and LCM to more than two natural numbers at a time. If $\{n_1, n_2, \ldots, n_k\} \subset \mathbb{N}$ is a finite subset of natural numbers, define $\text{GCD}(n_1, n_2, \ldots, n_k)$ to be the largest $d \in \mathbb{N}$ such that $d|n_i$ for all i, and define $\text{LCM}(n_1, n_2, \ldots, n_k)$ to be the smallest $m \in \mathbb{N}$ such that $n_i|m$ for all i. The following exercise asks you to explore several of the concepts introduced earlier in this section, to determine whether they might apply in this more general setting.

Exercise 4.17 There are a number of things to prove or refute in considering the extension of the concept of GCD and LCM to more than two natural numbers.

(a) Explain why the more general definitions of GCD and LCM given above are well-defined.
(b) Show that the most direct generalization of the "min" and "max" functions gives you a quick way to compute $\text{GCD}(n_1, n_2, \ldots, n_k)$ and $\text{LCM}(n_1, n_2, \ldots, n_k)$.
(c) Compute $\text{GCD}(20, 90, 150)$ and $\text{LCM}(6, 45, 200)$.
(d) Does

$$abc = \text{GCD}(a, b, c) \cdot \text{LCM}(a, b, c)$$

hold for all natural numbers a, b, and c? If not, can you characterize the triples for which it does hold?

Extending Theorem 4.14 to more than two numbers is the focus of Exercises 4.53 and 4.54.

4.5. Numbers and Closure

The set of natural numbers $\mathbb{N}$ has the very nice property that it is "closed under addition." This means that whenever $m, n \in \mathbb{N}$, then $m + n \in \mathbb{N}$ as well. For instance, 3 is a natural number and so is 17, as is 20, their sum. In general, we say that a set S of numbers is *closed under addition* if $m, n \in S$ implies that $m + n \in S$ as well. Similarly, a set S of numbers is *closed under multiplication* if $m, n \in S$ implies that $m \cdot n \in S$ as well. The sets $\mathbb{N}$, $\mathbb{Z}$, $\mathbb{Q}$, and $\mathbb{R}$ are all closed under addition and multiplication.

It is possible for finite sets of numbers to be closed under multiplication. For example, the set $\{-1, 0, 1\}$ is closed under multiplication. It is not, however, closed under addition, as $1 + 1 = 2$ and $2 \notin \{-1, 0, 1\}$.

Exercise 4.18 Is any non-empty finite set closed under addition?

Let $n \in \mathbb{N}$ be at least 2, and define $S_n \subset \mathbb{Z}$ to be the set of all integers that leave a remainder of 1 when divided by n:

$$S_n = \{nk + 1 \mid k \in \mathbb{Z}\}.$$

The set S_n is not closed under addition: the numbers 1 and 1 are both in S_n, but $1+1 = 2$ is not in S_n. However, these sets are closed under multiplication.

PROPOSITION 4.15. *Each of the sets S_n is closed under multiplication.*

PROOF. Fix $n \in \mathbb{N}$, and let a and b be elements of S_n. Then $a = nk+1$ and $b = nm+1$, for some $k, m \in \mathbb{Z}$. A bit of arithmetic shows that

$$a \cdot b = (nk + 1)(nm + 1) = n(nkm + k + m) + 1.$$

Since $nkm+k+m \in \mathbb{Z}$, it follows that $a \cdot b = nz+1$ for some $z \in \mathbb{Z}$, and so $a \cdot b \in S_n$. □

A set of numbers S is *closed under subtraction* if $m, n \in S$ implies that $m - n \in S$ as well. The numbers 3 and 17 provide just one of many possible counterexamples to the claim that $\mathbb{N}$ is closed under subtraction: 3 and 17 are both natural numbers, but $3 - 17 = -14$ is not. The larger set of all integers, $\mathbb{Z}$, is closed under subtraction.

The notion of being closed under division is perhaps the most complicated, as 0 is an exceptional number. Thus a set S is said to be *closed under division* if given $a, b \in S$ with $b \neq 0$, the number a/b is also in S. The sets $\mathbb{Q}$ and $\mathbb{R}$ are closed under division, while $\mathbb{Z}$ is not.

Exercise 4.19 Is $\mathbb{Q}_n$ closed under addition? subtraction? multiplication? division?

We end this section with a discussion of a set of numbers based on the golden ratio $\phi = (1 + \sqrt{5})/2$. We first discussed ϕ in Section 1.2, where we noted that it is a root of the polynomial $x^2 - x - 1$. In Exercise 4.36 we ask you to prove (in three ways!) that ϕ is irrational, a fact we will assume in this discussion.

Let $\mathbb{Z}[\phi]$ be the set of all integer combinations of 1 and ϕ:

$$\mathbb{Z}[\phi] = \{a + b\phi \mid a, b \in \mathbb{Z}\}.$$

$\mathbb{Z}[\phi]$ is a set of numbers satisfying $\mathbb{Z} \subset \mathbb{Z}[\phi] \subset \mathbb{R}$. (The inclusion $\mathbb{Z} \subset \mathbb{Z}[\phi]$ is proper as $\phi \notin \mathbb{Z}$; the claim that the inclusion $\mathbb{Z}[\phi] \subset \mathbb{R}$ is proper is intuitive but is a bit more delicate to establish.) This set is called "$\mathbb{Z}$ adjoined by ϕ" or "$\mathbb{Z}$ adjoined by the golden ratio."

PROPOSITION 4.16. *The set of numbers $\mathbb{Z}[\phi]$ is closed under addition, subtraction, and multiplication.*

PROOF. Let $x = a + b\phi$ and $y = c + d\phi$ be numbers in $\mathbb{Z}[\phi]$. Then

$$x + y = (a + c) + (b + d)\phi\,.$$

Since $a + c$ and $b + d$ are in $\mathbb{Z}$, it follows by the definition that $x + y \in \mathbb{Z}[\phi]$.

The same argument, with some minus signs replacing plus signs, shows that $\mathbb{Z}[\phi]$ is closed under subtraction.

In proving that $\mathbb{Z}[\phi]$ is closed under multiplication we will appeal to the fact that ϕ is a root of $x^2 - x - 1$, which implies that $\phi^2 = \phi + 1$. Thus we have

$$\begin{aligned} x \cdot y &= (a + b\phi)(c + d\phi) \\ &= ac + (ad + bc)\phi + bd\phi^2 \\ &= ac + (ad + bc)\phi + bd(\phi + 1) \\ &= (ac + bd) + (ad + bc + bd)\phi\,. \end{aligned}$$

Since $\mathbb{Z}$ is closed under addition and multiplication, $ac+bd \in \mathbb{Z}$ and $ad+bc+bd \in \mathbb{Z}$, hence $x \cdot y \in \mathbb{Z}[\phi]$. □

The set $\mathbb{Z}[\phi]$ is not closed under division, though. For example, 1 and $2 \in \mathbb{Z}[\phi]$, but $1/2$ is not in $\mathbb{Z}[\phi]$. To prove this, start by assuming to the contrary that $1/2 \in \mathbb{Z}[\phi]$. Thus $1/2 = a + b\phi$ with $a, b \in \mathbb{Z}$. If $b = 0$ then $a = 1/2$, which contradicts $a \in \mathbb{Z}$. However, if $b \neq 0$ then $1/2 = a + b\phi$ implies

$$\phi = \frac{\frac{1}{2} - a}{b} \in \mathbb{Q},$$

which contradicts the fact that $\phi \notin \mathbb{Q}$.

On the other hand, $\phi^{-1} = 1/\phi$ is in $\mathbb{Z}[\phi]$. We can take the identity $\phi^2 = \phi + 1$, divide every term by ϕ, and then rearrange the equation to show

$$\phi^{-1} = -1 + \phi \in \mathbb{Z}[\phi]\,.$$

Since $\mathbb{Z}[\phi]$ is closed under multiplication we immediately get the following result:

PROPOSITION 4.17. *The set*

$$\left\{\ldots, \frac{1}{\phi^2}, \frac{1}{\phi}, 1, \phi, \phi^2, \ldots\right\} = \{\phi^n \mid n \in \mathbb{Z}\}$$

is contained in $\mathbb{Z}[\phi]$.

We hope that the example of $\mathbb{Z}[\phi]$ inspires an inclination to ask if similar results hold when you extend the integers using other irrational numbers. The exercise below gives you the chance to explore a similar situation, this time focused on $\sqrt{2}$. You might even ask about using $\sqrt{-1}$, which appears in Project 11.4 on the Gaussian integers.

Exercise 4.20 The set of numbers $\mathbb{Z}[\sqrt{2}]$ is a subset of $\mathbb{R}$ consisting of all integer combinations of 1 and $\sqrt{2}$. That is,

$$\mathbb{Z}[\sqrt{2}] = \{a + b\sqrt{2} \mid a, b \in \mathbb{Z}\}.$$

This set is called "$\mathbb{Z}$ adjoined by $\sqrt{2}$." Determine whether $\mathbb{Z}[\sqrt{2}]$ is closed under addition, subtraction, and multiplication.

The sets of numbers S_n, $\mathbb{Z}[\sqrt{2}]$, and $\mathbb{Z}[\phi]$ are all closed under multiplication, and so you can ask about decompositions of given elements as products of the others. Is there a notion of "prime numbers" for these systems? Unique factorization? The cases of $\mathbb{Z}[\sqrt{2}]$ and $\mathbb{Z}[\phi]$ are beyond the scope of this book. They are often discussed in courses in abstract algebra and number theory, where you will find that sometimes adjoining square roots to the integers results in a system (called a "ring") with unique factorization, and sometimes unique factorization fails. Perhaps the easiest example to present comes from $\mathbb{Z}[\sqrt{-5}]$, where

$$6 = 2 \cdot 3 = (1 + \sqrt{-5})(1 - \sqrt{-5})$$

and none of the four factors shown can be further decomposed. Unique factorization in S_2, S_3, and S_4 is explored in Exercises 4.48, 4.49, and 4.50.

4.6. End-of-Chapter Exercises

Exercises you can work on after Section 4.1

4.21 Let $S \subset \mathbb{Z}$ be a subset of the integers that has a minimal element. Prove that S has the well-ordering property, that is, every non-empty subset of S contains a minimal element.

4.22 Prove that postage of $n = 6$ cents or more can be achieved by using only 2-cent and 7-cent stamps.

4.23 In this exercise you will construct two proofs of the fact that $7^n + 5$ is divisible by 3 for all $n \in \mathbb{N}$.

(a) Prove this result using a standard induction argument.
(b) Prove this result using a minimal criminal argument. As a hint, if $7^m + 5$ is the minimal criminal then 3 divides $7^{m-1} + 5$, and it also divides $7^m - 7^{m-1}$.

Modular arithmetic (Chapter 6) will offer another method for proving this fact.

4.24 Prove that

$$1 \cdot 2 + 2 \cdot 3 + \cdots + n(n+1) = \frac{(n)(n+1)(n+2)}{3}$$

for all $n \in \mathbb{N}$ using a minimal criminal argument.

4.25 Prove that

$$\frac{3}{4} + \frac{3}{16} + \cdots + \frac{3}{4^n} = 1 - \frac{1}{4^n}$$

for all $n \in \mathbb{N}$ using a minimal criminal argument.

4.26 Return to the induction exercises at the end of Chapter 2, and prove some of them using a minimal criminal argument instead.

Exercises you can work on after Section 4.2

4.27 Use one of the results of Exercise 4.6 and mathematical induction to prove that any two consecutive Fibonacci numbers f_n and f_{n+1} are relatively prime.

4.28 What would it mean to say that three integers a, b, and c are relatively prime? Try to come up with a reasonable mathematical definition for this concept. Make sure that your definition allows an example where the three integers a, b, and c are relatively prime as a triple, yet each of $\{a, b\}$, $\{a, c\}$, and $\{b, c\}$ is not a relatively prime pair of integers.

4.29 Let a, b, and c be natural numbers. Prove that if a and b are relatively prime and if a and c are also relatively prime, then a and bc are relatively prime.

4.30 The set of integer combinations of $1/2$ and $1/3$ is

$$S = \left\{ \frac{1}{2}a + \frac{1}{3}b \mid a, b \in \mathbb{Z} \right\}.$$

(a) Prove that the following inclusions are proper: $\mathbb{Z} \subset S \subset \mathbb{Q}$.

(b) Conjecture a relationship between S and

$$T = \left\{ \frac{n}{6} \mid n \in \mathbb{Z} \right\}.$$

(c) Prove your conjecture.

4.31 Let a and b be relatively prime. Prove that for any $k \in \mathbb{N}$ there are integers m and n such that

$$ma + nb = k.$$

Exercises you can work on after Sections 4.3 and 4.4

4.32 What proportion of the integers in $\{1, 2, \ldots, 100\}$ can be expressed as $2^m 3^n$ for non-negative integers m and n?

4.33 A natural number $n \in \mathbb{N}$ is *awesome* if whenever $n|b^2$, then $n|b$. Or, said a bit more carefully and using notation from Chapter 2, n is awesome if $\forall b \in \mathbb{N}, n|b^2 \Rightarrow n|b$.

(a) Show that 1 and all the prime numbers are awesome.
(b) Show by an example that not all natural numbers are awesome.
(c) Show by an example that there are non-prime numbers that are awesome.
(d) Formulate a conjecture that identifies in a simple and precise way exactly which numbers are awesome.
(e) Prove your conjecture.

4.34 In this exercise we return to the sets $\mathbb{Q}_n$ defined at the beginning of Chapter 3 and in Exercise 3.2.

(a) Characterize when $\mathbb{Q}_a \subseteq \mathbb{Q}_b$.
(b) Characterize when $\mathbb{Q}_a = \mathbb{Q}_b$.
(c) Characterize when $\mathbb{Q}_a \subset \mathbb{Q}_b$.

If it wasn't understood already, you should prove your characterizations are correct!

4.35 Recall that $\mathbb{P}$ is the set of primes.

(a) Explain why

$$\bigcup_{p \in \mathbb{P}} p\mathbb{N} = \mathbb{N} - \{1\}.$$

(b) Explain why

$$\bigcap_{p \in \mathbb{P}} p\mathbb{N} = \emptyset.$$

(c) Explain why

$$p_1\mathbb{N} \cap p_2\mathbb{N} \cap \cdots \cap p_k\mathbb{N}$$

is an infinite set if $k \in \mathbb{N}$ and $p_1, p_2, \ldots, p_k$ are any k primes. Then contrast this statement with the one in part (b).

4.36 Here we outline three proofs that the golden ratio ϕ is irrational. It is your job to fill in the gaps in these arguments.

Proof 1: A variation on the $\sqrt{2}$-argument presented in Section 1.7

(a) Modify the argument that proves $\sqrt{2}$ is irrational to show that $\sqrt{5}$ is irrational.
(b) Assume that ϕ is a rational number, and show that this implies $\sqrt{5}$ is also rational.
(c) Conclude that the assumption that ϕ is rational must be false.

Proof 2: Using the definition of ϕ

(a) Assume that $\phi = m/n$, in lowest terms. Show that $m/n = n/(m-n)$ by recalling that ϕ is related to dividing segments into "extreme and mean ratio."
(b) Conclude that the assumption that m/n is in lowest terms is false, and therefore that no such expression exists for ϕ.

Proof 3: Using the quadratic equation it satisfies

(a) Assume that $\phi = m/n$, in lowest terms. Show that the equation $\phi^2 = \phi + 1$ implies $m^2 = mn + n^2$.
(b) Conclude that $n|m^2$ and that this contradicts the assumption that m/n is in lowest terms. (Be careful here: see Exercise 4.33.)

4.37 Use the Fundamental Theorem of Arithmetic to prove that $\sqrt{k}$ is irrational for each positive integer k that is not a perfect square.

4.38 For which pairs of natural numbers a and b is $\sqrt{a} + \sqrt{b}$ rational?

4.39 Generalize the statement and proof in Exercise 4.37 to $\sqrt[n]{k}$.

4.40 Find all pairs of natural numbers a and b with $\text{GCD}(a,b) = 14$ and $\text{LCM}(a,b) = 210$.

4.41 Let m and n be natural numbers where $m|n$. What is $\text{GCD}(m,n)$? What is $\text{LCM}(m,n)$?

4.42 Let m and n both divide k. Prove that $\text{LCM}(m,n)$ also divides k.

4.43 Let m and n be natural numbers with $\text{GCD}(m,n) = d$. Let e be a divisor of both m and n. Prove that $\text{GCD}(m/e, n/e) = d/e$.

4.44 Prove that if m and n are relatively prime natural numbers and their product mn is a perfect square, then m and n must both be perfect squares.

4.45 Write a rational number $q \in \mathbb{Q}$ in lowest terms and take the product of its numerator and denominator. For how many rational numbers strictly between 0 and 1 will 10! be the resulting product?

4.46 Is the following formula true?

$$\text{LCM}(a,b,c) = \text{LCM}\,(\text{LCM}(a,b), c)\,?$$

Either prove it or find a counterexample.

Exercises you can work on after Section 4.5

The Fundamental Theorem of Arithmetic is so well known that people might take it for granted. The following two exercises show that we shouldn't be so cavalier.

4.47 Consider the set $\mathbb{E}$ of even integers.

(a) Show that $\mathbb{E}$ is closed under addition and multiplication.

(b) Call a number $n \in \mathbb{E}$ an *even-prime* if $n \in \mathbb{N}$ and n cannot be expressed as a product of two elements in $\mathbb{E}$. Prove that an even number is an even-prime if and only if it can be written as $2(2m - 1)$ for some $m \in \mathbb{N}$.

(c) Prove that every positive $n \in \mathbb{E}$ is an even-prime or can be expressed as a product of even-primes.

(d) Show by an example that there are numbers $n \in \mathbb{E}$ that can be expressed in two different ways as a product of even-primes.

(e) Theorem 4.11 proves the uniqueness of prime decompositions for all $n \in \mathbb{N}$ with $n \geq 2$. By your work above you have shown that this proof cannot apply to $\mathbb{E}$. Can you find a claim in the proof of Theorem 4.11 that is false when applied to $\mathbb{E}$?

4.48 Let S_4 be the set of all integers giving a remainder of 1 when divided by 4. Thus

$$S_4 = \{\ldots, -3, 1, 5, 9, 13, 17, \ldots\} = \{4k + 1 \mid k \in \mathbb{Z}\}.$$

Proposition 4.15 shows that S_4 is closed under multiplication, and so it is reasonable to discuss the ways in which elements in S_4 can be written as products of elements in S_4. For example, working in S_4 we can write $45 = 9 \cdot 5$, but we cannot write $45 = 3 \cdot 15$ because 3 and 15 are not elements of S_4. Similarly, 9 has no non-trivial factorizations within S_4, as the prime decomposition of 9 is $3 \cdot 3$, and $3 \notin S_4$.

(a) Call a positive integer $p \in S_4$ a *4-prime* if its only positive factors in S_4 are 1 and p. Show that if p is an actual prime number in $\mathbb{N}$, and p is also in S_4, then p is a 4-prime as well.

(b) Show that 21 and $33 \in S_4$ are 4-primes.

(c) Show that 693 can be written as a product of 4-primes in two distinct ways.

If you found this exercise interesting, you might look at the article "Unique factorization in multiplicative systems" by James and Niven [**JN54**].

4.49 Let $\mathbb{O}$ be the set of odd natural numbers.

(a) Prove that $\mathbb{O}$ is closed under multiplication.

(b) Describe a notion of primes within $\mathbb{O}$, and prove that $\mathbb{O}$ has the same unique factorization property as $\mathbb{Z}$.

4.50 Recall that for a given integer $n \geq 2$, S_n consists of those integers that leave a remainder of 1 when divided by n. In Exercise 4.49 you proved that S_2 (the odd natural numbers) has the unique factorization property. In Exercise 4.48 you proved that S_4 does not. This leaves an obvious question: does S_3 have the unique factorization property? Develop a conjecture, and then prove it!

4.51 We could extend the definition presented in Exercise 4.20 by letting $\mathbb{Q}[\sqrt{2}]$ consist of all *rational* combinations of 1 and $\sqrt{2}$. That is,

$$\mathbb{Q}[\sqrt{2}] = \{a + b\sqrt{2} \mid a, b \in \mathbb{Q}\}.$$

Determine whether "$\mathbb{Q}$ adjoined by $\sqrt{2}$" is closed under addition, subtraction, multiplication, and division by non-zero elements.

4.52 Given your experience with $\mathbb{Z}[\phi]$ and $\mathbb{Z}[\sqrt{2}]$, you might be tempted to define "$\mathbb{Z}$ adjoin $\sqrt[3]{2}$" to be:

$$\{a + b\sqrt[3]{2} \mid a, b \in \mathbb{Z}\}.$$

(a) Show by an example that this set is not closed under multiplication.

(b) Prove that the set

$$\{a + b\sqrt[3]{2} + c\sqrt[3]{4} \mid a, b, c \in \mathbb{Z}\}$$

is closed under multiplication.

More Exercises!

4.53 Let a, b, and c be real numbers. Any sum of the form

$$al + bm + cn\,,$$

with l, m, and $n \in \mathbb{Z}$, is called an *integer combination* of a, b, and c. Is it true that $\mathrm{GCD}(a, b, c)$ is the smallest positive integer combination of the natural numbers a, b and c? Is every linear combination of a, b, and c a multiple of $\mathrm{GCD}(a, b, c)$?

4.54 Extend the ideas and questions in Exercise 4.53 from three terms to n terms.

4.55 Let $P_n(x)$ represent the set of polynomials in the variable x of degree less than or equal to n. For example, $x^2 + 3x - \sqrt{3}$ is in $P_3(x)$, but $x^4 - 1$ is not. Determine whether $P_n(x)$ is closed under addition, subtraction, and multiplication.

5 Functions

In this chapter we explore the concept of a mathematical function. Undoubtedly, functions have been prevalent in your previous mathematical experiences. You may have spent a lot of time working with functions given by an explicit formula, such as $f(x) = x^2 \sin(2x)$. In this chapter we develop a broader understanding of functions, presenting a commonly used and intuitive (but incomplete) definition in Section 5.1, followed by a variety of related terminology and concepts, and then a rigorous definition in Section 5.5.

5.1. What is a Function?

The notion of a function is broader than numerical functions that are given by explicit formulas.

DEFINITION 5.1 (Initial Version). Let A and B be sets. A *function from A to B* is a rule that assigns to each $a \in A$ a unique associated element $b \in B$.

There are two standard ways to represent "f is a function from A to B." One is to write

$$f : A \to B,$$

and the other is

$$A \xrightarrow{f} B\,.$$

Both notations are supposed to indicate that the function f takes an input element $a \in A$ and outputs an element $b \in B$, and we express this by writing $f(a) = b$. In the sections that follow, you will often read statements like "Let $f : A \to B$." This is a short but complete sentence stating that f is a function from A to B.

For example, consider the function g defined in a piecewise fashion for real numbers x:

$$g(x) = \begin{cases} -x^2 & \text{if } x \leq 0, \\ x^4 & \text{if } x > 0. \end{cases}$$

The formulas $-x^2$ and x^4 both give real numbers if x is a real number, so we could use the notation $g : \mathbb{R} \to \mathbb{R}$ or $\mathbb{R} \xrightarrow{g} \mathbb{R}$ because both A and B equal $\mathbb{R}$. As another example, you might have encountered parametric curves which trace out patterns in the plane. One of these gives a circle parametrized by

$$\gamma(t) = (\sin(t), \cos(t))\,,$$

as illustrated in Figure 1. Here the input is a real number $t \in \mathbb{R}$ and the output is a point $(x, y) \in \mathbb{R}^2$, so we write $\gamma : \mathbb{R} \to \mathbb{R}^2$ or $\mathbb{R} \xrightarrow{\gamma} \mathbb{R}^2$.

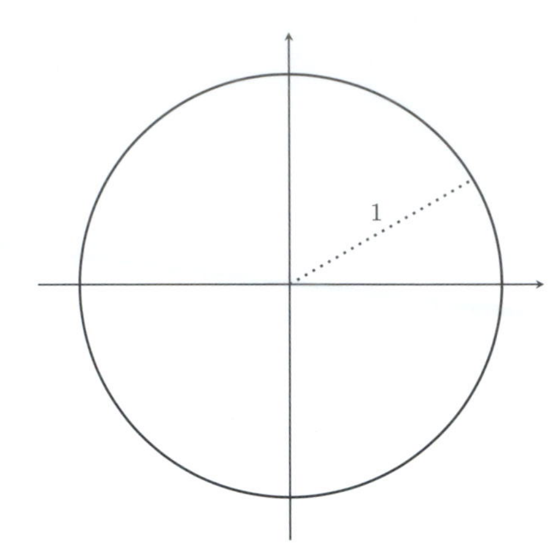

Figure 1. The image of the function $\gamma(t) = (\sin(t), \cos(t))$ is a circle.

Exercise 5.1 Define two functions from $\mathbb{R}$ to $\mathbb{Z}$.

The examples in this section emphasize that functions are not restricted to inputs and outputs involving real numbers. In particular, many functions do not have natural graphical representations. As a first example, we can construct a function f from $\mathbb{Z}$ to the set Parity = {even, odd}. We use the rule $f(n) =$ even when n is an even integer and $f(n) =$ odd when n is odd, so

$$f(0) = f(4) = \text{even}, \quad \text{while} \quad f(-3) = \text{odd}.$$

Formal languages are collections of strings of characters from a given alphabet. A particular example is the set of binary strings, consisting of all finite strings of 0s and 1s, including the empty string, which we denote ε. We call the set of all finite binary strings Σ^*:

$$\Sigma^* = \{\varepsilon, 0, 1, 00, 01, 10, 11, 000, 001, 010, 011, 100, 101, \ldots\}.$$

The notion of *length* provides a function from Σ^* to $\mathbb{Z}$. Using a slightly different style of function notation than we have seen so far, we denote the length of a string by $\|\cdot\|$, so that we have $\|010\| = 3$, $\|1101\| = 4$, and $\|\varepsilon\| = 0$. This function counts the total number of 0s and 1s in the given string. There is also a function $\|\cdot\|_0$ that counts only the 0s, where for example

$$\|1101\|_0 = 1.$$

Similarly there is a function $\|\cdot\|_1$ that counts only the 1s, so that

$$\|1101\|_1 = 3.$$

Exercise 5.2 Prove the following two elementary facts about $\|\cdot\| : \Sigma^* \to \mathbb{Z}$.

(a) Let ω be formed by concatenating the two strings ω' and ω'', that is $\omega = \omega'\omega''$. Prove that $\|\omega\| = \|\omega'\| + \|\omega''\|$.
(b) Prove that $\|\omega\| = \|\omega\|_0 + \|\omega\|_1$ for any string $\omega \in \Sigma^*$.

We can also construct a function

$$\Sigma^* \setminus \{\varepsilon\} \xrightarrow{\phi} \{0, 1\}$$

by defining $\phi(\omega)$ to be the first character in the string ω, reading left-to-right; for instance, $\phi(11010) = 1$. Notice that we need to remove ε from consideration because ε has no first character.

Our next examples use the set of polynomials with integer coefficients. We denote this set by $\mathbb{Z}[x]$, notation that is purposely reminiscent of examples from Section 4.5. The set $\mathbb{Z}[x]$ contains $p(x) = 7 - 6x + 3x^4$, and it also contains linear functions like $p(x) = 2 - x$ and constant functions like $p(x) = 11$.

We can define $f : \mathbb{Z}[x] \to \mathbb{Z}$ by

$$f(p(x)) = p(1).$$

For example, $f(7 - 6x + 3x^4) = 4$. We can also define a function $g : \mathbb{Z}[x] \to \mathbb{Z}[x]$ by

$$g(p(x)) = xp(x).$$

Thus $g(7 - 6x + 3x^4) = 7x - 6x^2 + 3x^5$.

As a final introductory example, we can define a function κ from $\mathbb{Z}[x]$ to Σ^*, where a polynomial $p(x)$ is mapped to the shortest possible string that has a 1 in each position i where the coefficient of x^{i-1} is non-zero; other positions are filled in with 0s. For instance,

$$\kappa(7 - x + x^3) = 1101$$

has 1s in positions 1, 2, and 4, and

$$\kappa(-5x + 17x^4 + 6x^6) = 0100101.$$

To make the definition of our function κ precise, we set $\kappa(p(x)) = \varepsilon$ when $p(x) = 0$, the zero polynomial; for all other $p(x) \in \mathbb{Z}[x]$, the right-most position of $\kappa(p(x))$ is a 1.

REMARK 5.2. Because functions are ubiquitous, there are a number of different ways to speak of them. For example, in many situations the word "map" is synonymous with "function." If $f(x) = y$, which is sometimes written $x \mapsto y$, you will hear both "f maps x to y" and "f sends x to y."

Exercise 5.3 It is time for you to construct some functions involving the set of polynomials $\mathbb{Z}[x]$.

(a) Construct a function $f : \mathbb{N} \to \mathbb{Z}[x]$.
(b) Construct a more interesting function[1] from $\mathbb{N}$ to $\mathbb{Z}[x]$ than you did for part (a).
(c) Construct an interesting function $g : \mathbb{Z} \to \mathbb{Z}[x]$.
(d) Construct an interesting function $g : \mathbb{Z} \times \text{Parity} \to \mathbb{Z}[x]$.
(e) Construct an interesting function $\mathbb{Z}[x] \xrightarrow{h} \mathbb{N} \times \mathbb{Z}$.
(f) Construct an interesting function $\Sigma^* \xrightarrow{h} \mathbb{Z}[x]$.

1 As an example, you might have defined your function from $\mathbb{N}$ to $\mathbb{Z}[x]$ by $f(n) = n$. That is valid, but not interesting. The function $f(n) = 1 + 2x + \cdots + nx^{n-1}$ strikes us as being more interesting. We hope your answer to (b) is even more interesting than this. And please don't ask us to define "interesting"!

5.2. Domain, Codomain, and Range

There are a number of words that are used quite frequently when discussing functions, and you need to absorb them.

DEFINITION 5.3. If f is a function from X to Y, then the *domain* of f is X; the *codomain* is Y. The *image* of an element $x \in X$ is the unique element $y \in Y$ where $f(x) = y$. The *range* of f is the subset of Y consisting of all images:

$$\text{Range}(f) = \{f(x) \mid x \in X\}.$$

Sometimes the range of f is called the *image of the function* f.

EXAMPLE 5.4. Let $\gamma : \mathbb{R} \to \mathbb{R}^2$ be the parametric curve defined in the previous section. The domain of γ is $\mathbb{R}$, the codomain is $\mathbb{R}^2$, and as is illustrated in Figure 1, the range of γ is the unit circle:

$$\text{Range}(\gamma) = \{(x, y) \in \mathbb{R}^2 \mid x^2 + y^2 = 1\}.$$

This highlights that the range can be significantly smaller than the specified codomain.

Exercise 5.4 Determine the ranges of the functions $\|\cdot\|$, ϕ, and κ defined in Section 5.1.

EXAMPLE 5.5. Here are some additional examples of functions.

(a) For any set A, the identity function $f : A \to A$ has $f(a) = a$ for every $a \in A$. It is denoted 1_A, which we often use, and sometimes Id_A. The domain, codomain, and range all equal A.

(b) The floor function $\lfloor x \rfloor$ sends $x \in \mathbb{R}$ to the greatest integer less than or equal to x, and the ceiling $\lceil x \rceil$ sends x to the least integer greater than or equal to x. For example, $\lfloor \pi \rfloor = 3$ and $\lceil \pi \rceil = 4$. These are both functions from $\mathbb{R}$ to $\mathbb{Z}$. In both cases the range is all of the codomain $\mathbb{Z}$.

(c) An infinite sequence of real numbers $(a_1, a_2, a_3, \ldots)$ is just another way to describe a function $f : \mathbb{N} \to \mathbb{R}$, where the correspondence is $a_n = f(n)$. In a later chapter we will ask whether such a function can ever have its range equal to the codomain $\mathbb{R}$.

(d) You can map a set to its power set by $f : A \to \mathcal{P}(A)$ with $f(a) = \{a\}$ for every $a \in A$. You can also use complements to construct another map from A to $\mathcal{P}(A)$: let $g : A \to \mathcal{P}(A)$ with $g(a) = A - \{a\}$. As an example, consider the set

$$A = \{\text{Ole, Lena, Sven}\}.$$

Then $f(\text{Lena}) = \{\text{Lena}\}$ and $g(\text{Lena}) = \{\text{Ole, Sven}\}$. The range of f is

$$\{\{\text{Ole}\}, \{\text{Lena}\}, \{\text{Sven}\}\},$$

while the range of g is

$$\{\{\text{Ole, Lena}\}, \{\text{Lena, Sven}\}, \{\text{Sven, Ole}\}\}.$$

(e) The function $f : A \times B \to A$ with $f((a, b)) = a$ is often called a *projection*, in this case a projection onto the first coordinate. The function $g(a, b) = b$ is the projection onto the second coordinate;[2] its range is B.
(f) The function $f : A \to A \times A$ with $f(a) = (a, a)$ is often called a *diagonal*. To see why this name is appropriate, set $A = [0, 1]$ and sketch the points in the subset $\{(a, a) \mid a \in [0, 1]\}$ of the plane $\mathbb{R}^2$.

Exercise 5.5 Determine which of the following are functions with the given domains and codomains.

(a) $f : \mathbb{Z} \to \mathbb{N}$ with $f(x) = |x|$.
(b) $f : \mathbb{Z} \to \mathbb{N}$ with $f(x) = |x| + 1$.
(c) $f : \mathbb{R} \to \mathbb{R}$ with $f(x) = \sqrt{x^2 - x + 1}$.
(d) $f : \mathbb{Z} \to \mathbb{R}$ with $f(x) = \sqrt{x^2 - x + 1}$.
(e) $f : \mathbb{Z} \to \mathbb{Z}$ with $f(x) = x^2 - x + 1$.
(f) $f : \mathbb{Z} \to \mathbb{N}$ with $f(x) = x^2 - x + 1$.
(g) $f : \mathbb{R} \to \mathbb{R}^+$ with $f(x) = x^2 - x + 1$, where $\mathbb{R}^+$ is the set of positive real numbers.

5.3. Injective, Surjective, and Bijective

The core vocabulary of functions includes the adjectives injective, surjective, and bijective.

DEFINITION 5.6. A function $f : X \to Y$ is *injective* if for all $x_1, x_2 \in X, f(x_1) = f(x_2)$ implies $x_1 = x_2$.

Put another way, f is injective if no two distinct xs have the same image. An injective function is also called *one-to-one*. The claim that a function is injective is sometimes encoded by the symbol $\hookrightarrow$, as in $f : X \hookrightarrow Y$ and $X \overset{f}{\hookrightarrow} Y$.

EXAMPLE 5.7. The function $f : \mathbb{Z} \to \mathbb{Z}$ given by $f(x) = 3x$ is injective for the following reason: for any $x_1, x_2 \in \mathbb{Z}$, having $f(x_1) = f(x_2)$ means that $3x_1 = 3x_2$, which gives $x_1 = x_2$ upon dividing both sides by 3.

EXAMPLE 5.8. Let $p : \Sigma^* \to \Sigma^*$ be the function that appends a 1 to the end of any given string. As examples, $p(1101) = 11011$ and $p(\varepsilon) = 1$. To prove that this function is injective, let ω_1 and ω_2 be two strings in Σ^*. If $p(\omega_1) = p(\omega_2)$, then the string $\omega_1 1$ and the string $\omega_2 1$ are identical. This means that ω_1 and ω_2 are identical strings as well.

Exercise 5.6 Determine whether the following functions are injective.

(a) $f : \mathbb{R}^+ \to \mathbb{R}^+$ given by $f(x) = \sqrt{x}$, where $\mathbb{R}^+$ is the set of positive real numbers.
(b) $g : \mathbb{R} \setminus \{0\} \to \mathbb{R} \setminus \{0\}$ given by $g(x) = 1/x$.
(c) $h : \mathbb{R} \setminus \{0\} \to \mathbb{R} \setminus \{0\}$ given by $h(x) = 1/x^2$.
(d) $m : \Sigma^* \to \mathbb{Z}[x]$ given by $m(\omega) = \|\omega\|_0 + x^{\|\omega\|}$. As an example, $m(10011) = 2 + x^5$.

2 The notation $g(a, b)$ is technically incorrect; the input element is the ordered pair (a, b), so the output should be denoted $g((a, b))$, as we did in the previous sentence with f. But, in fact, notation like $g(a, b)$ is quite standard, to reduce the number of parentheses in a context where the meaning is clear.

DEFINITION 5.9. A function $f : X \to Y$ is *surjective* if for each $y \in Y$ there exists an $x \in X$ such that $f(x) = y$.

Put another way, f is surjective if each element in Y is hit by some element in X. A surjective function is also called *onto*. The claim that a function is surjective is sometimes encoded by the symbol $\twoheadrightarrow$, as in $f : X \twoheadrightarrow Y$ and $X \overset{f}{\twoheadrightarrow} Y$.

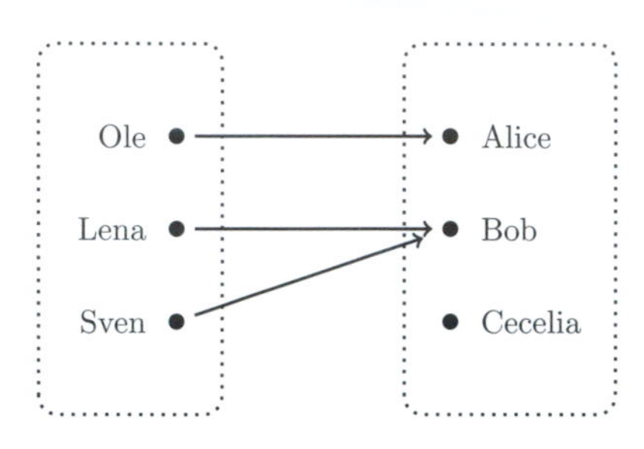

Figure 2. A function between two finite sets.

In Figure 2 we illustrate a function going from the set {Ole, Lena, Sven} to the set {Alice, Bob, Cecelia}, where each person in the first set is sent to his or her favorite partner when playing games. This function is not injective, as Lena and Sven both like playing with Bob. It is also not surjective: Cecelia is lonely, not being the preferred partner of anyone in the first set.

EXAMPLE 5.10. The following three examples provide some insight into how you prove that a given function is or is not surjective.

(a) The function $f : \mathbb{Z} \to \mathbb{Z}$ given by $f(x) = 3x$ is not surjective for the following reason: 1 is an element of (the codomain) $\mathbb{Z}$, but the equation $f(x) = 3x = 1$ is not satisfied by any element of (the domain) $\mathbb{Z}$.
(b) The function $p : \Sigma^* \to \Sigma^*$ given by appending a 1 to the end of any string is also not surjective. To prove this you could first note that the process of appending a 1 increases the length of any string ω by 1. Since ε has length 0, $p(\omega) = \varepsilon$ is not satisfied by any $\omega \in \Sigma^*$.
(c) Let $g : \Sigma^* \to \mathbb{Z}$ be defined by $g(\omega) = \|\omega\|_0 - \|\omega\|_1$. This function is surjective; we can also say that it is "onto $\mathbb{Z}$." To prove this, we need to show that the range of g is all of $\mathbb{Z}$, which we do by considering three cases:
    (i) Because $g(\varepsilon) = 0$, we know 0 is in the range of g.
    (ii) If $n > 0$, define ω_n to be the string consisting of n 0s. Then $g(\omega_n) = n$, and thus all positive integers are in the range of g.
    (iii) If $n < 0$, define $\overline{\omega}_n$ to be the string consisting of $|n|$ 1s. Then $g(\overline{\omega}_n) = n$, and thus all negative integers are in the range of g.

Exercise 5.7 Prove the following claims:

(a) The function in Example 5.5 (a) is both surjective and injective for any set A.
(b) The functions in Example 5.5 (b) are surjective but not injective.
(c) The function in Example 5.5 (f) is injective, and it is surjective if and only if A contains at most one element.

DEFINITION 5.11. A function is *bijective* if it is both injective and surjective.

There is a familiar graphical interpretation of injective, surjective, and bijective when $f : \mathbb{R} \to \mathbb{R}$:

(a) f is injective if each horizontal line intersects the graph of f in at most one point.
(b) f is surjective if each horizontal line intersects the graph of f in at least one point.
(c) f is bijective if each horizontal line intersects the graph of f in one and only one point.

You need to be a bit careful with this graphical interpretation: it applies when the codomain is $\mathbb{R}$, but not necessarily in other circumstances. For example, in Section 5.1 we defined the function γ that traces out a circle in the plane. You might be tempted to say that horizontal lines can cross a circle twice, and thus γ is not injective. The x-axis, for example, intersects the circle at $(-1, 0)$ and $(1, 0)$. But this confuses the path traced out by γ with the graph of a function. To prove that γ is not injective, you can point out that $\gamma(0) = \gamma(2\pi) = (0, 1)$, and thus two real numbers are mapped to the same point in $\mathbb{R}^2$. If you restrict the domain of γ to the half-open interval $[0, 2\pi)$, then γ is indeed an injective function, even though the image of γ is still a circle.

Bijective functions, called *bijections*, are very important because they provide matchings between the elements in two sets A and B. They are often referred to as *one-to-one correspondences*, emphasizing this fact.

Exercise 5.8 Find bijections $f : A \to B$ for the given open intervals of real numbers.

(a) $A = (0, 1)$ and $B = (0, 2)$
(b) $A = (0, 1)$ and $B = (-3, 5)$
(c) $A = (0, 1)$ and $B = (2, \infty)$

Exercise 5.9 Let $f : A \to B$ be an injection. Prove that the function $\hat{f} : A \to \text{Range}(f)$, defined by $\hat{f}(a) = f(a)$ for all $a \in A$, is a bijection.

When X and Y are finite sets, we can count the number of functions $f : X \to Y$ of different types. For example, if $X = \{x_1, x_2\}$ and $Y = \{y_1, y_2, y_3\}$, there are nine different functions, one for each column in Table 1. As we have indicated, six of these functions are injective; none are surjective. Exercises 5.35, 5.36, and the following exercise help you further explore the number of injective, surjective, bijective, and otherwise unrestricted functions between two finite sets.

Table 1 *Six of the nine possible functions from* $\{x_1, x_2\}$ *to* $\{y_1, y_2, y_3\}$ *are injective.*

	f_1	f_2	f_3	f_4	f_5	f_6	f_7	f_8	f_9
$x_1 \mapsto$	y_1	y_1	y_1	y_2	y_2	y_2	y_3	y_3	y_3
$x_2 \mapsto$	y_1	y_2	y_3	y_1	y_2	y_3	y_1	y_2	y_3
Injection?	×	✓	✓	✓	×	✓	✓	✓	×

Exercise 5.10 Consider all of the different functions from a set A with three elements to a set B with five elements. Try to answer the following questions without creating the full table of functions as we did in the smaller case above.

(a) How many functions are there?
(b) How many of these functions are injections?
(c) How many of these functions are bijections?
(d) How do your answers change if $|A| = 5$ and $|B| = 3$?

5.4. Composition

The function $\cos(3x^2+2)$ is a composition of the "outer" cosine function and the "inner" polynomial $3x^2 + 2$. This concept of function composition is not, however, restricted to functions that can be presented by formulas.

DEFINITION 5.12. If $f : A \to B$ and $g : B \to C$ are functions, then their *composition* is written

$$g \circ f : A \to C$$

and is defined by sending each $a \in A$ to the element $g(f(a))$. That is, $(g \circ f)(a) = c$ if $f(a) = b$ and $g(b) = c$.

The notation $A \xrightarrow{f} B$ is particularly helpful for composition, as you can visualize $g \circ f$ by

$$A \xrightarrow{f} B \xrightarrow{g} C\,.$$

Similarly a diagram like the one shown in Figure 3 can be used to visualize function composition.[3] This figure depicts two functions $f : A \to B$ and $g : B \to C$, where $A = \{1, 2, 3\}$, $B = \{a, b, c, d\}$, and $C = \{x, y, z\}$. Following the arrows shows, for example, that $(g \circ f)(2) = g(b) = z$.

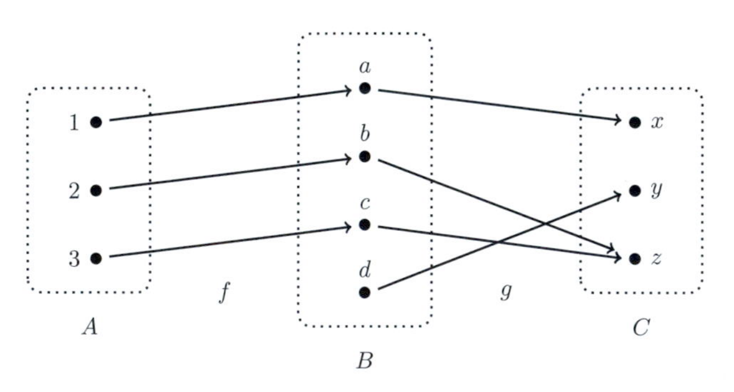

Figure 3. Two functions $f : A \to B$ and $g : B \to C$.

3 Yes, the order of f and g in the name $g \circ f$ is opposite to the order presented in $A \xrightarrow{f} B \xrightarrow{g} C$ and in Figure 3. This is due to standard function notation like $f(x)$ having the function name on the left, while mathematical text and visuals such as ours are naturally read left-to-right.

EXAMPLE 5.13. Let $f : \mathbb{R} \to \mathbb{R}$ be given by $f(x) = e^x$ and $g : \mathbb{R} \to \mathbb{R}$ by $g(x) = x - 1$. In this case, since $A = B = C = \mathbb{R}$, both $f \circ g$ and $g \circ f$ are legitimate functions from $\mathbb{R}$ to $\mathbb{R}$. However, $(g \circ f)(x) = e^x - 1$, while $(f \circ g)(x) = e^{x-1}$.

EXAMPLE 5.14. Let f and g be the functions shown in Figure 3. The function f is injective but not surjective because $d \in B$ is not in the range of f. The function g is surjective but not injective because $g(b) = g(c)$ but $b \neq c$. The composition $g \circ f$ is neither injective nor surjective.

Exercise 5.11 Referring to the previous example, what is $f \circ g$, if anything?

We end this section with a handful of important, foundational facts about the adjectives injective, surjective, and bijective in the context of composition of functions. But first we offer a couple of cautionary examples. You might expect that if $g \circ f$ is injective, then g and f must be injective. However, consider the example where $f(x) = e^x$ and $g(x) = x^2$, both having domain and range equal to $\mathbb{R}$. Then the composition $(g \circ f)(x) = (e^x)^2$ is injective, but the same is not true for g, as you should now verify.

Exercise 5.12

(a) Verify that if $f(x) = e^x$ and $g(x) = x^2$, then g is not injective, but the composition $g \circ f$ is injective.
(b) Find functions f and g where g is surjective and $g \circ f$ is surjective, but f is not surjective.

These examples should help you appreciate the following proposition.

PROPOSITION 5.15. *Let $f : A \to B$ and $g : B \to C$.*

(a) *If $g \circ f$ is injective, then f is injective.*
(b) *If $g \circ f$ is surjective, then g is surjective.*

PROOF. To prove part (a), assume to the contrary that f is not injective. Then there must be distinct elements a and a' in A such that $f(a) = f(a')$. But then it must also be the case that $g(f(a)) = g(f(a'))$. This contradicts the hypothesis that $g \circ f$ is injective.

To prove part (b), note that because $g \circ f$ is surjective it must be the case that given any $c \in C$ there is an $a \in A$ such that $(g \circ f)(a) = c$. But if we let $b = f(a)$, we know that $g(b) = c$. And so given any $c \in C$ there is a $b \in B$ such that $g(b) = c$. □

Thankfully, the properties that are held by both f and g also hold for their composition.

PROPOSITION 5.16. *Let $f : A \to B$ and $g : B \to C$.*

(a) *If f and g are both surjective, then so is $g \circ f$.*
(b) *If f and g are both injective, then so is $g \circ f$.*
(c) *If f and g are both bijective, then so is $g \circ f$.*

Exercise 5.13 Prove Proposition 5.16. As a hint for (a), you might start with the statement "Let c be any element of C. Since g is surjective, ..." As a hint for (b), you might start with the statement "Suppose that a_1 and a_2 are any two elements of A such that

$(g \circ f)(a_1) = (g \circ f)(a_2)$, which we can also write as $g(f(a_1)) = g(f(a_2))$. Since g is injective, ..." Finally, as a hint for (c), make sure you've done parts (a) and (b)!

REMARK 5.17. Our definition of function composition requires that the codomain of f is the same as the domain of g. This is more restrictive than is necessary. If $f : A \to B$ and $g : D \to C$, where $B \subseteq D$, then you can still define the composition $g \circ f : A \to C$ by $a \mapsto c = g(f(a))$. See Exercises 5.45 and 5.69 at the end of this chapter if you would like to explore this modest extension of the definition.

5.5. What is a Function? Redux!

While Definition 5.1 likely aligns well with your prior knowledge of functions and has allowed us to first explore some of their fundamental concepts, mathematicians are often uncomfortable defining functions as "rules" taking elements from one set to another. One reason for this discomfort is because the notion of a rule is somewhat vague. A related reason is that "rule" is often misinterpreted as meaning "formula," which is too restrictive. However, if that interpretation is too restrictive, just what is it we mean when we use "rule" in the definition?

An alternate and ultimately preferable approach is to define functions using the set theory we developed in Chapter 3. The motivation comes from our experience with polynomials, trigonometric functions, and other functions that are quite closely identified with their graphs. In the case of a function $f : \mathbb{R} \to \mathbb{R}$, the graph of f is the set of points $\{(x, f(x)) \mid x \in \mathbb{R}\}$, which is a subset of $\mathbb{R} \times \mathbb{R}$. We are used to this idea in the context of real numbers, but we can define the "graph" of a function $f : A \to B$ regardless of whether it is feasible to visualize it in the plane $\mathbb{R}^2$.

DEFINITION 5.18. Let f be a function from a set A to a set B, both non-empty. Then the *graph* of f is the subset of $A \times B$ described by

$$\text{graph of } f = \{(a, f(a)) \mid a \in A\}\,.$$

If f is a function from A to B, then the following property holds for the graph of f: given any $a \in A$, there is a unique element (a, b) in the graph of f. This statement is just a direct generalization of the standard vertical line test for the case when A and B are subsets of $\mathbb{R}$. But we can reverse our way of thinking to see that any subset $S \subseteq A \times B$ that satisfies this property can be used to *define* a function from A to B: you simply say that the function associated to S maps $a \in A$ to $b \in B$ when $(a, b) \in S$. Put another way, the subset of $A \times B$ becomes the "rule."

DEFINITION 5.19 (Unambiguous Version). Let A and B be sets. A *function from A to B* is a subset $f \subseteq A \times B$ such that for each element $a \in A$ there is a unique element $(a, b) \in f$.

Notice that we are defining a function as a *set*, even though we are giving the set a lower-case name, "f."

EXAMPLE 5.20. Let $A = \{1, 2\}$ and $B = \{1, 2, 3, 4, 5\}$. Then $f = \{(1, 1), (2, 4)\}$ is a function from A to B, as is $\{(1, 3), (2, 3)\}$. On the other hand, none of the following subsets of $A \times B$ are functions.

(a) $f = \{(1,1),(2,4),(2,5)\}$
(b) $f = \{(2,4)\}$
(c) $f = \{(1,1),(2,6)\}$

The first does not have a unique ordered pair with leading entry equal to 2; the second has no ordered pair with leading entry 1; and the third is not contained in $A \times B$.

Exercise 5.14 Suppose that $A = B = \{1,2,3,4\}$ and the function f contains $(1,1)$, $(2,3)$, and $(4,1)$. Must f contain any other elements?

We emphasize that the functions we have seen in the first sections of this chapter are indeed functions according to Definition 5.19: the "rule" mapping a to b is simply the condition that (a,b) is in the set f. For example, the functions in Figure 3 are

$$f = \{(1,a),(2,b),(3,c)\} \quad \text{and} \quad g = \{(a,x),(b,z),(c,z),(d,y)\}.$$

And the function κ from Section 5.1 is an infinite subset of $\mathbb{Z}[x] \times \Sigma^*$, two of its elements being $(1-3x^3+x^4, 10011)$ and $(x-x^2, 011)$.

Exercise 5.15 If you define the function $\|\cdot\| : \Sigma^* \to \mathbb{Z}$ from Section 5.1 as a subset $\|\cdot\| \subseteq \Sigma^* \times \mathbb{Z}$, what elements of $\|\cdot\|$ have the form $(\omega, 3)$ for some $\omega \in \Sigma^*$?

We now revisit some of the terminology from the previous sections. If f is a function from A to B, then the *domain* of f is A; the *codomain* is B. The *image* of an element $a \in A$ is the unique $b \in B$ such that $(a,b) \in f$. By defining $f(a)$ to be this element b, all of our previous definitions translate seamlessly. For example, the *range* of f is the subset of B consisting of all images:

$$\text{Range}(f) = \{f(a) \mid a \in A\} = \{b \mid (a,b) \in f\}.$$

It is important that you become comfortable with both types of function notation, as each is useful in different settings.

We can restate the definition of the composition of functions using the unambiguous definition of functions: if $f \subseteq A \times B$ and $g \subseteq B \times C$ are functions, then their *composition* is the subset of $A \times C$ defined by

$$g \circ f = \{(a,c) \mid \text{there exists } b \in B \text{ such that } (a,b) \in f \text{ and } (b,c) \in g\}.$$

For example, the composition of f and g in Figure 3 is $g \circ f = \{(1,x),(2,z),(3,z)\}$, with a, b, and c acting as the intermediate elements in B.

Exercise 5.16 Restate the definitions of *injective* and *surjective* in a way that utilizes the unambiguous definition of a function f as a subset of $X \times Y$.

We finish this section with a proof of the Pigeonhole Principle, a lemma that is widely used throughout mathematics. It derives its name from a common, informal phrasing: if m pigeons fly into n coops, with $m > n$, then at least one coop gets at least two pigeons. Stated in the terminology of functions, the Pigeonhole Principle is:

LEMMA 5.21 (The Pigeonhole Principle). *Let $f : A \to B$, where A and B are finite, non-empty sets. If $|A| > |B|$, then f is not injective.*

The proof of the Pigeonhole Principle fits well with the unambiguous definition of functions presented in this section.

PROOF. Our proof is by induction on the size of the set B.

Base Case: If $|B| = 1$, then $|A| \geq 2$, so A contains at least two distinct elements, a_1 and a_2. This means that f contains both (a_1, b) and (a_2, b), where b is the only element of B, so f is not injective.

Inductive Step: Assume the lemma is true for all finite sets B whose size is less than or equal to n for some $n \in \mathbb{N}$. Now consider a function $f \subseteq A \times B$ where $|A| = m$ is greater than $|B| = n + 1$. For this argument we fix an element $b \in B$.

Case 1: If A contains at least two distinct elements, a_1 and a_2, such that f contains both (a_1, b) and (a_2, b), then f is not injective.

Case 2: If A contains no element a such that $(a, b) \in f$, then f is also a subset of $A \times (B - \{b\})$. Since $|B - \{b\}| = n$ and $|A| > |B - \{b\}|$, we know that f is not an injection, by our induction hypothesis.

Case 3: By the previous two cases, we may now assume there is a unique $a \in A$ such that $(a, b) \in f$. We can then define

$$\hat{f} = f - \{(a, b)\} \subseteq (A - \{a\}) \times (B - \{b\}).$$

We know that for every $\hat{a} \in A - \{a\}$ there is a unique $(\hat{a}, \hat{b}) \in \hat{f}$, because f being a function implies that $\hat{f}$ is also a function. Notice that

$$|A - \{a\}| = m - 1 > n = |B - \{b\}|,$$

and so we may apply our our induction hypothesis to $\hat{f}$, proving that $\hat{f}$ is not injective. It follows that f cannot be injective either. □

Exercise 5.17 Explain two quick steps in the proof of the Pigeonhole Principle above: if f is a function, so is $\hat{f}$; and if $\hat{f}$ is not injective, neither is f.

The Pigeonhole Principle is the focus of Project 11.5. We highly recommend that you look at this project, which begins with a number of accessible results that follow from the Pigeonhole Principle.

GOING BEYOND THIS BOOK. The definition of a function introduced in this section is the result of a long evolution in the history of mathematics; see the nice survey of this history in Israel Kleiner's "Evolution of the function concept: A brief survey" [**Kle89**].

5.6. Inverse Functions

Two functions are said to be inverses of each other if one function "undoes" the other, in a manner described more carefully near the end of this section. You have encountered the idea of inverse functions before, perhaps in the process of solving equations. For example, if you have reduced a problem to $x^3 = -8$, then you would take the cube root of both sides to determine that $x = -2$. This works because the functions x^3 and $\sqrt[3]{x}$ are inverses. Similarly if you meet $e^{2x} = 9$, you can apply the natural logarithm to both sides to determine that $2x = \ln(9)$, and therefore $x = \frac{1}{2}\ln(9) = \ln(9^{1/2}) = \ln(3)$. Here we used the fact that the natural logarithm function $\ln(x)$ and the exponential function e^x are inverses.

The definition of a function f as a subset of $A \times B$ is particularly convenient for defining the inverse of f.

DEFINITION 5.22. If $f \subseteq A \times B$ is a function, define the *inverse* f^{-1} as a subset of $B \times A$:

$$f^{-1} = \{(b,a) \in B \times A \mid (a,b) \in f\}.$$

Let f be a function from a subset of $\mathbb{R}$ to $\mathbb{R}$. Because the inverse of f is created by simply reversing the order of elements, f^{-1} contains all points of the form $(f(x), x)$. This is equivalent to reflecting the graph of f across the line $y = x$; the example of $\ln(x)$ and e^x is illustrated in Figure 4.

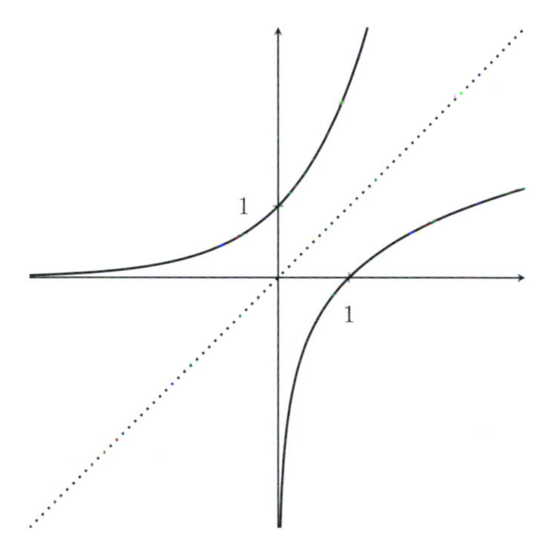

Figure 4. Plotting a function and its inverse – in this case e^x and $\ln(x)$ – reveals symmetry about the line $y = x$.

Exercise 5.18 Carefully plot $f(x) = x^3$ and $g(x) = \sqrt[3]{x}$, determining all of the points where they intersect.

There are many situations where f^{-1} is not a function. Consider the example of $f(x) = x^4 - 4x^2$, shown along with the subset f^{-1} of $\mathbb{R} \times \mathbb{R}$ in Figure 5. Looking at the graph of f you can see that there are four values of x where $f(x) = -1$, and via the quadratic formula you can determine that the four solutions of $x^4 - 4x^2 = -1$ are $\pm\sqrt{2 \pm \sqrt{3}}$. Because multiple values are taken to -1 by f, you cannot determine the value of x just from knowing that $f(x) = -1$.

There are many vertical lines in Figure 5, including the line $x = -1$, that pass through several points of f^{-1}. Thus f^{-1} does not satisfy the definition of a function. Specifically, it is not the case that for each element $x \in \mathbb{R}$ there is a *unique* element $(x, y) \in f^{-1}$. There is another problem with thinking of f^{-1} as a function from $\mathbb{R}$ to $\mathbb{R}$. Because there is no real-valued solution to $x^4 - 4x^2 = -5$, there is no element $(-5, y) \in f^{-1}$. So for some values of $x \in \mathbb{R}$ there is no element $(x, y) \in f^{-1}$.

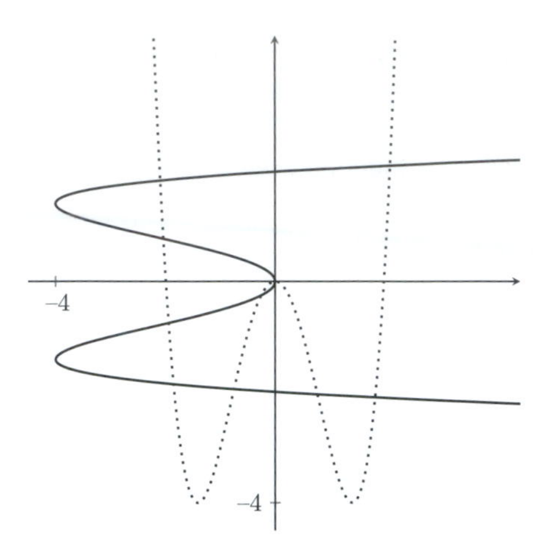

Figure 5. The function $f(x) = x^4 - 4x^2$ is shown with a dotted curve. Viewing f as a subset of $\mathbb{R} \times \mathbb{R}$, you can form its formal inverse, shown as a solid curve, which is not a function.

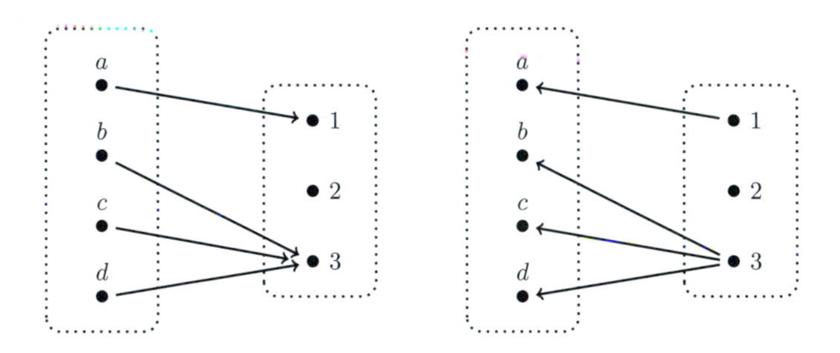

Figure 6. The function f is shown on the left, with f^{-1} illustrated on the right.

The same fundamental issues are illustrated by the function from $\{a, b, c, d\}$ to $\{1, 2, 3\}$ shown in Figure 6. In this case, you can visualize f^{-1} by reversing the arrows, producing

$$f^{-1} = \{(1, a), (3, b), (3, c), (3, d)\}.$$

As the figure highlights, the number 2 is not the first element of an ordered pair in f^{-1} (notice that f is not surjective), and the number 3 is the first element of three ordered pairs in f^{-1} (notice that f is not injective).

These examples motivate the need for a characterization of what makes f^{-1} a function.

THEOREM 5.23. *Let $f : A \to B$ be a function. Then f^{-1} defines a function from B to A if and only if f is a bijection.*

PROOF. Assume first that f is a bijection, that is, f is both injective and surjective. Since f is surjective, given any $b \in B$ there is an $a \in A$ such that $(a, b) \in f$. This means that given any $b \in B$ there is an ordered pair $(b, a) \in f^{-1}$. To see that this pair is unique given any $b \in B$, and thus that f^{-1} is a function, assume to the contrary that there is also a pair $(b, a') \in f^{-1}$, with $a \neq a'$. Then (a, b) and $(a', b) \in f$, contradicting the fact that f is injective.

If f is not a bijection, then f is either not surjective or not injective. If f is not surjective then there is a $b \in B$ that does not appear as the second element of an ordered pair in f. Thus there is no $a \in A$ such that $(b, a) \in f^{-1}$, so f^{-1} is not a function. If f is not injective, then there are distinct elements $a, a' \in A$ and an element $b \in B$ such that (a, b) and (a', b) are both in f. Thus (b, a) and (b, a') are both in f^{-1}, which contradicts the requirement that the ordered pair beginning with b is unique. □

Exercise 5.19 Let's examine the logical structure of the proof for Theorem 5.23. If P is the proposition "f is a bijection," then P is stating that f is both injective and surjective. Thus P is equivalent to $P_1 \wedge P_2$, where P_1 is "f is surjective" and P_2 is "f is injective."

Let Q be the proposition "f^{-1} is a function." In our proof of Theorem 5.23 we showed:

$$\underbrace{((P_1 \wedge P_2) \Rightarrow Q)}_{\text{first paragraph}} \wedge \underbrace{(\sim P_1 \Rightarrow \sim Q) \wedge (\sim P_2 \Rightarrow \sim Q)}_{\text{second paragraph}} .$$

Verify that this expression is logically equivalent to what is claimed in the theorem: $(P_1 \wedge P_2) \Leftrightarrow Q$.

If $f : A \to B$ is a bijection, we can now say why f^{-1} "undoes" f, and vice versa, in the common notation of functions. Since f is a function, for every $a \in A$ there is a unique $b \in B$ such that $(a, b) \in f$, and thus $(b, a) \in f^{-1}$. Since f^{-1} is also function, we can write both $f(a) = b$ and $f^{-1}(b) = a$, so

$$(f^{-1} \circ f)(a) = f^{-1}(f(a)) = f^{-1}(b) = a$$

for all $a \in A$. Moreover, since f is a bijection, for every $b \in B$ there is a unique element $a \in A$ such that $(a, b) \in f$, and thus $(b, a) \in f^{-1}$. As before, we can again write $f(a) = b$ and $f^{-1}(b) = a$, so

$$(f \circ f^{-1})(b) = f(f^{-1}(b)) = f(a) = b$$

for all $b \in B$. Put another way, we now know that $f^{-1} \circ f = 1_A$ and $f \circ f^{-1} = 1_B$. (Recall that the functions 1_A and 1_B are the identity functions mentioned in Example 5.5.)

Exercise 5.20 Let $f : A \to B$ and $g : B \to A$. Prove that if $f \circ g = 1_B$ and $g \circ f = 1_A$, then f and g are bijections, and $g = f^{-1}$.

5.7. Functions and Subsets

There are many confusing things about inverses and how they are used in practice. For one, because you often want to know what elements in the domain are mapped by f to a particular element in the range, it is not uncommon to use f^{-1} even when f^{-1} is not a function. It is also not uncommon to apply a function not just to single elements but to a subset of the domain. In this section we explore these notationally challenging practices.

5.7.1. Where did this come from? Let $f : A \to B$ be a function. Given some $b \in B$, you might wonder what elements of A are taken to b by f. For example, the length function $\|\cdot\|$ takes a binary string in Σ^* and returns a non-negative integer: we find that $\|110\| = 3$, and $\|001\| = 3$ as well. As shown in Exercise 5.15, there are eight strings of length 3:

$$\{\omega \in \Sigma^* \mid \|\omega\| = 3\} = \left\{ \begin{array}{cccc} 000, & 001, & 010, & 011, \\ 100, & 101, & 110, & 111 \end{array} \right\}.$$

If $f : A \to B$, then the *pre-image* of $b \in B$ is defined to be

$$\text{pre-image of } b = \{a \in A \mid f(a) = b\},$$

or stated in terms of ordered pairs,

$$\text{pre-image of } b = \{a \in A \mid (a, b) \in f\}.$$

Notice that the pre-image of an element b may be a single element in A or it may be a subset of elements. In particular, if b is not in the range of f, then the pre-image of b is the empty set.

In the previous section we examined $f(x) = x^4 - 4x^2$. Looking at the graph in Figure 5 you see that f takes four distinct real numbers to -3, and with a bit of algebra you can show that

$$\text{pre-image of } -3 = \left\{\pm 1, \pm\sqrt{3}\right\}.$$

And since -5 is not in the range of $f(x) = x^4 - 4x^2$, the pre-image of -5 is $\emptyset$.

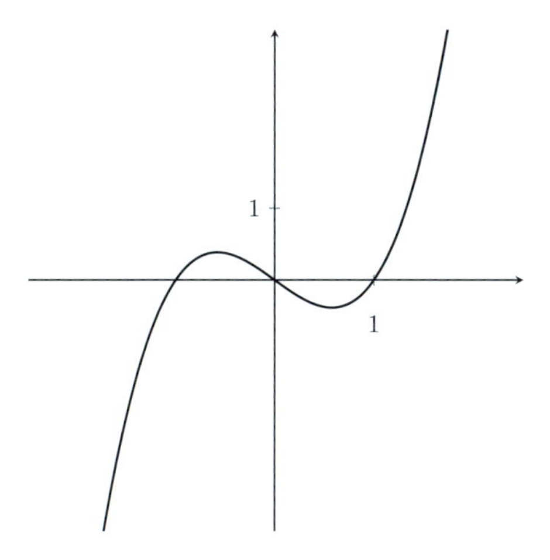

Figure 7. The cubic function for Exercise 5.21.

Exercise 5.21 Let $f : \mathbb{R} \to \mathbb{R}$ be the function $f(x) = x^3 - x$, shown in Figure 7.

(a) What is the pre-image of 0?
(b) What is the pre-image of 6?
(c) Which real numbers have a pre-image consisting of exactly two numbers? (You may need a bit of calculus to help with this part.)

5.7.2. From Elements to Subsets. Continuing our examination of $f(x) = x^3 - x$ from Exercise 5.21, it is not uncommon to want to understand where a function like f takes an entire subset of real numbers. For example, we might like to apply f to the closed interval $[0, 1]$ and write

$$f([0,1]) = \left[\frac{-2}{3\sqrt{3}}, 0\right],$$

by which we mean

$$\{x^3 - x \mid x \in [0,1]\} = \left[\frac{-2}{3\sqrt{3}}, 0\right].$$

The lower limit, $-2/3\sqrt{3}$, is the minimum for f over the interval $[0, 1]$.

Any function $f : A \to B$ can be used to describe an associated function from the power set of A to the power set of B. There is no commonly accepted notation for this "induced" function, and we will temporarily denote it by $f_{\mathcal{P}}$. The rule that defines $f_{\mathcal{P}}$ is this:

$$\text{If } S \subseteq A, \text{ then } f_{\mathcal{P}}(S) = \{b \in B \mid b = f(s) \text{ for some } s \in S\}.$$

That is, $f_{\mathcal{P}}(S)$ is the set of all images of elements in $S \subseteq A$.

As an example, let f be the floor function, $f(x) = \lfloor x \rfloor$. Then

(a) $f_{\mathcal{P}}(\{\ln(3), e, \pi\}) = \{1, 2, 3\}$,
(b) $f_{\mathcal{P}}([1, 3.5]) = \{1, 2, 3\}$,
(c) $f_{\mathcal{P}}(\{1/2, 1/3, 1/4, 1/5, \ldots\}) = \{0\}$,
(d) $f_{\mathcal{P}}(\mathbb{R}) = \mathbb{Z}$.

Exercise 5.22 Let $f : \mathbb{R} \to \mathbb{Z}$ be the function

$$f(x) = \lfloor x^2 \rfloor.$$

(a) Verify that $f(\pi) = 9$ and $f_{\mathcal{P}}(\{\pi\}) = \{9\}$.
(b) What is $f_{\mathcal{P}}(\{-1, 0, 2\})$?
(c) What is $f_{\mathcal{P}}(\emptyset)$?
(d) What is $f_{\mathcal{P}}([2, \pi])$?
(e) What is $f_{\mathcal{P}}(\mathbb{R})$?

5.7.3. Subsets of the Range. If $f(x) = x^3 - x$, then the pre-image of 0 is the set $\{-1, 0, 1\}$. Because f is not injective, we know that f^{-1} is not itself a function. However, it is tempting to write

$$f^{-1}(0) = \{-1, 0, 1\}.$$

Such a statement is in essence claiming that f^{-1} can be thought of as a function, but one where the image of an element might be a subset of $\mathbb{R}$, the domain of the function f. That is, the range of f^{-1} is in the power set $\mathcal{P}(\mathbb{R})$.

We can combine the notions in the previous two sections – working with pre-images even when f^{-1} is not a function and the construction of $f_{\mathcal{P}}$ from f – to construct a function from subsets of B to subsets of A. Given a function $f : A \to B$, there is an induced function from $\mathcal{P}(B)$ to $\mathcal{P}(A)$. It is defined for each $T \subseteq B$ by

$$f_{\mathcal{P}}^{-1}(T) = \{a \in A \mid f(a) \in T \subseteq B\}.$$

In particular, if $f(x) = x^3 - x$ then we can write

$$f_{\mathcal{P}}^{-1}(\{0\}) = \{-1, 0, 1\}.$$

Exercise 5.66 asks you to verify that $f_{\mathcal{P}}$ and $f_{\mathcal{P}}^{-1}$ are indeed functions. As with $f_{\mathcal{P}}$, our use of the notation $f_{\mathcal{P}}^{-1}$ is temporary.

We return to the function $f(x) = x^4 - 4x^2$ to illustrate this construction. For this f we have

(a) $f_{\mathcal{P}}^{-1}(\{0\}) = \{-2, 0, 2\}$,
(b) $f_{\mathcal{P}}^{-1}(\{-1\}) = \{\pm\sqrt{2 \pm \sqrt{3}}\}$,
(c) $f_{\mathcal{P}}^{-1}(\{-5\}) = \emptyset$,
(d) $f_{\mathcal{P}}^{-1}([0, 45]) = [-3, -2] \cup \{0\} \cup [2, 3]$.

As another example, consider $\pi : \mathbb{N} \to \mathbb{N}$ defined by

$$\pi(x) = \text{ the smallest prime greater than or equal to } x.$$

Then each of the following statements is true:

(a) $\pi(1) = 2$ and $\pi_{\mathcal{P}}(\{1\}) = \{2\}$,
(b) $\pi_{\mathcal{P}}(\{1, 2, 3, 4, 5\}) = \{2, 3, 5\}$,
(c) $\pi_{\mathcal{P}}^{-1}(\{11\}) = \{8, 9, 10, 11\}$,
(d) $\pi_{\mathcal{P}}^{-1}(\{5, 101\}) = \{4, 5, 98, 99, 100, 101\}$.

Exercise 5.23 We can consider the trigonometric functions sine and cosine to have domain $\mathbb{R}$ and codomain $[-1, 1]$. What are $\sin([0, \pi])$ and $\cos_{\mathcal{P}}^{-1}(\{1\})$?

Exercise 5.24 Consider the following two functions from $\mathbb{R}$ to $\mathbb{R}$:

$$f(x) = \begin{cases} 1 & \text{if } x \in \mathbb{Q}, \\ 0 & \text{otherwise,} \end{cases}$$

$$g(x) = \begin{cases} x & \text{if } x \in \mathbb{Q}, \\ 0 & \text{otherwise.} \end{cases}$$

Plot enough points for each function so that you have a decent impression of its graph. Then compute the following.

(a) $f(7/3)$ and $g(7/3)$
(b) $f(\sqrt{2})$ and $g(\sqrt{2})$
(c) $f_{\mathcal{P}}([-1, 2])$
(d) $g_{\mathcal{P}}(\{\sqrt{2}, \sqrt{3}, \sqrt{4}\})$
(e) $f_{\mathcal{P}}(g_{\mathcal{P}}^{-1}(\mathbb{Q}))$
(f) $f_{\mathcal{P}}^{-1}(g_{\mathcal{P}}(S))$, where $S = \{\frac{1}{n+1} \mid n \in \mathbb{N}\}$

5.7.4. Abuse of Notation. Given $f : A \to B$ we have now introduced three associated concepts:

f^{-1}, which may or may not be a function from B to A;
$f_{\mathcal{P}}$, which is a function from $\mathcal{P}(A)$ to $\mathcal{P}(B)$;
$f_{\mathcal{P}}^{-1}$, which is a function from $\mathcal{P}(B)$ to $\mathcal{P}(A)$.

We also mentioned that there is no standard notation for $f_{\mathcal{P}}$ and $f_{\mathcal{P}}^{-1}$. In practice, the symbols f and f^{-1} are used, or perhaps we should say reused or overused, in their places. The multiple meanings for the same notation is potentially confusing, but in context it is usually clear what is being described. Thus if $f(x) = x^3 - x$, then you will indeed see expressions like

$$\begin{aligned} &f(1) = 0\,, \\ &f([-1,1]) = [-2/3\sqrt{3}, 2/3\sqrt{3}]\,, \\ &f^{-1}(0) = \{-1,0,1\}\,. \end{aligned}$$

In each case, you can determine what is being stated without the burden of additional notation.

Using notation in an imprecise fashion, or using the same notation for two related but distinguishable constructions, is referred to as an *abuse of notation*. As the overuse of the symbols f and f^{-1} is a rather common abuse of notation, for the remainder of the book we will drop the subscript "$\mathcal{P}$" and follow the convention of using the symbols f and f^{-1} in all of the contexts we have described in this section.

5.8. A Few Facts About Functions and Subsets

Let $f : A \to B$ be a bijection and let f^{-1} be its inverse function. We have seen that $f^{-1}(f(a)) = a$ for all $a \in A$ and $f(f^{-1}(b)) = b$ for all $b \in B$. When you apply functions and inverses to subsets instead of elements, similar but weaker statements hold.

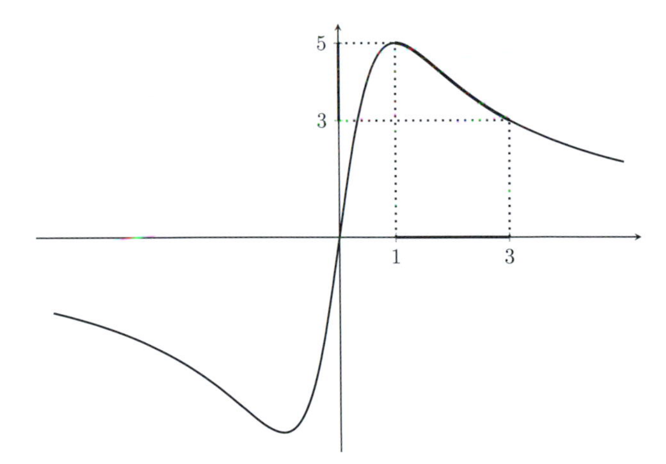

Figure 8. The function $f(x) = 10x/(1 + x^2)$ takes the closed interval $[1,3]$ to $[3,5]$.

It will be helpful to begin with an example. Consider the function

$$f(x) = \frac{10x}{1 + x^2}$$

whose graph is plotted in Figure 8, highlighting the fact that f takes the closed interval $[1,3]$ to $[3,5]$. You can see in the graph of f that there are additional values of x that map into the closed interval $[3,5]$. In particular, a bit of algebra shows that $f(1/3) = 3$, which is the key fact in determining that $f^{-1}([3,5]) = [1/3,3]$. Thus

$$[1/3,3] = f^{-1}(f([1,3])) \neq [1,3].$$

The original interval is not, however, completely unrelated to $[1/3,3]$: it is a subset of it.

PROPOSITION 5.24. *Let $f : A \to B$ be a function.*

(a) *If $S \subseteq A$, then $S \subseteq f^{-1}(f(S))$.*
(b) *If $T \subseteq B$, then $f(f^{-1}(T)) \subseteq T$.*

PROOF. We first prove part (a). For each $s \in S$ we know that $f(s) \in f(S)$ by the discussion in Section 5.7.2. By the discussion in Section 5.7.3, we know that $s \in f^{-1}(f(S))$. Hence $S \subseteq f^{-1}(f(S))$.

To prove part (b), simply notice that

$$f(f^{-1}(T)) = f(\{x \in A \mid f(x) \in T\}) \subseteq T.$$

□

Exercise 5.25 Let $f : A \to B$. Prove or find a counterexample to each of the following claims.

(a) If f is injective, then for all subsets $T \subseteq B, f(f^{-1}(T)) = T$.
(b) If f is surjective, then for all subsets $T \subseteq B, f(f^{-1}(T)) = T$.
(c) If f is bijective, then for all subsets $T \subseteq B, f(f^{-1}(T)) = T$.

The behavior of functions applied to unions and intersections is similarly aligned with intuition, but it is once again weaker than what you might at first expect.

PROPOSITION 5.25. *Let $f : A \to B$ be a function and $X, Y \subseteq A$. Then*

(a) $f(X \cup Y) = f(X) \cup f(Y)$,
(b) $f(X \cap Y) \subseteq f(X) \cap f(Y)$.

PROOF. We first prove that $f(X \cup Y) \subseteq f(X) \cup f(Y)$. Let $w \in f(X \cup Y)$. Then there is a $z \in X \cup Y$ with $f(z) = w$. Since $z \in X \cup Y$ we have $z \in X$ or $z \in Y$. Thus $f(z) \in f(X)$ or $f(z) \in f(Y)$, hence $w = f(z) \in f(X) \cup f(Y)$.

Conversely, let $w \in f(X) \cup f(Y)$. Then $w \in f(X)$ or $w \in f(Y)$, so $w = f(x)$ for some $x \in X \subseteq X \cup Y$ or $w = f(y)$ for some $y \in Y \subseteq X \cup Y$. Either way we conclude that $w \in f(X \cup Y)$. This establishes part (a).

To prove part (b) we begin with an element $w \in f(X \cap Y)$. Because w is in the image of $X \cap Y$, there must be some $z \in X \cap Y$ with $f(z) = w$. Since $z \in X$, $w = f(z) \in f(X)$, and similarly since $z \in Y$, $w = f(z) \in f(Y)$. Thus $w \in f(X) \cap f(Y)$. □

Exercise 5.26 Construct an example showing that $f(X \cap Y) = f(X) \cap f(Y)$ doesn't always hold.

5.9. End-of-Chapter Exercises

Exercises you can work on after Sections 5.1 and 5.2

5.27 The following questions ask about the domain and range of some standard trigonometric functions.

(a) What is the range of $\sin(x)$?
(b) What is the domain of $\tan(x)$?
(c) What is the range of $\sec(x)$?
(d) What is the domain of $\csc(x)$?

5.28 Determine the range of the following functions.

(a) The function $f : \mathbb{R} \to \mathbb{R}$ defined by $f(x) = 1/(1 + x^2)$.
(b) The function $f : \Sigma^* \to \mathbb{Q}_2$ defined by

$$f(\omega) = \frac{\|\omega\|_0}{2^{\|\omega\|_1}}.$$

5.29 Determine the range of the following functions that involve $\mathbb{Z}[x]$, the set of all polynomials with integer coefficients.

(a) The function $f : \mathbb{Z}[x] \to \mathbb{Z} \times \mathbb{Z}$ defined by $f(p(x)) = (p(0), p(1))$.
(b) The function $g : \mathbb{Z}[x] \to \mathbb{Z}[x]$ defined by $g(p(x)) = xp(x)$.

5.30 Determine which of the following are functions with the given domains and codomains.

(a) $f : \mathbb{N} \to \mathbb{N}$ with $f(n) = (n + 1)^2 - 2$.
(b) $f : \mathbb{Q} \to \mathbb{Z}$ with $f(a/b) = a + b$.
(c) For any finite set A, $f : \mathcal{P}(A) \to \mathbb{Z}$ with $f(S) = |S|$.
(d) $f : \mathbb{Q}_2 \to \mathbb{Q}_2$ with $f(x) = x^2$.
(e) $f : \mathbb{Q}_2 \setminus \{0\} \to \mathbb{Q}_2 \setminus \{0\}$ with $f(x) = 1/x$.
(f) For any set A, $f : \mathcal{P}(A) \to \mathcal{P}(A)$ with $f(S) = A \setminus S$.

5.31 Define the function $g : \mathbb{Z} \to \mathbb{Z}$ by

$$g(x) = \begin{cases} x - 5 & \text{if } x \text{ is odd,} \\ 3x & \text{if } x \text{ is even.} \end{cases}$$

(a) What is Range(g)?
(b) Now consider the same definition for $g(x)$ but as a function $g : \mathbb{N} \to \mathbb{Z}$ with a different domain. What is Range(g) in this case?

Exercises you can work on after Section 5.3

5.32 Define a function $f : \mathbb{R} \to \mathbb{R}$ by

$$f(x) = \begin{cases} x + 1 & \text{if } \lceil x \rceil \text{ is odd,} \\ x - 1 & \text{if } \lceil x \rceil \text{ is even,} \end{cases}$$

where $\lceil x \rceil$ is the ceiling function. Is f injective or surjective? Or neither?

5.33 Define the function $f : \mathbb{Z} \to \mathbb{Z}[x]$ by

$$f(n) = \begin{cases} 1 - x^{-n} & \text{if } n < 0, \\ 1 & \text{if } n = 0, \\ 1 + x^n & \text{if } n > 0. \end{cases}$$

Prove that f is injective but not surjective.

5.34 Let $f : \mathbb{R} \to \mathbb{R}$ be the function $f(x) = ax + b$, where a and b are elements of $\mathbb{R} \setminus \{0\}$. For any such pair of coefficients, must f be ...

(a) ... injective?
(b) ... surjective?
(c) ... bijective?

5.35 Consider the following continuation of Exercise 5.10 for finite sets A and B.

(a) Determine the number of different bijections from A to B if $|A| = |B|$.
(b) Determine the number of different injections from A to B if $|A| \leq |B|$.

5.36 Consider the following continuation of Exercises 5.10 and 5.35.

(a) Determine the number of different surjections from A to B if $|A| = 4$ and $|B| = 3$.
(b) Determine the number of different surjections from A to B if $|A| = n$ and $|B| = k$, with $k \leq n$. As a hint, Stirling numbers of the second kind, discussed in Section 3.8, might prove helpful.

5.37 Construct an injective map $f : \mathbb{N} \hookrightarrow \mathbb{N} \times \mathbb{N}$. Then try your best to construct an injective map $f : \mathbb{N} \times \mathbb{N} \hookrightarrow \mathbb{N}$.

5.38 Let $f : A \to C$ and $g : B \to D$, and define $h : A \times B \to C \times D$ by

$$h(a, b) = (f(a), g(b)) .$$

(a) Prove that if f and g are injections, then so is h.
(b) Prove that if f and g are surjections, then so is h.

5.39 Determine which of the following functions are bijections.

(a) $f : \mathbb{Z} \to \mathbb{Z}$ given by $f(x) = x - 2$.
(b) $f : \mathbb{R} \setminus \{0\} \to \mathbb{R} \setminus \{0\}$ given by $f(x) = 1/x$.
(c) For a set A, $f : \mathcal{P}(A) \to \mathcal{P}(A)$ given by $f(S) = S^c = A - S$.

5.40 Given sets A and B, recall the function $f : A \times B \to A$ that projects onto the first coordinate: $f(a, b) = a$. What conditions do you need to put on the sets A and B to insure that this function is ...

(a) ... injective?
(b) ... surjective?
(c) ... bijective?

5.41 Determine whether there is a bijection $f : A \to B$ for the following given sets A and B.

(a) $A = \mathbb{R}$ and $B = (0, \infty)$.

(b) $A = [0, 1]$ and $B = [0, 1)$.

Don't let the single missing point in part (b) deceive you: accommodating it might prove to be quite challenging!

5.42 Try to find bijections $f : A \to B$ for the following given sets A and B. We believe the first one is not so hard, the next is harder, and so on. Some bijections might not be possible; this topic will reappear in Chapter 7.

(a) $A = \mathbb{N}$ and $B = \mathbb{N} \cup \{\text{frog}\}$.
(b) $A = \mathbb{Z}$ and $B = \mathbb{N}$.
(c) $A = \mathbb{N}$ and $B = \mathbb{Z}$.
(d) $A = \mathbb{N}$ and $B = \mathbb{N} \times \mathbb{N}$.
(e) $A = \mathbb{N}$ and $B = \mathbb{R}$.

Caution: The last one is REALLY hard.

Exercises you can work on after Section 5.4

5.43 Suppose that A, B, and C are finite sets that each contain five elements. Do there exist functions $f : A \to B$ and $g : B \to C$ that satisfy the following conditions? Prove that your answers are correct.

(a) $|\text{Range}(f)| = 4$ and $g \circ f$ is injective.
(b) $|\text{Range}(f)| = 4$ and $g \circ f$ is surjective.
(c) $|\text{Range}(g)| = 3$ and $|\text{Range}(g \circ f)| = 2$.
(d) $|\text{Range}(g)| = 2$ and $|\text{Range}(g \circ f)| = 3$.

5.44 Let $f : \mathbb{Z} \to \mathbb{Z}[x]$ be the function defined in Exercise 5.33. Let $g : \mathbb{Z}[x] \to \mathbb{Z}[x]$ be defined by

$$g(p(x)) = \frac{1}{2}\left(p(x) + p(-x)\right) .$$

For example,

$$g(x^2 + 2x - 1) = \frac{1}{2}\Big((x^2 + 2x - 1) + ((-x)^2 + 2(-x) - 1)\Big) = x^2 - 1 .$$

Describe the composition $g \circ f : \mathbb{Z} \to \mathbb{Z}[x]$.

5.45 The arcsine function, $\arcsin(x)$, is a function from $[-1, 1]$ to $[-\pi/2, \pi/2]$, where $\arcsin(x) = \theta$ if and only if $\sin(\theta) = x$. In this exercise we ask you to describe the functions $\sin \circ \arcsin$ and $\arcsin \circ \sin$: for each composition, justify your answers to the following questions:

(a) What is the domain of the composition?
(b) What is the range of the composition?
(c) Is the composition injective?
(d) Is the composition surjective?
(e) Is the composition bijective?

Note that the range of arcsine is properly contained in the domain of the sine function. As we remarked at the end of Section 5.4, a composition like $\sin \circ \arcsin$ still makes sense, even though the range of $\arcsin(x)$ is not identical to the domain of $\sin(x)$.

5.46 Prove that function composition is associative. That is, if

$$A \xrightarrow{f} B \xrightarrow{g} C \xrightarrow{h} D,$$

then

$$(h \circ g) \circ f = h \circ (g \circ f).$$

5.47 Consider the function $h : \mathbb{R} \to \mathbb{Q}$ defined by

$$h(x) = \begin{cases} \frac{1}{n} & \text{if } x \in \mathbb{Q} \setminus \{0\}, \text{ and } x = \frac{m}{n} \text{ in lowest terms,} \\ 0 & \text{otherwise.} \end{cases}$$

For example, $h(12/5) = 1/5$, $h(-4/6) = 1/3$ and $h(\sqrt{2}) = 0$.

(a) What is the range of h?

(b) We have heard this function called the *Popcorn Function* as well as *Stars Over Babylon*. Why are those appropriate names? Plot at least 12 points on the graph of this function using rational x-values with $-1 \le x \le 2$.

(c) A function f is an *idempotent* if $f \circ f$ and f are the same function. Show that $h(x)$ is an idempotent. Two simpler idempotent functions can be found in Exercise 5.24.

(d) Comment on Zippy in Figure 9.

©Bill Griffith, Pinhead Productions

Figure 9. Idempotent functions are defined in Exercise 5.47.

5.48 Define a function $f : \mathbb{R} \to \mathbb{R}$ by

$$f(x) = \begin{cases} x & \text{if } x \in [0, 1), \\ x - 1 & \text{otherwise.} \end{cases}$$

It is easy to see that the only values of x for which $f(x) = x$ are those with $x \in [0, 1)$. In this exercise you will extend this insight to iterated versions of f, namely the functions

$$f^n = \underbrace{f \circ f \circ \cdots \circ f}_{n \text{ terms}}.$$

It may be helpful to sketch the graphs of f and $f^2 = f \circ f$, before starting these problems.

(a) Which values of x satisfy $f(x) = f^2(x)$?
(b) Which values of x satisfy $f(x) = f^3(x)$?
(c) Which values of x satisfy $f^2(x) = f^3(x)$?
(d) Let $m, n \in \mathbb{N}$. Which values of x satisfy $f^m(x) = f^n(x)$?
(e) What is Range(f)? Range(f^3)?

Exercises you can work on after Section 5.5

5.49 Determine which of the following are functions with the given domains X and codomains Y. For those that are functions, determine which are injective, surjective, or bijective.

(a) $X = Y = \mathbb{Z}$ and $f = \{(x, -3x) \mid x \in X\}$.
(b) $X = [-1, 0]$, $Y = [-1, 1]$, and $f = \{(x, y) \in X \times Y \mid x^2 + y^2 = 1\}$.
(c) $X = [-1, 1]$, $Y = [-1, 0]$, and $f = \{(x, y) \in X \times Y \mid x^2 + y^2 = 1\}$.
(d) $X = [-1, 0]$, $Y = [-1, 0]$, and $f = \{(x, y) \in X \times Y \mid x^2 + y^2 = 1\}$.
(e) $X = Y = \mathbb{N}$ and

$$f = \{(m, n+1) \in \mathbb{N} \times \mathbb{N} \mid m \text{ has exactly } n \text{ distinct prime factors}\}.$$

For example, $(80, 3) \in f$ because $80 = 2^4 \cdot 5$ has two distinct prime factors.
(f) $Y = \{1, 2, 3, 4\}$, $X = \mathcal{P}(Y)$, and

$$f = \{(x, y) \in X \times Y \mid \text{the smallest element of } x \text{ is } y\}.$$

For example, $(\{1, 2, 4\}, 1) \in f$.

5.50 Prove that the composition of two functions is a function, using the ordered pairs definition of functions.

5.51 The functions $\|\cdot\|$ and κ were defined in Section 5.1. Does either $\|\cdot\| \circ \kappa$ or $\kappa \circ \|\cdot\|$ make sense? If so, describe the effect of the resulting function(s).

5.52 Let $f \subseteq A \times B$ be a function.

(a) Prove that f is injective if and only if

$$|f \cap (A \times \{b\})| \leq 1$$

for all $b \in B$.
(b) Prove that f is surjective if and only if

$$|f \cap (A \times \{b\})| \geq 1$$

for all $b \in B$.

5.53 Let $f \subseteq A \times B$ be a function. In many situations you may want to restrict the domain of f or expand its range.

(a) If $C \subseteq A$, then define the *restriction of f to C* as

$$f|_C = (C \times B) \cap f.$$

Prove that $f|_C$ is a function from C to B.

(b) If D is any set that contains B, then you can use f to define a function with domain A and codomain D, using the fact that

$$f \subseteq A \times B \subseteq A \times D.$$

Prove that f is still a function in this setting.

5.54 Give two examples of infinite subsets $A \subset \mathbb{R}$ where $\sin|_A$, the restriction of the sine function, is an injective function. (The restriction of a function is defined in Exercise 5.53.)

5.55 Prove that the restriction of sine to the integers, $\sin|_{\mathbb{Z}}$, is injective. (The graph of $\sin|_{\mathbb{Z}}$, for integers between -20 and 20, is shown in Figure 10.) We recommend using the trigonometric identity

$$\sin(a) - \sin(b) = 2\sin\left(\frac{a-b}{2}\right)\cos\left(\frac{a+b}{2}\right).$$

You may assume that $\pi \notin \mathbb{Q}$, a fact that you can prove in Section 8.6.

Was $\mathbb{Z}$ one of your answers to Exercise 5.54?

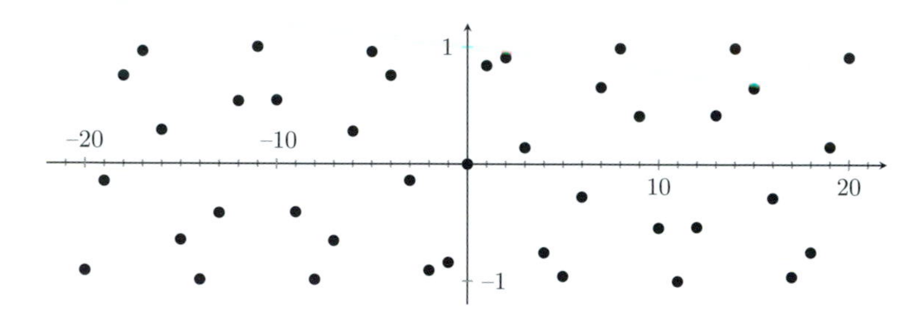

Figure 10. The sine function restricted to the integers.

5.56 In a class of n students, some students are friends and others are not. We assume that being friends is a symmetric relationship, so that if Alice is friends with Bob, then Bob is also friends with Alice. Prove that there must be two students in the class with the same number of friends. Here's a hint: can Alice have no friends in the class while at the same time Bob is friends with $n-1$ students?

5.57 Let

$$S = \{3n+1 \mid 0 \le n \le 33\} = \{1, 4, 7, \ldots, 97, 100\}.$$

Prove that if 19 distinct elements are chosen from S, then two of those integers must sum to 104.

Exercises you can work on after Section 5.6

5.58 The function $f(x) = \frac{1}{5}(x^3 + x)$ and its inverse are shown in Figure 11.

(a) What is $f^{-1}(2)$?
(b) What is $f^{-1}(2/5)$?
(c) What is $f^{-1}(6)$?
(d) $f^{-1}(1)$ is between which two integers?

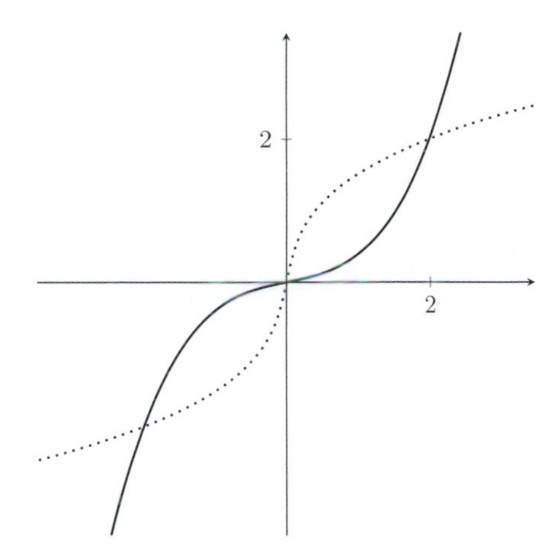

Figure 11. The function $f(x) = \frac{1}{5}(x^3 + x)$ is shown with a solid curve and its inverse is represented by the dotted curve.

5.59 Define $f : \mathbb{Z} \to \mathbb{Z}$ by

$$f(x) = \begin{cases} x + 3 & \text{if } x \text{ is odd,} \\ x - 1 & \text{if } x \text{ is even.} \end{cases}$$

Prove that f is a bijection, and describe f^{-1}.

5.60 For any set S, let $\cdot^c$ be the function from $\mathcal{P}(S)$ to $\mathcal{P}(S)$ taking a subset of S to its complement in S.

(a) Prove that $\cdot^c$ is a bijection.
(b) Prove that $\cdot^c$ is its own inverse.

5.61 Let $f : A \to B$ be a bijection. By Theorem 5.23, $f^{-1} : B \to A$ is also function; moreover, by Exercise 5.20 it is a bijection. Since $f^{-1} : B \to A$ is a bijection, it follows that it has an inverse function, $(f^{-1})^{-1} : A \to B$. Prove that f and $(f^{-1})^{-1}$ are the same function.

Exercises you can work on after Sections 5.7 and 5.8

5.62 Let $f(x) = x^4 - 4x^2$, as shown in Figure 5.

(a) What is the pre-image of 5?
(b) For what values of x does the set $f^{-1}(x)$ have exactly two elements? If it's helpful, the minimum values of $f(x)$ occur at the points $(\pm\sqrt{2}, -4)$.

5.63 Let $r : \mathbb{Z}[x] \to \mathcal{P}(\mathbb{R})$ be defined by

$$r(p) = \{x \in \mathbb{R} \mid p(x) = 0\}.$$

In other words, r takes a polynomial to its set of real roots.

(a) Verify that $r(x^2 - 4x + 3) = \{1, 3\}$, and determine $r(x^3 - x)$.
(b) Find a polynomial $p(x) \in \mathbb{Z}[x]$ where $r(p) \cap \mathbb{Q} = \emptyset$.
(c) Show that $r^{-1}(\emptyset)$ is not empty.

5.64 Let $f : A \to B$ and $X \subseteq A$. Then $f(X) \subseteq B$. Show by an example that it is possible to have $f(a) \in f(X)$ without having $a \in X$.

5.65 Determine $f(S)$ and $f^{-1}(T)$ for the following functions and subsets.

(a) $f : \mathbb{R} \to \mathbb{R}$ given by $f(x) = x^2 - 1$, with $S = (-2, 1]$ and $T = (0, 1]$.
(b) $f : \mathbb{R} \to \mathbb{Z}$ given by $f(x) = \lfloor x \rfloor + \lceil x \rceil$, with $S = (3/2, 4]$ and $T = \{4, 5, 8\}$.
(c) $f : \mathbb{Q}^+ \to \mathbb{Z}$ given by $f(x) = m + n$ if $x = m/n$ is a fraction in lowest terms, with $S = \{1/2, 4/12, 3/4, 3, 6\}$ and $T = \{3, 5\}$.

5.66 Given a function f from a set A to a set B, we defined $f_{\mathcal{P}} : \mathcal{P}(A) \to \mathcal{P}(B)$ and $f_{\mathcal{P}}^{-1} : \mathcal{P}(B) \to \mathcal{P}(A)$ in Section 5.7.

(a) Verify that the “function” $f_{\mathcal{P}}$ is actually a function.
(b) Verify that the “function” $f_{\mathcal{P}}^{-1}$ is actually a function.

5.67 Let $f(x) = x^4 - 4x^2$, as shown in Figure 5.

(a) What is $f([-3, 0])$?
(b) What is $f(f^{-1}([0, 45]))$?
(c) What is $f^{-1}(f([0, 2]))$?

5.68 Let $f : A \to B$ be a function, and let $\{A_i \mid i \in I\}$ be an indexed family of subsets of A. Prove the following.

(a) $f\left(\bigcup_{i \in I} A_i\right) = \bigcup_{i \in I} f(A_i)$

(b) $f\left(\bigcap_{i \in I} A_i\right) \subseteq \bigcap_{i \in I} f(A_i)$

More Exercises!

5.69 Let $f : A \to B$ and $g : D \to C$.

(a) Assuming that $B \subseteq D$, construct a definition of the composition $g \circ f : A \to C$ (as described in Remark 5.17) using the unambiguous definition of a function.
(b) Give an unambiguous definition of $g \circ f$ which only requires

$$\text{Range}(f) \subseteq D,$$

and explain why this is different from the situation in part (a).
(c) If it is not the case that $\text{Range}(f) \subseteq D$, what is the maximal subset $X \subset A$ such that you can define the composition $g \circ f|_X : X \to C$? Here $f|_X$ denotes the restriction of f to X, as defined in Exercise 5.53.

5.70 Given a partition $\mathcal{B}$ of the set $D = \{1, 2, \ldots, 8, 9\}$, let $p : D \to \mathbb{N}$ be the function where $p(a)$ is the size of the block of $\mathcal{B}$ that contains a. Prove that for any two partitions of D, $\mathcal{B}_1$ and $\mathcal{B}_2$, there are distinct numbers $a, b \in D$ such that $p_1(a) = p_1(b)$ and $p_2(a) = p_2(b)$.

5.71 We have seen that a bijection $f : A \to B$ has an inverse function $f^{-1} : B \to A$, with $f \circ f^{-1} = 1_B$ and $f^{-1} \circ f = 1_A$. If $f : A \to B$ is any function (not necessarily a bijection) then we say f has a *left inverse* if there is a function $g : B \to A$ such that $g \circ f = 1_A$.

(a) Let $f : \mathbb{R} \to \mathbb{R}$ be defined by

$$f(x) = \begin{cases} x & x < 0, \\ x+1 & x \geq 0. \end{cases}$$

Find a left inverse for f.

(b) Find another left inverse for f, thereby showing that left inverses are not unique.

(c) Prove that if $f : A \hookrightarrow B$ is an injection, then f has a left inverse.

5.72 Construct a similar exercise to Exercise 5.71 that defines and explores *right inverses* instead of left inverses.

5.73 The goal of this exercise is to prove the following claim:

> *Let $n \in \mathbb{N}$, and define $S_n = \{1, 2, 3, \ldots, 2n\} \subset \mathbb{N}$. Let A be any subset of S_n with $|A| \geq n+1$. Then A must contain x and y, where x is a proper divisor of y.*

Our approach draws on ideas from Chapter 4, particularly the Fundamental Theorem of Arithmetic (Theorem 4.11).

(a) Use the Fundamental Theorem of Arithmetic (Theorem 4.11) to prove that every natural number has a unique expression as $n = 2^k \cdot n_{\text{odd}}$ where k is a non-negative integer and n_{odd} is an odd natural number.

(b) Define a function $\mathcal{O} : \mathbb{N} \to \mathbb{N}$ via $\mathcal{O}(n) = n_{\text{odd}}$, and prove that the restriction of $\mathcal{O}$ to the set A is not injective.

(c) Conclude that there must be an x and $y \in A$ where x is a proper divisor of y.

(d) Explain why we really do have to assume that $|A| \geq n + 1$, regardless of the value of n.

5.74 We use $\mathcal{F}_{A \to B}$ to denote the set of functions from A to B. Let $g : A \to A$ be a bijection. Define $\Delta_g : \mathcal{F}_{A \to B} \to \mathcal{F}_{A \to B}$ by

$$\Delta_g(f) = f \circ g,$$

or if you prefer

$$\Delta_g = \{(f, f \circ g) \mid f \in \mathcal{F}_{A \to B}\}.$$

Prove that Δ_g is a bijective function.

If You Have Studied Calculus

5.75 Find a condition on the coefficients that characterizes when a cubic polynomial $p(x) = ax^3 + bx^2 + cx + d$ is bijective, and prove that your characterization is correct.

5.76 Recall that $\mathbb{Z}[x]$ is the set of all polynomials with integer coefficients.

(a) Let $d : \mathbb{Z}[x] \to \mathbb{Z}[x]$ be the derivative. For example, $d(3x^3 - 4x + 1) = 9x^2 - 4$ and $d(2x + 18) = 2$. Explain why d is a function.

(b) Is d injective? Give a proof or find a counterexample.

(c) Is d surjective? Give a proof or find a counterexample.

5.77 Let $f(x) = \ln(x) + 2x$, with the domain of f equal to the set of positive real numbers.

(a) Explain why you know that $f(x)$ is injective, and has an inverse.
(b) What is $f^{-1}(2)$?
(c) $f^{-1}(4)$ is between what two integers?

5.78 Let $\mathbb{Q}[x]$ be the set of all polynomials with *rational* coefficients.

(a) Construct a similar set of questions to Exercise 5.76 in the context of integration. Take care with "+C"!
(b) Answer your questions.

6 Relations

In this chapter we introduce the concept of relations on a given set, quickly moving to equivalence relations and their key properties. Modular arithmetic is one of the best known and widely applicable examples of an equivalence relation, and this is the focus of the second half of the chapter.

6.1. Introduction to Relations

Sometimes elements of a set are related by a mathematical property. For example, we might want to emphasize that the integers come in two types, even and odd, so we could say that $a, b \in \mathbb{Z}$ are related if they have the same parity. The idea of "less than" is another relation, where we could write $-1 < 5$ but not $5 < 4$. As in the case of functions, we can capture this very general notion using ordered pairs.

DEFINITION 6.1. A *relation* R on a set S is a subset of $S \times S$.

Here are nine example relations, in three groups of three.

EXAMPLE 6.2. These are three relations on $\mathbb{R}$:

$$A = \{(x, y) \in \mathbb{R} \times \mathbb{R} \mid x = y\},$$
$$B = \{(x, y) \in \mathbb{R} \times \mathbb{R} \mid x \leq y\},$$
$$C = \{(x, y) \in \mathbb{R} \times \mathbb{R} \mid |x - y| \leq 1\}.$$

Because $\pi = \pi$, $\pi \leq \pi$, and $|\pi - \pi| = 0 \leq 1$, we know that (π, π) is in all three of these relations. You should check that (e, π) is in exactly two of them, and (π, e) is in only one.

Exercise 6.1 In the plane, sketch the set of points (x, y) that are in each of the relations in Example 6.2.

EXAMPLE 6.3. Here are three relations on $\mathbb{N}$:

$$D = \{(a, b) \in \mathbb{N} \times \mathbb{N} \mid a + b \text{ is even}\},$$
$$E = \{(a, b) \in \mathbb{N} \times \mathbb{N} \mid a|b\},$$
$$F = \{(a, b) \in \mathbb{N} \times \mathbb{N} \mid 6|(a - b)\}.$$

You should check that $(3, 9)$ is in all three of these relations, $(4, 22)$ is in exactly two of them, and $(3, 6)$ is in only one.

EXAMPLE 6.4. The following are three relations on $S = \{\text{Sue, Fred, Bob, Lena}\}$:

$G = \{(\text{Sue, Fred}), (\text{Sue, Bob}), (\text{Lena, Sue})\}$,
$H = \{(\text{Sue, Sue}), (\text{Bob, Bob})\}$,
the empty set, $\emptyset$.

Relation G might correspond to a concept like "admiration," where Sue admires both Fred and Bob, while Bob admires no one. But just as functions are best defined without "rules," relations don't need descriptions or meanings in order to exist: the definition says that *any* subset of $S \times S$ is a relation on S.

Also like functions, there is a useful notation for relations that often replaces ordered pairs: it is common to write xRy to mean that $(x, y) \in R$, or to connect x and y with a symbol like $\preceq$ or $\sim$. Thus $(1, \sqrt{2}) \in B$ could also be expressed as $1\,B\,\sqrt{2}$, but most often you will just see $1 \leq \sqrt{2}$.

Just as adjectives like "injective" and "surjective" are part of the language of functions, the following adjectives describe certain relations.

DEFINITION 6.5. A relation R on a set S is

(a) *reflexive* if for all $x \in S$, xRx,
(b) *symmetric* if for all $x, y \in S$, xRy implies yRx,
(c) *transitive* if for all $x, y, z \in S$, xRy and yRz implies xRz.

Consider the relation E on $\mathbb{N}$ defined in Example 6.3. It is a reflexive relation because $n|n$ for any $n \in \mathbb{N}$. It is not a symmetric relation: it's true that $2|4$, but it's not true that $4|2$. Finally, the transitive property follows from part (c) of Exercise 1.10.

Exercise 6.2 Determine whether each of the relations in Example 6.2 is reflexive, symmetric, and/or transitive.

Exercise 6.3 In this exercise you should construct a relation with specified properties.

(a) Flip a coin three times to determine if your relation should be reflexive, symmetric, and/or transitive. Then construct a relation on a non-empty set of your choice that agrees with your list of properties.
(b) Repeat the process, re-flipping the coin if needed to get a different list of properties.

Exercise 6.4 Explain why any function $f : S \to S$ is a relation on S. Then give a function $g : \mathbb{Z} \to \mathbb{Z}$ that is neither reflexive, symmetric, nor transitive.

6.2. Partial Orders

Many well-known relations are orderings. The real numbers are ordered by $\leq$ and $\geq$, and subsets of a set S are ordered by $\subseteq$ and $\supseteq$. All four of these relations are reflexive and transitive. These relations are not symmetric, though, and in fact the notion of symmetry runs counter to the notion of an ordering.

DEFINITION 6.6. A relation R is *antisymmetric* if whenever aRb and bRa, then $a = b$.

For example, the less than or equal relation $\leq$ on $\mathbb{R}$ is antisymmetric, since $a \leq b$ and $b \leq a$ implies that $a = b$.

DEFINITION 6.7. A relation is a *partial order* if it is reflexive, antisymmetric, and transitive. A *partially ordered set*, or *poset*, consists of a set S and a partial order $\preceq$ on S. Posets are often represented as pairs, $(S, \preceq)$, so as to be clear about the set and the ordering under discussion.

By the results of Exercise 6.2, you already know that $\leq$ is a reflexive and transitive relation on $\mathbb{R}$, so $(\mathbb{R}, \leq)$ is a poset. In Exercise 6.37, you are asked to show that $\subseteq$ is an antisymmetric relation on $\mathcal{P}(S)$, the power set of any set S, which along with Exercise 6.33 shows that $(\mathcal{P}(S), \subseteq)$ is a poset.

Exercise 6.5 In the previous section we noted that divisibility $|$ is a reflexive and transitive relation on $\mathbb{N}$. Prove that divisibility is antisymmetric, and therefore it is a partial order on $\mathbb{N}$, which we can write $(\mathbb{N}, |)$.

Let $\preceq$ be a partial order on a finite set S. If a and b are distinct elements with $b \preceq a$, and there is no third element such that $b \preceq c \preceq a$, then a is said to *cover* b. Just as we saw in Chapter 3 with set inclusion, a Hasse diagram is a method for representing a partial order on a finite set, where line segments indicate covering. Figure 1 is a Hasse diagram for the set $[12] = \{1, 2, \ldots, 12\}$, where the partial order is given by divisibility; Exercise 6.38 asks you to verify that $([n], |)$ is a poset for any natural number n. Figure 1 in Chapter 3 shows a Hasse diagram for the power set of $\{a, b, c\}$ ordered by set inclusion.

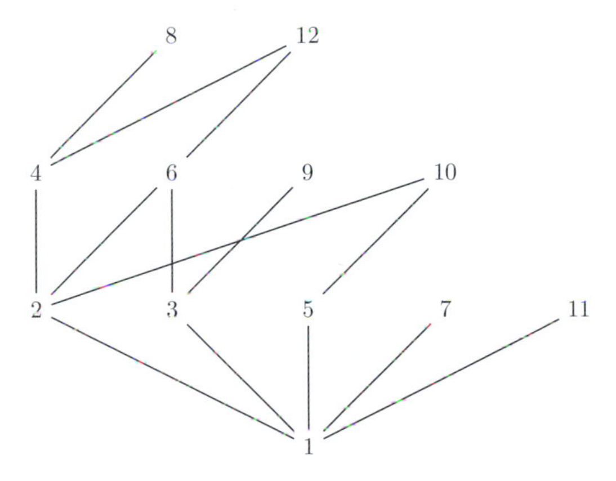

Figure 1. A Hasse diagram for the partial order given by divisibility on the set of integers $[12] = \{1, 2, \ldots, 12\}$.

Exercise 6.6 Can you redraw the Hasse diagram in Figure 1 to avoid crossing line segments?

DEFINITION 6.8. A *maximal element* in a poset $(S, \preceq)$ is an element M such that if $M \preceq a$, then $a = M$. Similarly a *minimal element* is an element m where if $a \preceq m$, then $a = m$.

The poset $(\mathcal{P}(\{a, b, c\}), \subseteq)$ has $\emptyset$ as the unique minimal element and $\{a, b, c\}$ as the unique maximal element. Figure 1 shows that the natural number 1 is the unique minimal element in the poset $([12], |)$, while there are six maximal elements: $7, 8, 9, 10, 11$ and 12. Notice that this is the second time we have defined minimal element, the first occurring in Section 4.1 in the context of $\mathbb{R}$ and $\leq$. Exercise 6.41 asks you to show that the earlier definition is consistent with Definition 6.8.

Exercise 6.7 Recalling the introduction to Fibonacci numbers in Section 1.2, let S_n be the set of rabbit pairs born in the first n months, which becomes a poset $(S_n, \preceq)$ if we define $x \preceq y$ when rabbit pair x either is the same as rabbit pair y or is a descendant of rabbit pair y. Explain why $(S_n, \preceq)$ is a poset, and draw the Hasse diagram for $(S_5, \preceq)$. What are the minimal and maximal elements of $(S_5, \preceq)$? Feel free to give the rabbits cute or useful names.

Exercise 6.8 Let $\mathcal{I}$ be the set of all closed intervals $[a, b]$ where $0 \leq a \leq b \leq 1$. For example, $\mathcal{I}$ contains $[0.5, 0.79]$, $[0, 1]$, and the point $0.87 = [0.87, 0.87]$. Order the elements of $\mathcal{I}$ by saying $[a, b] \preceq [c, d]$ if $[a, b] = [c, d]$ or $b < c$.

(a) Explain why $[0, 1/2] \preceq [3/4, 1]$, but $[1/6, 1/3] \not\preceq [1/4, 1/2]$.
(b) Verify that $(\mathcal{I}, \preceq)$ is a poset.
(c) Prove that the maximal elements in this poset are the intervals $[a, 1]$ where $0 \leq a \leq 1$.
(d) Prove that the minimal elements in this poset are the intervals $[0, b]$ where $0 \leq b \leq 1$.
(e) Show by example that it is possible for an element to be both maximal and minimal.

DEFINITION 6.9. In a poset, we say that two elements x and y are *comparable* if $x \preceq y$ or $y \preceq x$. Figure 1 shows that not all pairs of elements in a poset need to be comparable; for example, neither $5|7$ nor $7|5$ is true.

A poset is *totally ordered*[1] if every pair of elements is comparable. For example, the poset $(\mathbb{R}, \leq)$ is a totally ordered set: if a and b are any two real numbers, then $a \leq b$ or $b \leq a$.

A poset is *well ordered*[2] if it is a totally ordered set such that every non-empty subset contains a minimal element. This definition should remind you of the Well-Ordering Principle for $\mathbb{N}$ (Theorem 4.3), which states that $(\mathbb{N}, \leq)$ is well ordered.

Exercise 6.9 Which of the following posets are totally ordered?

(a) $(\mathbb{Z}, \leq)$
(b) $(\mathbb{N}, \leq)$
(c) $(\mathbb{N}, |)$
(d) $(T, |)$, where $T = \{2^k \mid k \in \mathbb{N}\}$

Exercise 6.10 Which of the totally ordered sets from Exercise 6.9 are well ordered?

Many end-of-chapter exercises help you further explore the structure of posets, including the problem of determining the largest totally ordered subsets in a poset, its maximal chains. This leads naturally to the study of antichains and Mirsky's Theorem in Project 11.6. Also, Exercise 6.76 explains how certain partial orders can be refined to total orders.

1 For some reason, no one calls it a "toset" ...
2 ... or a "woset."

6.3. Equivalence Relations

There are many contexts in mathematics where you might think of two mathematical objects as being "equivalent," even when they are not identical. For example, two triangles in the plane are congruent if their edges have the same lengths. Another example is similarity: two triangles in the plane are similar if they have the same angles. In both cases the triangles share important features, and in that sense they are equivalent. This informal notion is captured by equivalence relations.

DEFINITION 6.10. A relation R on a set S is an *equivalence relation* if it is reflexive, symmetric, and transitive. Equivalence relations are often denoted with the symbol $\sim$ or some other variation on an equals sign, like $\approx, \simeq, \cong$, or $\equiv$.

Out of the six relations described in Examples 6.2 and 6.3, three are equivalence relations: A, D, and F. It shouldn't be surprising that A is an equivalence relation; it's hard to be more "equivalent" than being equal! Having mentioned this, we want to immediately comment that the concept of equivalence relation is intended to allow different elements to be related without necessarily being equal, even if equality is part of the definition of the relation. As an example, let $\sim_{\mathrm{s}}$ be the relation on $\mathbb{R}$ that says $x \sim_{\mathrm{s}} y$ when $\sin(x) = \sin(y)$. We can have $x \sim_{\mathrm{s}} y$ without $x = y$; as examples, $0 \sim_{\mathrm{s}} \pi$ and $\pi/4 \sim_{\mathrm{s}} -5\pi/4$.

PROPOSITION 6.11. *The relation* $\sim_s$ *is an equivalence relation on* $\mathbb{R}$.

PROOF. We need to establish that $\sim_{\mathrm{s}}$ has three properties.

Reflexive: This follows since $\sin(x) = \sin(x)$ for all $x \in \mathbb{R}$.
Symmetric: If $x \sim_{\mathrm{s}} y$, then $\sin(x) = \sin(y)$. But then it is also true that $\sin(y) = \sin(x)$, so $y \sim_{\mathrm{s}} x$.
Transitive: If $x \sim_{\mathrm{s}} y$ and $y \in \sim_{\mathrm{s}} z$, then

$$\sin(x) = \sin(y) \quad \text{and} \quad \sin(y) = \sin(z).$$

Thus $\sin(x) = \sin(z)$, so $x \sim_{\mathrm{s}} z$. □

A generalization of $\sim_{\mathrm{s}}$ and Proposition 6.11 is explored in Exercise 6.57 at the end of this chapter.

Exercise 6.11 In Example 6.3 we defined the relation D on the natural numbers, $\mathbb{N}$. Consider the extension of this relation to the integers:

$$\equiv_e \text{ is the relation } \{(a, b) \in \mathbb{Z} \times \mathbb{Z} \mid a + b \text{ is even}\}.$$

Prove that $\equiv_e$ is an equivalence relation on $\mathbb{Z}$.

If R is an equivalence relation on a set S, there is an important subset of S associated to each $a \in S$:

DEFINITION 6.12. If R is an equivalence relation on a set S, the *equivalence class* of $a \in S$ is the set of all elements related to a:

$$[a] = \{b \in S \mid aRb\}.$$

For example, you have just shown that $\equiv_e$ is an equivalence relation on $\mathbb{Z}$. The equivalence class of 0 is

$$[0] = \{\ldots, -2, 0, 2, 4, 6, \ldots\}$$

and the equivalence class of 1 is

$$[1] = \{\ldots, -3, -1, 1, 3, 5, \ldots\}.$$

Figure 2 shows that every integer is in exactly one of these two equivalence classes.

$\cdots$ ● ○ ● ○ ● ○ ● $\cdots$

−3 −2 −1 0 1 2 3

Figure 2. The two equivalence classes for $\equiv_e$. The class [0] is indicated by hollow circles, and [1] has filled circles.

Exercise 6.12 Let $\approx$ be the relation on $\mathbb{R}$ defined by having $x \approx y$ when x and y have the same integer part, that is, when $\lfloor x \rfloor = \lfloor y \rfloor$. Prove that this is an equivalence relation, and prove that the equivalence class of each element is a half-open interval of the form $[n, n+1)$, where $n \in \mathbb{Z}$.

With $\equiv_e$ and $\approx$ from the previous two exercises, you might have noticed that the equivalence classes of two elements are either the same or they are disjoint. This is not a coincidence, as you may begin to believe after working through the next exercise.

Exercise 6.13 Consider the equivalence relation $\sim_s$ on $\mathbb{R}$ from Proposition 6.11.

(a) Show that the equivalence class of $\pi/2$ is

$$\left[\frac{\pi}{2}\right] = \{\ldots, -\frac{3\pi}{2}, \frac{\pi}{2}, \frac{5\pi}{2}, \ldots\} = \{\frac{\pi}{2} + n \cdot 2\pi \mid n \in \mathbb{Z}\}.$$

(b) Show that the equivalence class of $\pi/6$ is

$$\left[\frac{\pi}{6}\right] = \{\frac{\pi}{6} + n \cdot 2\pi \mid n \in \mathbb{Z}\} \cup \{\frac{5\pi}{6} + n \cdot 2\pi \mid n \in \mathbb{Z}\}.$$

(c) Prove that for any two equivalence classes $[a]$ and $[b]$ for $\sim_s$, either $[a] = [b]$ or $[a] \cap [b] = \emptyset$.

What you proved in the exercise above is a special case of the following general result.

THEOREM 6.13. *If R is an equivalence relation on a set S and $a, b \in S$, then $[a] = [b]$ or $[a] \cap [b] = \emptyset$.*

Before presenting a proof of this result, we should think about the logic of the theorem. The statement of the theorem is of the form $P \Rightarrow Q_1 \vee Q_2$, namely

$$\underbrace{R \text{ is an equivalence relation}}_{P} \Rightarrow \underbrace{[a] = [b]}_{Q_1} \vee \underbrace{[a] \cap [b] = \emptyset}_{Q_2}.$$

This is equivalent to $P \wedge (\sim Q_2) \Rightarrow Q_1$. (You might have proven this logical equivalence if you worked on Exercise 2.46.) This restatement of what needs to be proven makes intuitive sense: if you are being asked to show that at least one of two possibilities is true, and one of the possibilities is false, then the other one had better be true! This is the approach we take in the proof.

PROOF OF THEOREM 6.13. Let R be an equivalence relation on S, and assume there are elements $a, b \in S$ with $[a] \cap [b] \neq \emptyset$. Thus there must be some $c \in S$ that is in both $[a]$ and $[b]$. So aRc and bRc. Since R is symmetric, we also know cRb. Since R is transitive, aRc and cRb imply that aRb; the symmetric property then implies that bRa as well.

Now that we know that a and b are related, we show that $[a] = [b]$ by proving that each equivalence class contains the other. If d is any element of $[b]$, then bRd. Since aRb, the transitive property shows that aRd, and thus $d \in [a]$. So we know that $[b] \subseteq [a]$. A similar argument starting with $d \in [a]$ shows that $[a] \subseteq [b]$, and thus $[a] = [b]$. □

The result of Theorem 6.13, that $[a] = [b]$ or $[a] \cap [b] = \emptyset$, suggests a relationship between equivalence classes and partitions, which were defined in Section 3.7. The blocks of a partition of a set S are non-overlapping, and their union is all of S, just as we have seen with examples of equivalence classes. The following theorem says that for every set S there is a bijection between the set of equivalence relations on S and the set of partitions of S.

THEOREM 6.14. *Let S be a set. If $\mathcal{B} = \{S_i\}$ is a partition of S, then the relation defined by*

$$x \sim_{\mathcal{B}} y \text{ when } x \text{ and } y \text{ are in the same block } S_i$$

is an equivalence relation on S. Conversely, if $\sim$ is an equivalence relation on S, then the equivalence classes induced by $\sim$ form a partition of S.

PROOF. If you look back at the statement of Theorem 6.13 and the definition of partition, you will see that we have already proven that equivalence relations give rise to partitions: Theorem 6.13 says that two distinct equivalence classes cannot intersect (requirement (a) for a partition), and requirement (b) follows because every element is in an equivalence class, namely $a \in [a]$.

Now, let $\mathcal{B}$ be a partition of S, and let $\sim_{\mathcal{B}}$ be the relation given in the statement of the theorem. Is this relation ...

... *reflexive*? Yes, because $x \sim_{\mathcal{B}} x$ simply states that x and x are in the same block of $\mathcal{B}$.

... *symmetric*? Yes, because $x \sim_{\mathcal{B}} y$ means that x and y are in the same block of the partition. But then y and x are in the same block of the partition, so $y \sim_{\mathcal{B}} x$.

... *transitive*? Yes, because if $x \sim_{\mathcal{B}} y$ and $y \sim_{\mathcal{B}} z$, then x and y are in the same block S_i of the partition, and y and z are in the same block S_j. Since $y \in S_i$ and $y \in S_j$, we know that $S_i \cap S_j \neq \emptyset$. Since these are blocks in a partition, it follows that $S_i = S_j$. Thus this block contains both x and z, meaning that $x \sim_{\mathcal{B}} z$. □

Exercise 6.14 If S is any set, then the set of singleton sets $\{\{s\} \mid s \in S\}$ forms a partition. Describe the corresponding equivalence relation on S.

Exercise 6.15 How many equivalence relations are there on the set $\{a, b, c\}$?

6.4. Modulo m

In Sections 6.1 and 6.3 you encountered the equivalence relation F, which was originally defined on $\mathbb{N}$ but can be extended to $\mathbb{Z}$:

$$\equiv_6 \text{ is the relation } \{(a, b) \in \mathbb{Z} \times \mathbb{Z} \mid 6|(a - b)\}\,.$$

You have also encountered a number of earlier exercises where the concept of even and odd mattered, and other exercises where it matters if an integer n is of the form $3m$, $3m + 1$, or $3m + 2$. In 1801, having had a number of similar things occur earlier in his work, Carl Friedrich Gauss introduced the idea of *congruence modulo m*.

DEFINITION 6.15. If $a, b \in \mathbb{Z}$ and $m \in \mathbb{N}$, we say *a is congruent to b modulo m* when $m|(a - b)$. The notation is

$$a \equiv b \ (\text{MOD } m) \text{ whenever } m|(a - b)\,.$$

The extra "(MOD m)" in the relation notation may seem cumbersome at first, but its usage is common. You may also see $a \equiv b$ (MOD m) denoted by the more compact notation $a \equiv_m b$. The equivalence class of $a \in \mathbb{Z}$ is denoted $[a]_m$. The number m is called the *modulus*, and when the value of the modulus is clear, the subscript is sometimes dropped from these notations.

PROPOSITION 6.16. *For any $m \in \mathbb{N}$, the* MOD *m relation is an equivalence relation on $\mathbb{Z}$.*

PROOF. We need to establish the three required properties.

Reflexive: For any $a \in \mathbb{Z}$, $a - a = 0$. Since $m|0$, it follows that $a \equiv a$ (MOD m).
Symmetric: Assume that $a \equiv b$ (MOD m). By the definition this means $m|(a - b)$. But $(b - a) = -(a - b)$, and so $m|(b - a)$ as well. Hence $b \equiv a$ (MOD m).
Transitive: If $a \equiv b$ (MOD m) and $b \equiv c$ (MOD m), then $a - b = km$ and $b - c = lm$ for some integers k, l. Since $a - c = a - b + b - c = km + lm = (k + l)m$ implies $m|(a - c)$, we know that $a \equiv c$ (MOD m). □

Exercise 6.16 What are the equivalence classes of $\mathbb{Z}$ with respect to the equivalence relation MOD 2? What are the equivalence classes for MOD 3?

For each $m \in \mathbb{N}$ define the *set of remainders* (or the *residues*) to be

$$\text{RE}_m = \{0, 1, \ldots, m - 1\}\,.$$

For example, $\text{RE}_3 = \{0, 1, 2\}$.

PROPOSITION 6.17. *For any $m \in \mathbb{N}$, let RE_m be as defined above. Then for every $a \in \mathbb{Z}$ there is a unique $r \in \text{RE}_m$ such that $a \equiv r$ (MOD m).*

Exercise 6.17 Use the Division Algorithm from Section 1.6 to prove Proposition 6.17.

COROLLARY 6.18. *For a and $b \in \mathbb{Z}$, $a \equiv b$ (MOD m) if and only if $a = pm + r$ and $b = qm + r$ for some $p, q \in \mathbb{Z}$ and a common $r \in \text{RE}_m$.*

PROOF. Assume first that $a = pm + r$ and $b = qm + r$. Then $a - b = (p - q)m$, so $m|(a - b)$, which implies that $a \equiv b \ (\text{MOD } m)$.

Now assume that $a \equiv b \ (\text{MOD } m)$. We know that $m|(a - b)$, so $a - b = km$ for some $k \in \mathbb{Z}$. We also know by Proposition 6.17 that there is an $r \in \text{RE}_m$ such that $b \equiv r(\text{MOD } m)$, so $b = qm + r$ for some $q \in \mathbb{Z}$. Thus

$$a = km + b = km + (qm + r) = pm + r$$

if we set $p = k + q$. □

Exercise 6.18 Prove that $a \equiv b \ (\text{MOD } m)$ if and only if $b = a + km$ for some $k \in \mathbb{Z}$.

Exercise 6.19 Prove that the set of equivalence classes of $\mathbb{Z}$ with respect to congruence MOD m is $\{[0]_m, [1]_m, \ldots, [m-1]_m\}$.

GOING BEYOND THIS BOOK. It is known that there are true statements of arithmetic that cannot be proven. In **[Con13]** John Conway describes a number of examples of elementary arithmetic functions, whose definitions are based on remainders modulo an integer m, that appear to be true but are not yet proven. The functions are interesting in their own right, and the possibility that they lead to elementary true but not provable statements is certainly intriguing.

6.5. Modular Arithmetic

In this section we will explore modular arithmetic. Let $\mathbb{Z}_m$ denote the set of equivalence classes of $\mathbb{Z}$ modulo m. For example, $\mathbb{Z}_3 = \{[0]_3, [1]_3, [2]_3\}$, which is also equal to $\{[0]_3, [4]_3, [-16]_3\}$ but not $\{[0]_3, [4]_3, [16]_3\}$.

The fact that $[1]_3 = [4]_3$ means that you need to use great care when defining operations or functions in terms of equivalence classes. Any element of an equivalence class is called a *representative* of that class, so 1, 4, and -2 are just three of infinitely many representatives of the class $[1]_3$. This is similar to the familiar case of fractions like $3/7$ and $9/21$ both representing the same value. In the case of fractions, we know that any representative is equally valid for use in computations. We would, however, be concerned if someone defined a function $f : \mathbb{Q} \to \mathbb{Z}$ by saying $f(a/b) = a$, as then $f(3/7) \neq f(9/21)$, but $3/7$ and $9/21$ represent the same rational number.

A CAUTIONARY TALE: Derek is working at the board, and he's written down what he thinks is a function from $\mathbb{Z}_4$ to $\mathbb{Z}_3$:

$$f([a]_4) = [a^2]_3 .$$

As an aside he's written

$$f([2]_4) = [2^2]_3 = [4]_3 = [1]_3.$$

John, however, is made a bit nervous by this, and he says "Wait a minute, Derek. The numbers 2 and 6 are in the same equivalence class modulo 4, so $[2]_4 = [6]_4$. Why shouldn't it be the case that

$$f([2]_4) = f([6]_4) = [6^2]_3 = [36]_3 = [0]_3 \text{ ?"}$$

Exercise 6.20 Explain why this poses a problem for f being a function from $\mathbb{Z}_4$ to $\mathbb{Z}_3$.

MORAL OF THE STORY: If you define a mathematical object in terms of representatives of an equivalence class, you need to make sure that your definition does not depend on which representative you pick. Mathematicians say that an object defined on equivalence classes is *well defined* when it has the same value regardless of which representative is used from a given equivalence class.

Our Cautionary Tale partly explains why we need to present a proof of the following theorem, stating that addition and multiplication can be defined on $\mathbb{Z}_m$ in terms of representatives of the equivalence classes, and that these operations are well defined.

THEOREM 6.19. *If* $a \equiv a'$ (MOD m) *and* $b \equiv b'$ (MOD m), *then*

(a) $a + b \equiv a' + b'$ (MOD m),
(b) $ab \equiv a'b'$ (MOD m).

PROOF. We know that if $a \equiv a'$, then $a' = a + km$ for some $k \in \mathbb{Z}$; similarly we know that $b' = b + lm$ for some $l \in \mathbb{Z}$. Thus $a' + b' = a + b + (k + l)m$, so by Exercise 6.18 we have $a + b \equiv a' + b'$. Further, $a'b' = (a + km)(b + lm) = ab + (kb + la + klm)m$, so $ab \equiv a'b'$ as well. □

We can now define *modular arithmetic*: There are functions $+$ and $\cdot$ from $\mathbb{Z}_m \times \mathbb{Z}_m$ to $\mathbb{Z}_m$ which are defined by

$$[a]_m + [b]_m = [a + b]_m \quad \text{and} \quad [a]_m \cdot [b]_m = [a \cdot b]_m .$$

We emphasize that Theorem 6.19 says that the choice of representatives doesn't affect the result of addition or multiplication, so each of these is a legitimate function from $\mathbb{Z}_m \times \mathbb{Z}_m$ to $\mathbb{Z}_m$: each element of $\mathbb{Z}_m \times \mathbb{Z}_m$ is mapped to a unique element of $\mathbb{Z}_m$.

Because of the connection between equivalence classes and equivalence relations, we have two ways of expressing addition modulo m. In terms of equivalence relations we can make a statement like:

$$3 + 2 \equiv 1 \text{ (MOD 4)}.$$

What we have shown is that we can also make the exact same claim using the language of equivalence classes by writing:

$$[3]_4 + [2]_4 = [1]_4 .$$

Notice that in this second statement we are discussing sets (the equivalence classes) and so we should use the symbol $=$ in this context, while we should use $\equiv$ (MOD 4) when discussing individual numbers. The same situation arises for multiplication modulo m.

Exercise 6.21 The equivalence classes [0] and [1] in $\mathbb{Z}_m$ have some of the fundamental properties of 0 and 1 in $\mathbb{Z}$.

(a) Prove that $0 + a \equiv a$ (MOD m) for any integer a and any modulus $m \in \mathbb{N}$.
(b) Prove that $1 \cdot a \equiv a$ (MOD m) for any integer a and any modulus $m \in \mathbb{N}$.

Because of the properties above, [0] is said to be an additive identity in $\mathbb{Z}_m$ and [1] is a multiplicative identity in $\mathbb{Z}_m$.

$+$	$[0]_3$	$[1]_3$	$[2]_3$
$[0]_3$	$[0]_3$	$[1]_3$	$[2]_3$
$[1]_3$	$[1]_3$	$[2]_3$	$[0]_3$
$[2]_3$	$[2]_3$	$[0]_3$	$[1]_3$

$\cdot$	$[0]_3$	$[1]_3$	$[2]_3$
$[0]_3$	$[0]_3$	$[0]_3$	$[0]_3$
$[1]_3$	$[0]_3$	$[1]_3$	$[2]_3$
$[2]_3$	$[0]_3$	$[2]_3$	$[1]_3$

Figure 3. Addition and multiplication for the integers modulo 3.

It is common to display the results of modular arithmetic in tables. For example, the tables for arithmetic in $\mathbb{Z}_3$ are shown in Figure 3. The connection with $(\text{MOD } m)$ notation is easy to see. For example, since $5 \equiv 2\,(\text{MOD } 3)$ and $8 \equiv 2\ (\text{MOD } 3)$, we know that $5 \cdot 8 \equiv 2 \cdot 2 \equiv 1\ (\text{MOD } 3)$, as displayed in the lower-right entry of the multiplication table.

Exercise 6.22 It is time for you to practice modular arithmetic.

(a) Create the addition and multiplication tables for $\mathbb{Z}_5$.
(b) Create the addition and multiplication tables for $\mathbb{Z}_6$.
(c) You should now be able to check that the product of non-zero entries in $\mathbb{Z}_5$ is never zero, just like in ordinary arithmetic. But in $\mathbb{Z}_6$, the product of non-zero entries can be zero. Why do you think this happens in $\mathbb{Z}_6$ but not in $\mathbb{Z}_5$?

Exercise 6.23 State a version of Theorem 6.19 that involves n terms, and use mathematical induction to prove it.

Here is one example of the utility of being able to do arithmetic modulo a number m. Trying to find the remainder of 9^{34} divided by 11 sounds quite hard. But $9 \equiv -2\ (\text{MOD } 11)$, so $9^{34} \equiv (-2)^{34}\ (\text{MOD } 11)$. Further, since $(-2)^5 = -32$ and $-32 \equiv 1\ (\text{MOD } 11)$, we have

$$(-2)^{34} \equiv (-2)^4(-2)^{30} \equiv 16 \cdot 1^6 \equiv 5\ (\text{MOD } 11).$$

Thus the remainder is 5.

Here is another example of the utility of modular arithmetic. Exercise 4.23 asks you to use induction and a minimal criminal argument to prove that, for all $n \in \mathbb{N}$, $7^n + 5$ is divisible by 3. Using modular arithmetic, this is now a computation. Being divisible by 3 is equivalent to being congruent to 0 modulo 3. And we can establish the result by applying the fact that $7 \equiv 1\ (\text{MOD } 3)$ and $5 \equiv 2\ (\text{MOD } 3)$:

$$7^n + 5 \equiv 1^n + 2 \equiv 1 + 2 \equiv 0\ (\text{MOD } 3)\,.$$

Exercise 6.24 What's the last digit of 9^{99}? Notice that if you just want the last digit, you can work with the integers modulo 10.

In our Cautionary Tale, we attempted to define a function from $\mathbb{Z}_4$ to $\mathbb{Z}_3$ by squaring representatives. The definition did not actually make sense, as it was not consistent across all elements in an equivalence class. Given that example, it may seem a bit daft for us to now consider the following proposed definition for a function from $\mathbb{Z}_6$ to $\mathbb{Z}_3$:

$$f([a]_6) = [a^2]_3\,.$$

To see if this function is well defined, we need to check if the answer is the same regardless of the representative we choose for any given equivalence class. By Exercise 6.18 we know that $a \equiv b$ (MOD 6) if and only if $b = a + 6k$ for some $k \in \mathbb{Z}$. Notice that

$$f([a+6k]_6) = [(a+6k)^2]_3 = [a^2 + 12ak + 36k^2]_3 \,.$$

Since 3 divides 12 and 36 we know $12ak \equiv 0$ (MOD 3) and $36k^2 \equiv 0$ (MOD 3). Theorem 6.19 allows us to replace these terms by zeros, yielding

$$a^2 + 12ak + 36k^2 \equiv a^2 \text{ (MOD 3)} \,.$$

Thus $f([a]_6) = f([b]_6)$ whenever $a \equiv b$ (MOD 3), so the function is well defined.

Exercise 6.25 Find a step in the argument given above for $f : \mathbb{Z}_6 \to \mathbb{Z}_3$ that fails when applied to the supposed function in the Cautionary Tale.

Exercise 6.26 Show that the following functions are well defined, where $m, n \in \mathbb{N}$.

(a) $f : \mathbb{Z}_8 \to \mathbb{Z}_4$ given by $f([a]_8) = [a]_4$.
(b) $f : \mathbb{Z}_{mn} \to \mathbb{Z}_n$ given by $f([a]_{mn}) = [a]_n$.
(c) $f : \mathbb{Z}_{12} \to \mathbb{Z}_6$ given by $f([a]_{12}) = [a^2]_6$.
(d) $f : \mathbb{Z}_{mn} \to \mathbb{Z}_n$ given by $f([a]_{mn}) = [a^2]_n$.

REMARK 6.20. In section 5.7.4 we discussed a common abuse of notation that occurs in the context of functions. There are many places in mathematics where one abuses notation by giving a single symbol multiple meanings or by simplifying notation. The expectation is that the context makes clear what the notation does not.

There are many abuses of notation that occur in the context of modular arithmetic. For example, when the modulus is clear or unimportant, you will often see $a \equiv b$ instead of the more complete $a \equiv b$ (MOD m). Similarly in denoting equivalence classes you might see $[a]$ instead of $[a]_m$. Perhaps even more confusing, one often drops the square bracket notation and lets a stand for $[a]_m$. This occurs most commonly in addition and multiplication tables, where the square brackets make for a rather cluttered diagram.

6.6. Invertible Elements

This section makes use of material from Sections 4.4 and 4.5. Theorem 6.19 shows that addition and multiplication make sense in $\mathbb{Z}_m$; it is also the case that subtraction is well defined.

THEOREM 6.21. *If $a \equiv a'$ (MOD m) and $b \equiv b'$ (MOD m), then $a - b \equiv a' - b'$ (MOD m).*

Exercise 6.27 Prove Theorem 6.21.

It is less clear that you can divide in $\mathbb{Z}_m$, and given that we are working with integers, you might guess that this is not always possible. Let's experiment with $m = 15$ and see if we can solve

$$2x \equiv 1 \text{ (MOD 15)} \,.$$

If we can, then whatever x might be it deserves to be considered as "1/2" within $\mathbb{Z}_{15}$. Having no better plan available to us, we form the row of the multiplication table for

$\mathbb{Z}_{15}$ corresponding to $[2]_{15}$. We'll also start to abuse notation and write k instead of $[k]_{15}$.

$\cdot$	0	1	2	3	4	5	6	7	**8**	9	10	11	12	13	14
2	0	2	4	6	8	10	12	14	**1**	3	5	7	9	11	13

The highlighted column shows us that $2 \cdot 8 \equiv 1 \pmod{15}$, so 8 plays the role of $1/2$ in $\mathbb{Z}_{15}$. We can use this fact to solve an equation like

$$2x + 7 \equiv 12 \pmod{15}$$

as follows:

$$2x + 7 \equiv 12 \Rightarrow 2x \equiv 5 \Rightarrow 8 \cdot 2x \equiv 8 \cdot 5 \Rightarrow x \equiv 40 \equiv 10 \pmod{15}.$$

In this calculation, we used both parts of Theorem 6.19. part (a) to add -7 to both sides of a congruence, and part (b) to multiply both sides by 8. We can verify that we did not make an error by plugging our answer of $x \equiv 10 \pmod{15}$ back into the original congruence:

$$2 \cdot 10 + 7 \equiv 27 \equiv 12 \pmod{15}.$$

Exercise 6.28 What element of $\mathbb{Z}_{15}$ deserves to be called "1/4"?

DEFINITION 6.22. If $a + b \equiv 0 \pmod{m}$, we say that a and b are *additive inverses*.

If $ab \equiv 1 \pmod{m}$, we say that a and b are *multiplicative inverses* modulo m, and $[a]$ and $[b]$ are multiplicative inverses in $\mathbb{Z}_m$.

We can express $[b]$ as $[a]^{-1}$, notation that is similar to what we used for an inverse function in Chapter 5. But, as with functions, the concept of inverse requires great care.

Since $a + (-a) \equiv 0 \pmod{m}$ we know that $[-a]$ is an additive inverse of $[a]$, and so additive inverses always exist in $\mathbb{Z}_m$. The situation is more complicated for multiplicative inverses. You have seen that [8] is a multiplicative inverse of [2] in $\mathbb{Z}_{15}$, and you should have discovered that [4] is its own multiplicative inverse in $\mathbb{Z}_{15}$, so you can write $[8] = [2]^{-1}$ and $[4] = [4]^{-1}$ in $\mathbb{Z}_{15}$. Having now realized that at times you can do division in $\mathbb{Z}_m$, an unreasonably optimistic individual might conjecture that every non-zero element in $\mathbb{Z}_{15}$ has a multiplicative inverse. The truly careful person would say, "Let's try to find multiplicative inverses for other values, like [3], before we get too excited." To do this, we once again produce the relevant row in the multiplication table for $\mathbb{Z}_{15}$:

$\cdot$	0	1	2	3	4	5	6	7	8	9	10	11	12	13	14
3	0	3	6	9	12	0	3	6	9	12	0	3	6	9	12

The number 1 does not appear anywhere in this row, so there is no number a such that $3a \equiv 1 \pmod{15}$. You might now be sad, since [3] does not have a multiplicative inverse in $\mathbb{Z}_{15}$. But instead you really ought to be intrigued. *Why* does [2] have an inverse in $\mathbb{Z}_{15}$ but not [3]? Is there a satisfactory condition that describes when an element of $\mathbb{Z}_m$ has a multiplicative inverse?

Exercise 6.29 Show that in $\mathbb{Z}_{14}$ the elements [2] and [10] do not have multiplicative inverses, but [3] and [5] do.

The following theorem should sound plausible, given your results from working on the exercise above.

THEOREM 6.23. *Let* $[k]$ *be a non-zero element of* $\mathbb{Z}_m$. *Then* $[k]$ *has a multiplicative inverse in* $\mathbb{Z}_m$ *if and only if* $\mathrm{GCD}(k, m) = 1$.

PROOF. We have to establish two implications, both of which rely on the characterization of relatively prime integers given in Theorem 4.8.

($\Leftarrow$) If k and m are relatively prime, then there are integers s and t such that

$$sk + tm = 1\,.$$

But then $sk \equiv sk + tm \equiv 1 \pmod{m}$, so $[s]$ is a multiplicative inverse of $[k]$ in $\mathbb{Z}_m$.

($\Rightarrow$) If $[k]$ has a multiplicative inverse, then there exists an $s \in \mathbb{Z}$ such that $sk \equiv 1 \pmod{m}$. But this means that there is some $t \in \mathbb{Z}$ such that $sk = 1 + tm$, and therefore $sk + (-t)m = 1$. Thus $\mathrm{GCD}(k, m) = 1$. □

DEFINITION 6.24. Define $\mathcal{U}_m$ to be the subset of $\mathbb{Z}_m$ consisting of equivalence classes whose representatives are relatively prime to m:

$$\mathcal{U}_m = \{[a] \in \mathbb{Z}_m \mid \mathrm{GCD}(a, m) = 1\}\,.$$

In the study of arithmetic systems, elements that have multiplicative inverses are called *units*. The notation $\mathcal{U}_m$ comes from the fact that we are looking at $\mathcal{U}$nits in $\mathbb{Z}_m$.

Exercise 6.30 Show that if $[a] \in \mathcal{U}_m$, then $[a]$ has exactly one multiplicative inverse.

The set of units in $\mathbb{Z}_{15}$ is $\mathcal{U}_{15} = \{1, 2, 4, 7, 8, 11, 13, 14\}$, where we are abusing notation by dropping the brackets. The fact that all of these elements are units is verified by the multiplication table for $\mathcal{U}_{15}$ shown in Figure 4, as every row contains a 1.

·	1	2	4	7	8	11	13	14
1	1	2	4	7	8	11	13	14
2	2	4	8	14	1	7	11	13
4	4	8	1	13	2	14	7	11
7	7	14	13	4	11	2	1	8
8	8	1	2	11	4	13	14	7
11	11	7	14	2	13	1	8	4
13	13	11	7	1	14	8	4	2
14	14	13	11	8	7	4	2	1

Figure 4. The multiplication table for $\mathcal{U}_{15}$.

Exercise 6.31 What elements are in $\mathcal{U}_{14}$? Once you've found them, make a multiplication table for $\mathcal{U}_{14}$.

The following proposition explains why the multiplication tables for $\mathcal{U}_m$ don't contain any other elements of $\mathbb{Z}_m$.

PROPOSITION 6.25. *The set* $\mathcal{U}_m$ *is closed under multiplication and taking multiplicative inverses.*

Exercise 6.32 Prove Proposition 6.25. To do this, you will have to show that if $[a]$ and $[b]$ are both in $\mathcal{U}_m$, then so is $[ab]$. You will also have to show that if $[a] \in \mathcal{U}_m$, then $[a]^{-1} \in \mathcal{U}_m$.

REMARK 6.26. Exercise 6.19 shows that $|\mathbb{Z}_m| = m$, but what can you say about $|\mathcal{U}_m|$, the number of units modulo m? The answer takes some work to discover and justify; it is the topic of Project 11.7 on Euler's totient function. If this is of interest, we recommend starting with a couple of warm-up problems that can be found in Exercise 6.71.

6.7. End-of-Chapter Exercises

Exercises you can work on after Section 6.1

6.33 Let A be a non-empty set and consider the relation $\subseteq$ on $\mathcal{P}(A)$.

(a) Prove that $\subseteq$ is reflexive.
(b) Prove that $\subseteq$ is transitive.
(c) Show that $\subseteq$ is not symmetric.

6.34 Example 6.3 presents three relations on $\mathbb{N}$. Since any relation on $\mathbb{N}$ is a subset of $\mathbb{N} \times \mathbb{N}$, you can visualize relations on $\mathbb{N}$ as a subset of the points in the first quadrant whose coordinates are natural numbers. For example, a portion of relation D from Example 6.3 is shown in Figure 5. Make similar illustrations for the relations E and F. You may need to include more points from $\mathbb{N} \times \mathbb{N}$ than we did in Figure 5 in order for your figure to genuinely illustrate these relations.

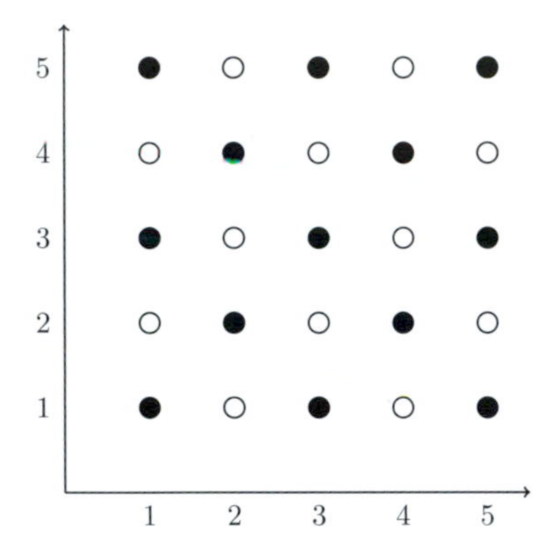

Figure 5. The elements of the relation D from Example 6.3 are illustrated by filled circles.

6.35 Since a relation on S is a subset of $S \times S$, you can often look at the "graph" of the relation, which is the obvious extension of the graph of a function. For example, in Figure 6 we illustrate the relation

$$R = \{(a,a),(a,b),(b,b),(b,d),(c,d),(d,b),(d,c)\}$$

on the set $S = \{a,b,c,d\}$.

(a) What geometric property of the graph of a relation corresponds to the property of being reflexive?
(b) What geometric property of the graph of a relation corresponds to the property of being symmetric?

6.36 Consider a finite set S with $|S| = n$.

(a) How many different relations exist on S?
(b) How many different reflexive relations exist on S?
(c) How many different symmetric relations exist on S?

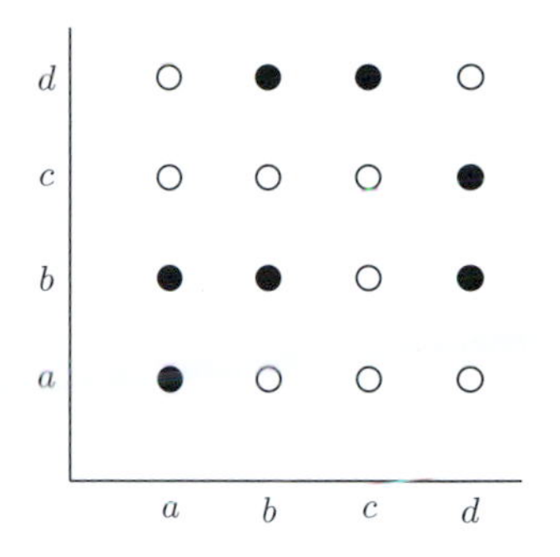

Figure 6. The elements of the relation R from Exercise 6.35 are illustrated by filled circles.

Exercises you can work on after Section 6.2

6.37 Continuing Exercise 6.33, prove that $\subseteq$ is an antisymmetric relation on $\mathcal{P}(S)$, and thus $(\mathcal{P}(S), \subseteq)$ is a poset for any set S.

6.38 Prove that for any $n \in \mathbb{N}$, $([n], |)$ is a poset, where $[n] = \{1, 2, \ldots, n\}$ and $|$ is divisibility.

6.39 Draw a Hasse diagram for the poset $([15], |)$. How does the diagram change if you add 0 as a new element to this set?

6.40 For any $n \in \mathbb{N}$, define $\mathcal{I}_n$ to be the partially ordered set whose elements are closed intervals in $[0, n]$ with integer endpoints, with $\preceq$ the same relation defined in Exercise 6.8.

(a) Draw a Hasse diagram for $\mathcal{I}_4$.
(b) What are the maximal elements of $\mathcal{I}_n$?
(c) What are the minimal elements of $\mathcal{I}_n$?

6.41 Show that the definition of minimal element given in Section 4.1 agrees with Definition 6.8 when applied to the totally ordered set $(\mathbb{R}, \leq)$.

6.42 Prove that any finite poset that is totally ordered is also well ordered.

6.43 Let R and R' be partial orders on a set S. Recalling that R and R' are subsets of $S \times S$, we may define R' to be a *refinement* of R if whenever $(a, b) \in R$, then $(a, b) \in R'$ as well.

(a) Consider the poset $(\mathcal{P}(\{1, 2\}), \subseteq)$. Create a refinement $\sqsubseteq$ of $\subseteq$ where $\{1\} \sqsubseteq \{2\}$.
(b) Consider the poset $(\mathcal{P}(\{1, 2, 3\}), \subseteq)$. Create a refinement $\sqsubseteq$ of $\subseteq$ where $\{1, 2\} \sqsubseteq \{1, 3\}$.

6.44 Show that there is a refinement of $\mathcal{P}(\{a, b, c\}, \subseteq)$ that is a total order. For a generalization of this result, see Exercise 6.76.

The remaining exercises in this section require the following definition.

DEFINITION 6.27. A *chain* in a poset $(S, \preceq)$ is a subset of elements where

$$s_1 \preceq s_2 \preceq \cdots \preceq s_k .$$

Put another way, a chain is a totally ordered subset of S. A *maximal chain* is one that is not properly contained in any other. The *height* of a poset is the number of elements in a largest chain, when such a value exists.

For example, in the poset $(\mathcal{P}(\{a,b,c,d\}), \subseteq)$, the subsets

$$\{a\} \subseteq \{a,c,d\} \subseteq \{a,b,c,d\}$$

form a chain, and

$$\emptyset \subseteq \{a\} \subseteq \{a,d\} \subseteq \{a,c,d\} \subseteq \{a,b,c,d\}$$

is a maximal chain. As can be seen in Figure 5 in Chapter 3, all of the maximal chains in this poset contain 5 elements, so the height of the poset is 5.

6.45 How would you rewrite Exercise 3.32 in terms of maximal chains?

6.46 Consider the poset on $[32] = \{1, 2, \ldots, 32\}$, ordered by divisibility.

(a) Verify that $\{1, 2, 6, 30\}$ is a maximal chain.
(b) Verify that $\{1, 3, 6, 12, 24\}$ is also a maximal chain.
(c) Prove that this poset has height 6.

6.47 Consider the poset of subsets of $[n] = \{1, 2, \ldots, n\}$ ordered by containment.

(a) Prove that this poset has height $n + 1$.
(b) Prove that this poset contains exactly $n!$ maximal chains.

6.48 Construct a poset that, for every $n \in \mathbb{N}$, has a maximal chain with n elements. That is, there is a maximal chain with only one element, there is a maximal chain with exactly two elements, and so on.

6.49 A *zigzag poset*, or *fence*, is a poset where the elements can be listed so that the order relation is alternating. An example of a zigzag poset on seven elements is shown in Figure 7. Using zigzag posets as an inspiration, prove that for any $n \in \mathbb{N}$ there is an infinite poset, where all the maximal chains have n elements.

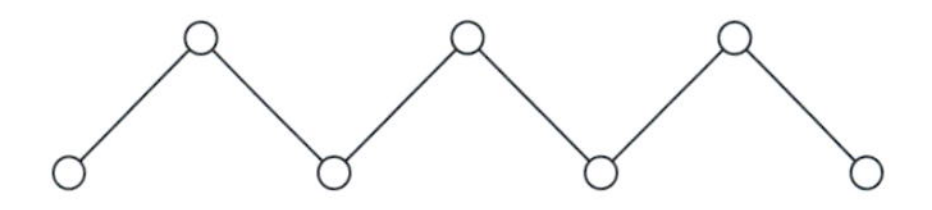

Figure 7. The Hasse diagram for a zigzag poset on seven elements.

Exercises you can work on after Section 6.3

6.50 Prove that *same magnitude*,

$$x \sim y \text{ if } x^2 = y^2,$$

is an equivalence relation on $\mathbb{R}$.

6.51 Define $\sim_{\text{bi}}$ by

$$\sim_{\text{bi}} = \{ (S_1, S_2) \mid \text{ there is a bijection } f : S_1 \to S_2 \},$$

for $S_1, S_2 \subseteq \mathbb{N}$.

(a) Prove that $\sim_{\text{bi}}$ is an equivalence relation on $\mathcal{P}(\mathbb{N})$.
(b) Describe two distinct equivalence classes for $\sim_{\text{bi}}$.

6.52 For any pair of subsets S_1 and S_2 of $\mathbb{N}$, define

$$S_1 \sim_{\triangle} S_2$$

when the symmetric difference $S_1 \triangle S_2$ is finite.

(a) Prove that $\sim_{\triangle}$ is an equivalence relation on $\mathcal{P}(\mathbb{N})$.
(b) Describe two distinct equivalence classes for $\sim_{\triangle}$.

6.53 Consider the following relation on $\mathbb{R}^2$. We say $(x, y) \sim (z, w)$ if $|x|+|y| = |z|+|w|$.

(a) Prove that $\sim$ is an equivalence relation.
(b) Which elements are in the equivalence class of $(2, -1)$? A graph may be useful here.

6.54 Determine which of the following relations are equivalence relations. For those that are, describe the elements of two different equivalence classes.

(a) $S = \mathbb{R}^2$, and $a \sim b$ if either $a = b$ or the line in $\mathbb{R}^2$ through a and b also goes through the origin.
(b) $S = \mathbb{R}^2$, and $a \sim b$ if either $a = b$ or the perpendicular bisector of the line segment joining a to b in $\mathbb{R}^2$ goes through the origin.
(c) S is the set of lines in $\mathbb{R}^2$, and $\ell_1 \parallel \ell_2$ when the lines ℓ_1 and ℓ_2 are parallel.

6.55 Determine which of the following relations are equivalence relations. For those that are, describe the elements of two different equivalence classes.

(a) $S = \mathbb{R}$ and $a \sim b$ if $a - b \in \mathbb{Z}$.
(b) $S = \mathbb{R}$ and $a \approx b$ if $a - b \in \mathbb{Q}$.
(c) $S = \mathbb{R}$ and $a \bowtie b$ if $a - b \in \mathbb{Q}^c$.

6.56 Let $\sim$ and $\approx$ be equivalence relations on a set S.

(a) Prove that the intersection $\sim \cap \approx$ is an equivalence relation.
(b) Give a counterexample to the following claim: The union $\sim \cup \approx$ is an equivalence relation.

Remember, both $\sim$ and $\approx$ are subsets of $S \times S$.

6.57 Let $f : A \to B$, and define $\sim_f$ to be the relation on A defined by

$$a_1 \sim_f a_2 \text{ whenever } f(a_1) = f(a_2).$$

(a) Prove that $\sim_f$ is an equivalence relation.
(b) Describe the equivalence classes of $\sim_f$ in the language of functions.

6.58 Let $A = \mathbb{N} \times \mathbb{N}$ and define a relation on A by $(a, b) \sim (c, d)$ when $ab = cd$. For example, $(2, 6) \sim (4, 3) \sim (1, 12)$.

(a) Show that $\sim$ is an equivalence relation.
(b) Find an equivalence class with exactly one element.
(c) Prove that for every $n \geq 2$ there is an equivalence class with exactly n elements.

(d) Prove there are no equivalence classes with infinitely many elements.
(e) Which equivalence classes contain exactly two elements?
(f) Which equivalence classes contain exactly three elements?

6.59 Let $B \subseteq A$, and let $\sim$ be the relation on $\mathcal{P}(A)$ defined by

$$S_1 \sim S_2 \Leftrightarrow S_1 \triangle S_2 \subseteq B,$$

where $\triangle$ is the symmetric difference. Prove that $\sim$ is an equivalence relation on $\mathcal{P}(A)$, and describe the equivalence class $[A]$.

6.60 Let $f : A \to A$ be a bijection. If $n \in \mathbb{N}$, we have already seen in Exercise 5.48 that

$$f^n : A \to A \text{ means } \underbrace{f \circ f \circ \cdots \circ f}_{n \text{ terms}}.$$

It is natural to similarly define

$$f^{-n} : A \to A \text{ as } \underbrace{f^{-1} \circ f^{-1} \circ \cdots \circ f^{-1}}_{n \text{ terms}},$$

and also

$$f^0 : A \to A \text{ as the identity function } 1_A .$$

(a) Explain why $f^m \circ f^n = f^{m+n}$ for any $m, n \in \mathbb{Z}$.
(b) If $a \in A$, then the *orbit* of a is

$$\text{ORB}(a) = \{f^n(a) \mid n \in \mathbb{Z}\}.$$

Prove that if $x \in \text{ORB}(a)$, then $\text{ORB}(x) = \text{ORB}(a)$.
(c) Prove that $\mathcal{B} = \{\text{ORB}(a) \mid a \in A\}$ is a partition of A.

Exercises you can work on after Sections 6.4 and 6.5

6.61 Solve the following for x, or prove that no solution exists.

(a) $3x + 5 \equiv 1 \ (\text{MOD } 7)$
(b) $x^2 \equiv 7 \ (\text{MOD } 9)$
(c) $2x - 7 \equiv 2 \ (\text{MOD } 8)$
(d) $4x^3 \equiv 3 \ (\text{MOD } 15)$

6.62 In Theorem 2.12, we used induction to show that, for all non-negative integers n, $7^n - 1$ is a multiple of 6.

(a) Use modular arithmetic to offer a simple and non-inductive proof of this theorem.
(b) Prove that, for all non-negative integers n, $6^n + 1$ is divisible by 7 if and only if n is odd.

6.63 Let $n \in \mathbb{N}$ be written in decimal notation as

$$n = a_k a_{k-1} \ldots a_1 a_0 ,$$

Use modular arithmetic to ...

(a) ... prove that n is divisible by 3 if and only if $\sum_{i=0}^{k} a_i$ is divisible by 3,
(b) ... prove that n is divisible by 9 if and only if $\sum_{i=0}^{k} a_i$ is divisible by 9.

(c) ...prove that n is divisible by 11 if and only if the alternating sum

$$\sum_{i=0}^{k}(-1)^i a_i$$

is divisible by 11.

If you did Exercise 1.47, compare your answer to that problem with your new solutions to parts (a) and (b) above.

6.64 Show that the following functions are not well defined.

(a) $f : \mathbb{Z}_4 \to \mathbb{Z}_8$ is given by $f([a]_4) = [a]_8$.
(b) $f : \mathbb{Z}_n \to \mathbb{Z}_{mn}$ is given by $f([a]_n) = [a]_{mn}$, where $m, n \in \mathbb{N}$ and $m > 1$.
(c) $f : \mathbb{Z}_6 \to \mathbb{Z}_{18}$ is given by $f([a]_6) = [a^2]_{18}$.

6.65 Show that $f : [a]_m \to [a^4]_5$ is well defined for any $m \in \mathbb{N}$.

6.66 For any integer $n \geq 2$, what is the value of the binomial coefficient $\binom{n}{2}$ modulo n?

Exercises you can work on after Section 6.6

6.67 The integers modulo 7, $\mathbb{Z}_7$, consists of the equivalence classes $\{[0], [1], [2], [3], [4], [5], [6]\}$. Which pairs of equivalence classes are additive inverses of each other? Which pairs of equivalence classes are multiplicative inverses of each other? Are there any equivalence classes that do not "pair" nicely? Explain!

6.68 Recalling that you found the multiplicative inverses of 3 and 5 in $\mathbb{Z}_{14}$ as part of Exercise 6.29, solve the following for x.

(a) $3x - 2 \equiv 5 \pmod{14}$
(b) $5x + 1 \equiv 0 \pmod{14}$

6.69 Recall that $\mathcal{U}_{12}$ is the set of units in $\mathbb{Z}_{12}$.

(a) What elements of $\mathbb{Z}_{12}$ are in $\mathcal{U}_{12}$?
(b) Create the multiplication table for $\mathcal{U}_{12}$.

6.70 Show that each element of $\mathcal{U}_m$ occurs once and only once in each row (or column) of the multiplication table for $\mathcal{U}_m$.

6.71 Determine the number of units in $\mathbb{Z}_m$ when m is prime, and when m is the square of a prime.

More Exercises!

6.72 In Exercise 4.11, you were asked to commit to memory an outline of the proof of the Fundamental Theorem of Arithmetic. Prove that you have!

6.73 Let R be a relation on any set S.

(a) Prove there is at least one equivalence relation on S that contains R.
(b) Let $\sim_R$ be the intersection of all equivalence relations on S that contain R. Prove that $\sim_R$ is itself an equivalence relation containing R.
(c) Prove that if $\approx$ is an equivalence relation containing R, then $\sim_R \subseteq \approx$.

The relation $\sim_R$ is often called the *equivalence relation generated by R.*

(d) Let T be the relation on $\mathbb{Q}$ given by qTq' when $q = 2q'$. Describe the equivalence relation $\sim_T$ generated by T, and describe the equivalence class $[1]$.

6.74 Here's the result you will ultimately want to prove in this exercise:

> *Let S be a set of n integers. Then there is a subset of S that sums to a number divisible by n.*

Here are some questions intended to help you develop a proof:

(a) Prove that the statement is true when $n = 2$. (This should be easy.)
(b) Prove that the statement is true when $n = 6$. (This might help you find an approach to the general result.)
(c) Prove that the statement is true when $n = 1001$. (This tests if your approach really works!)
(d) Prove the general result.

6.75 In this problem we outline how you can construct the rational numbers from the integers using an appropriate equivalence relation.

Let Q be the set $\mathbb{Z} \times \mathbb{N}$, and let $\sim$ be the relation

$$(a, b) \sim (c, d) \text{ whenever } ad = cb\,.$$

For example, $(18, 12) \sim (6, 4) \sim (3, 2)$.

(a) Prove that $\sim$ is an equivalence relation on the set Q.

Denote the equivalence class of (a, b) by $[a, b]$. We can define addition and multiplication on equivalence classes using the following formulas:

$$[a, b] + [c, d] = [ad + bc, bd] \quad \text{and} \quad [a, b] \cdot [c, d] = [ac, bd]\,.$$

Denote the set of equivalence classes by $\mathbb{Q}$.

(b) Verify that these formulas are well defined, that is, the answers do not depend on the choice of representatives from the equivalence classes.
(c) Show that $[0, 1]$ is an additive identity. That is, $[0, 1] + [a, b] = [a, b]$ for all $[a, b] \in \mathbb{Q}$.
(d) Show that $[1, 1]$ is a multiplicative identity, that is, $[1, 1] \cdot [a, b] = [a, b]$ for all $[a, b] \in \mathbb{Q}$.
(e) Show that $[a, b] + [-a, b] = [0, 1]$ for all $[a, b] \in \mathbb{Q}$.
(f) Show that if $[a, b] \neq [0, 1]$, there is a $[c, d] \in \mathbb{Q}$ such that $[c, d] \cdot [a, b] = [1, 1]$.

At this point you have probably already guessed that the set $\mathbb{Q}$ with these arithmetic operations is the same thing as the rational numbers. We have simply replaced the more standard notations of $\frac{a}{b}$ and a/b by $[a, b]$, and we have highlighted that there is an equivalence relation underlying the multiple ways you can write the same number via fractions.

6.76 The goal of this exercise is to construct a proof of the following theorem:

Theorem 6.28. *Every partial order on a finite set S can be refined to a total order on S.*

For the definition of refinement, see Exercise 6.43. The proof we outline uses induction on the number of pairs of incomparable elements. The base case is when there are no incomparable elements, meaning that the original partial order is already a total order on S.

For the inductive step, we assume that the theorem holds whenever there are n or fewer pairs of incomparable elements. Let R be a partial order on a finite set S, with $n+1$ incomparable pairs, and let a and b be incomparable elements. That is, $(a, b) \notin R$ and $(b, a) \notin R$. In the steps below you will be asked to prove there is a refinement of R that contains (a, b).

Define the set $A_\downarrow$ to consist of those elements that precede a with respect to the partial order R:

$$A_\downarrow = \{s \in S \mid sRa\}.$$

Similarly define the set $B_\uparrow$ as those elements that b precedes with respect to the partial order R:

$$B_\uparrow = \{s \in S \mid bRs\}.$$

(a) Prove that $A_\downarrow \cap B_\uparrow = \emptyset$.
(b) Define a new relation on S: $R' = R \cup \{A_\downarrow \times B_\uparrow\}$. Prove that R' is a reflexive relation.
(c) Prove that R' is antisymmetric. Note: This requires checking a number of cases, because a pair (x, y) can be in R' either because it is a pair in R or because it is in $A_\downarrow \times B_\uparrow$.

Proving that R' is transitive also requires the checking of a number of cases.

Transitivity: Assume that (x, y) and (y, z) are both in R'; we need to prove that this implies $(x, z) \in R'$. It cannot be the case that both pairs come from $A_\downarrow \times B_\uparrow$. If they did, then y would be in both $A_\downarrow$ and $B_\uparrow$ but we have already noted that $A_\downarrow \cap B_\uparrow = \emptyset$. If both pairs were already in R then the result follows since R is transitive.

Assume that $(x, y) \in R$ and $(y, z) \in A_\downarrow \times B_\uparrow$. Then from the definition of $A_\downarrow$ we know $(y, a) \in R$. Therefore $(x, a) \in R$, since R is transitive. Thus $x \in A_\downarrow$, and so

$$(x, z) \in A_\downarrow \times B_\uparrow \subset R'.$$

If instead we have $(x, y) \in A_\downarrow \times B_\uparrow$ and $(y, z) \in R$, then we know that $(b, y) \in R$ and so $z \in B_\uparrow$. Thus

$$(x, z) \in A_\downarrow \times B_\uparrow \subset R'.$$

(d) It follows from the work above that R' is a partial order on S. Verify that R' has n or fewer incomparable pairs, and so induction implies that R' can be refined to a total order R'' on S.
(e) Prove that R'' is also a total order that refines R.

Having now worked through the details of this argument, write up a proof of Theorem 6.28 that reads well from beginning to end.

7 Cardinality

These are among the marvels that surpass the bounds of our imagination, and that must warn us how gravely one errs in trying to reason about infinites by using the same attributes that we apply to finites.

—*Galileo Galilei*

7.1. The Hilbert Hotel, Count von Count, and Cookie Monster

Below are two stories. The first is well known within the mathematics community; the second is presented with two classic characters from children's television. In both cases, the stories highlight that our intuition about counting elements and determining the size of finite sets does not always transfer well to the setting of infinite sets.

7.1.1. The Hilbert Hotel. David Hilbert was a German mathematician who contributed transformative results in a number of areas of mathematics and provided valuable vision and leadership to the mathematics community. Because he introduced the following example in the 1920s, it is often described as the *Hilbert Hotel.* The Hilbert Hotel has many rooms. In fact, each room is numbered with a natural number, and there are just as many rooms as there are natural numbers. Thus there is a room #1, a room #2, ..., a room #n, ... and so on.

The Hilbert Hotel has acquired a nice reputation among tour group bus drivers, since it always seems to be able to accommodate very large tour groups. For instance, a bus drove up one evening that held infinitely many passengers: every passenger's shirt was labelled with a natural number, and every natural number appeared on just one shirt. This did not pose a problem for Hilbert, of course, because he made a single, precise announcement over his intercom system that told everyone where to go: "Go to the room whose number corresponds to the number on your shirt." So passenger #1 went to room #1, passenger #2 went to room #2, and so on.

Later that evening, a single person drove up in a car and asked for a room. Hilbert couldn't send the person to the first vacant room, since there were no vacancies. Instead, Hilbert simply made the following announcement over the intercom to everyone who already had a room: "I apologize for the inconvenience, but in order for us to accommodate a new guest, we need you to change rooms. If you are in room #k, please move to room #$(k + 1)$ for the night. Thank you! Breakfast will be served from 7 until 10 tomorrow morning." And then, of course, he sent the new arrival to room #1, which was free.

Exercise 7.1 Explain how Hilbert can accommodate any car that arrives with any finite number of passengers, even if all of the hotel's rooms are currently occupied.

Exercise 7.2 Hilbert faced another dilemma later that evening. A second bus showed up that, like the first, had a passenger for every natural number. Somehow, Hilbert was able to accommodate this group as well. How? Hilbert must follow standard hotel customer service rules; in particular, he must give an explicit command that will tell each person which specific room to go to.

GOING BEYOND THIS BOOK. In 1932 David Hilbert and Stephen Cohn-Vassen published a delightful and accessible book, *Anschauliche Geometrie*, which was translated and published in English as *Geometry and the Imagination* [**HCV52**]. In that text you find the following passage. "In mathematics ... we find two tendencies present. On the one hand, the tendency towards abstraction seeks to crystallize the logical relations inherent in the maze of materials ... being studied, and to correlate the material in a systematic and orderly manner. On the other hand, the tendency towards intuitive understanding fosters a more immediate grasp of the objects one studies, a live rapport with them, so to speak, which stresses the concrete meaning of their relations." This is a very appropriate insight for anyone reading this text.

7.1.2. Count von Count meets the Cookie Monster. In this story we reference a classic scene from *Sesame Street*. The first time Count von Count meets Cookie Monster, there is a plate of seven cookies sitting on a ledge. In a cooperative effort, the Count counts the cookies while Cookie Monster eats them.[1]

Had Count von Count and Cookie Monster been more ambitious, the story might have been a bit different. Instead of seven cookies, let's assume there are infinitely many cookies in front of the Count. We'll make them sugar cookies, with numbers written on them in frosting. Every cookie is frosted with a natural number, and every natural number appears on just one cookie.

Both the Count and Cookie Monster can act *very* quickly. The Count is able to count ten cookies and transfer them to Cookie Monster's plate in thirty seconds; he can count and transfer another ten cookies in a quarter-of-a-minute; the next ten cookies in an eighth-of-a-minute; and so forth. Cookie Monster is able to eat all the cookies on his plate just as rapidly, eating faster and faster as time goes on.

Assume the Count and Cookie Monster start counting and eating at two minutes before noon, and that they take turns so that the Count counts cookies and then Cookie Monster eats some of them. In the first thirty seconds the Count will count out the first ten cookies and put them on Cookie Monster's plate, and in the next thirty seconds Cookie Monster will eat *some* of these cookies. Then in the next fifteen seconds, Count will move the next ten cookies onto Cookie Monster's plate, and in the following fifteen seconds, Cookie Monster will eat *some* of the cookies on his plate. This pattern continues, with the time interval shrinking by half with each new round.

For example, it might be the case that in his first turn Cookie Monster eats cookie #1; in his second turn he eats cookie #2; in his third turn he eats cookie #3; and so on. Because the Count is piling up cookies much more quickly than Cookie Monster is eating them, the number of cookies on Cookie Monster's plate is constantly getting larger.

1 We encourage you to watch a clip of this first meeting. In a less cooperative later meeting, Cookie Monster displays a surprising affinity for apples.

At the end of his first turn, Cookie Monster has 9 cookies on his plate; after his second turn he has 18 cookies; after his third turn he has 27 cookies; and so on. And yet ...

Exercise 7.3 How many cookies remain on Cookie Monster's plate at noon?

Exercise 7.4 Which cookies will Cookie Monster have left on his plate at noon if ...

(a) ... in his first turn, Cookie Monster eats cookie #10; in his next turn he eats cookie #11; in his third turn he eats cookie #12; and so on. As in the initial scenario, at the end of his first turn Cookie Monster has 9 cookies; at the end of his second turn he has 18 cookies; and so on.
(b) ... in his first turn, Cookie Monster eats the odd-numbered cookies 1, 3, 5, 7 and 9; in his next turn he eats cookies 11, 13, 15, 17, and 19; in his third turn he eats cookies 21, 23, 25, 27 and 29; and so on.

Please note that in all of these situations, Cookie Monster eats infinitely many cookies, but the end results can be quite different.

GOING BEYOND THIS BOOK. The scenarios we imagine for the Count and Cookie Monster highlight what's known as the Ross–Littlewood paradox. Interested readers may also enjoy *Littlewood's Miscellany* [**Lit86**], a memoir and eclectic collection of anecdotes, puzzles, and paradoxes that also highlights the process of doing mathematics.

7.1.3. The Take Home Message. Much of our intuition about the size of sets is based on our experience with finite sets. In dealing with infinite sets, you need to be *very* careful. And so in the next section we begin a formal discussion of how to measure the size of infinite sets.

7.2. Cardinality

Here is a definition that in the context of finite sets seems to be a rather awkward way of saying two sets have the same size. We expect that by the end of this chapter you will have gained some appreciation for the beauty underlying this definition.

DEFINITION 7.1. Two sets, A and B, have the same *cardinality* if there is a bijection $f : A \to B$. We denote this by $A \asymp B$.

Exercise 7.5 Review your answers to Exercise 5.42, or work on that exercise for the first time, knowing that you are being asked to establish which of those pairs of infinite sets have the same cardinality. Some of those problems are hard, so feel free to put a question mark next to the ones for which you cannot find a proof in a reasonable amount of time.

LEMMA 7.2. *The relation $\asymp$ is an equivalence relation.*

PROOF. We need to show that $\asymp$ is reflexive, symmetric, and transitive.

Reflexive: If A is a set, then the identity function 1_A provides a bijection from A to A.

Symmetric: If $A \asymp B$, then there is a bijection $f : A \to B$. Theorem 5.23 shows that there is a bijection $f^{-1} : B \to A$, so $B \asymp A$.

Transitive: If $A \asymp B$ and $B \asymp C$, then we know there are bijections $f : A \to B$ and $g : B \to C$. By Proposition 5.16, the composition $g \circ f$ is then a bijection from A to C, hence $A \asymp C$. □

This approach to measuring the sizes of sets, particularly infinite sets, leads to some counterintuitive examples. We spend the rest of this section discussing two of these. The main purpose of this chapter is to build your intuition for cardinality to the point where these examples no longer seem strange.

Recall that open intervals in $\mathbb{R}$ are subsets of the form (a, b), (a, ∞), $(-\infty, b)$, or even $\mathbb{R} = (-\infty, \infty)$. Our first example is

PROPOSITION 7.3. *All non-empty open intervals in $\mathbb{R}$ have the same cardinality.*

In the exercise below you will establish the key lemmas that are used in proving Proposition 7.3.

Exercise 7.6 Prove each of the following claims.

(a) The exponential function e^x gives a bijection from $\mathbb{R}$ to $\mathbb{R}^+ = (0, \infty)$.
(b) The tangent function $\tan(x)$ gives a bijection from $(-\pi/2, \pi/2)$ to $\mathbb{R}$.
(c) If a and b are real numbers, with $a < b$, then $f(x) = (b - a)x + a$ is a bijection from $(0, 1)$ to (a, b).

PROOF OF PROPOSITION 7.3. First note that Exercise 7.6(c), combined with the fact that $\asymp$ is an equivalence relation, implies that $(a, b) \asymp (c, d)$ for any $a < b$ and $c < d$. Thus all of the bounded open intervals have the same cardinality.

Exercise 7.6(b) then shows that all of the bounded open intervals have the same cardinality as the entire real line $\mathbb{R}$.

It remains then to demonstrate that $\mathbb{R}$ has the same cardinality as any open interval of the form (a, ∞) or $(-\infty, b)$. First note that, for any $a \in \mathbb{R}$, the function $f(x) = x + a$ provides a bijection from $(0, \infty)$ to (a, ∞). Similarly, for any $b \in \mathbb{R}$, the function $f(x) = b - x$ is a bijection from $(0, \infty)$ to $(-\infty, b)$. Thus the intervals (a, ∞) and $(-\infty, b)$ have the same cardinality as $(0, \infty)$, and by Exercise 7.6(a) they therefore have the same cardinality as $\mathbb{R}$. □

Our second example was part of Exercise 5.42:

LEMMA 7.4. *The natural numbers $\mathbb{N}$ and the set of integers $\mathbb{Z}$ have the same cardinality.*

It is reasonable to believe that $\mathbb{N}$ is "smaller" than $\mathbb{Z}$. After all, $\mathbb{N}$ is a proper subset of $\mathbb{Z}$ that on first glance seems to be only "half" the size of the full set of integers. But, as you will show in the exercise below, the natural numbers can be put into a one-to-one correspondence with the integers.

Exercise 7.7 Convert the following chart into a proof of Lemma 7.4.

1	2	3	4	5	6	7	...
0	1	−1	2	−2	3	−3	...

7.3. Countability

Bijections provide matchings of the elements of one set to another, and since the natural numbers are also referred to as the counting numbers, the following definition seems reasonable.

DEFINITION 7.5. A set A is said to be *countably infinite* if there exists a bijection $f : \mathbb{N} \to A$. In other words, A is countably infinite if $\mathbb{N} \asymp A$. A set is *countable* if it is either finite or countably infinite.

In terms of this definition, Exercise 5.42 asks if each of the sets $\mathbb{N} \cup \{\text{frog}\}$, $\mathbb{Z}$, $\mathbb{N} \times \mathbb{N}$, and $\mathbb{R}$ are countably infinite. Since $f : \mathbb{N} \to \mathbb{N} \cup \{\text{frog}\}$ given by

$$\begin{array}{ccc} & f & \\ 1 & \to & \text{frog} \\ 2 & \to & 1 \\ 3 & \to & 2 \\ 4 & \to & 3 \\ \vdots & \vdots & \vdots \end{array}$$

is a bijection, we know that $\mathbb{N} \cup \{\text{frog}\}$ is countably infinite. Similarly Lemma 7.4 states that $\mathbb{Z}$ is countably infinite, with Exercise 7.7 suggesting the bijection $g : \mathbb{N} \to \mathbb{Z}$ given by

$$\begin{array}{ccr} & g & \\ 1 & \to & 0 \\ 2 & \to & 1 \\ 3 & \to & -1 \\ 4 & \to & 2 \\ 5 & \to & -2 \\ 6 & \to & 3 \\ 7 & \to & -3 \\ \vdots & \vdots & \vdots \end{array}$$

The displays of these two functions show that a countably infinite set A is one whose members can be labeled with the natural numbers so that each element of A is labeled with a single natural number, and no two elements of A receive the same label. Also, while it is certainly possible to write these bijections in a more "formulaic" manner such as

$$f(n) = \begin{cases} \text{frog} & \text{if } n = 1, \\ n - 1 & \text{otherwise} \end{cases}$$

and

$$g(n) = (-1)^n \lfloor n/2 \rfloor,$$

it is often not necessary to do so if it is perfectly clear that a listing of the elements is exhaustive and has no duplications. For example, the partial table shown in Figure 1 displays via subscripts a labeling of the elements of $\mathbb{N} \times \mathbb{N}$ with natural numbers that corresponds to a bijection $h : \mathbb{N} \to \mathbb{N} \times \mathbb{N}$, showing that $\mathbb{N} \asymp (\mathbb{N} \times \mathbb{N})$. Another proof of this fact, which was first noticed by Georg Cantor, appeals to the function $f : \mathbb{N} \times \mathbb{N} \to \mathbb{N}$ defined by $f(n, m) = 2^{n-1}(2m - 1)$.

Exercise 7.8 Verify that the function $f(n, m) = 2^{n-1}(2m - 1)$ is a bijection from $\mathbb{N} \times \mathbb{N}$ to $\mathbb{N}$. The Fundamental Theorem of Arithmetic may be useful to you.

$$
\begin{array}{cccccc}
\vdots & & & & & \\
(5,1)_{11} & \vdots & & & & \\
(4,1)_{7} & (4,2)_{12} & \vdots & & & \\
(3,1)_{4} & (3,2)_{8} & (3,3)_{13} & \cdots & & \\
(2,1)_{2} & (2,2)_{5} & (2,3)_{9} & (2,4)_{14} & \cdots & \\
(1,1)_{1} & (1,2)_{3} & (1,3)_{6} & (1,4)_{10} & (1,5)_{15} & \cdots
\end{array}
$$

Figure 1. The "running along diagonals" proof that $\mathbb{N} \times \mathbb{N}$ is countably infinite.

Exercise 7.9 We can connect our work in this section back to the Hilbert Hotel.

(a) How is the fact that $\mathbb{N} \cup \{\text{frog}\}$ is countable related to our Hilbert Hotel story?
(b) Write a follow-up event for the Hilbert Hotel story related to the fact that $\mathbb{N} \times \mathbb{N}$ is a countable set. Perhaps you might want to have multiple busses arriving at the same time?

Some of the sets described above, such as $\mathbb{Z}$ and $\mathbb{N} \times \mathbb{N}$, seem on first thought to be much larger than $\mathbb{N}$, despite the fact that we have just shown that their elements can be paired off with the elements of $\mathbb{N}$. Infinite sets often defy our initial (and secondary, and tertiary) intuition. For example, a deeper study of infinite sets would show that every infinite set S has a proper subset $S' \subset S$ such that there's a bijection $f : S \to S'$, a statement that is false for finite sets.

7.4. Key Countability Lemmas

Many common processes in mathematics preserve the property of being countable. Taking Cartesian products is one of them:

LEMMA 7.6. *If A and B are non-empty countable sets, then $A \times B$ is countable. Further, if either A or B is countably infinite, then $A \times B$ is countably infinite.*

PARTIAL PROOF. We will establish the most interesting case, where both A and B are countably infinite. In this case there is a bijection $\alpha : \mathbb{N} \to A$ and a bijection $\beta : \mathbb{N} \to B$. Combining these we can create a function $\gamma : \mathbb{N} \times \mathbb{N} \to A \times B$ defined by $\gamma(m, n) = (\alpha(m), \beta(n))$.

To prove that γ is injective, we note that if $\gamma(m, n) = \gamma(x, y)$, then $\alpha(m) = \alpha(x)$ and $\beta(n) = \beta(y)$. The fact that α and β are injections then shows that $m = x$ and $n = y$, hence $(m,n) = (x,y)$. To prove that γ is surjective we start with an arbitrary element $(a,b) \in A \times B$. Since α is a surjection, there is some $m \in \mathbb{N}$ such that $\alpha(m) = a$. Since β is a surjection, there is some $n \in \mathbb{N}$ such that $\beta(n) = b$. Thus $\gamma(m,n) = (a,b)$.

Because γ is a bijection, $(\mathbb{N} \times \mathbb{N}) \asymp (A \times B)$. Since $\mathbb{N} \asymp (\mathbb{N} \times \mathbb{N})$, the result follows by the transitivity of $\asymp$. □

Exercise 7.10 Prove the other cases of Lemma 7.6. You may first want to show that if S is any non-empty finite set, then $\mathbb{N} \asymp (S \times \mathbb{N})$.

LEMMA 7.7. *If S is a subset of $\mathbb{N}$, then S is countable. In particular, if S is an infinite subset of $\mathbb{N}$, then S is countably infinite.*

PROOF. If S is a finite subset of $\mathbb{N}$, then S is countable by definition.

Assume that $S \subseteq \mathbb{N}$ and that S is infinite. By the Well-Ordering Principle for $\mathbb{N}$ (Theorem 4.3), there must be a least element of S, which we denote s_1. And then there must be a least element of $S \setminus \{s_1\}$, which we call s_2. Continuing in this fashion we list all the elements

$$S = \{s_1, s_2, s_3, \ldots\}$$

in increasing order (so $s_1 < s_2 < s_3 < \cdots$). We know that our list is complete, as if $n \in S$ then n must appear among the first n terms in the list we have constructed.

The subscripts of the entries in this ordered list define a bijection $\mathbb{N} \to S$, given by $n \mapsto s_n$. □

This result leads to the following very useful corollary:

COROLLARY 7.8. *If A is countably infinite and there is an injection $f : B \hookrightarrow A$, then B is countable. In particular, if B is an infinite subset of A, then B is countably infinite.*

PROOF. If B is finite, then B is countable.

If B is infinite, we need to show that it is countably infinite. Since A is countably infinite, there is a bijection $g : \mathbb{N} \to A$, with an inverse bijection $g^{-1} : A \to \mathbb{N}$. Thus, the composition $h = g^{-1} \circ f : B \to \mathbb{N}$ is an injection. By Exercise 5.9 we know that the $\hat{h} : B \to \text{Range}(h)$ that has $\hat{h}(b) = h(b)$ for all $b \in B$ is a bijection from B to $\text{Range}(h) = \text{Range}(\hat{h})$.

The final step of the proof is left for you as Exercise 7.11. □

Exercise 7.11 Finish Corollary 7.8 by explaining why $\text{Range}(\hat{h})$, and thus B, is countably infinite.

With Corollary 7.8 in hand, we have a potentially simpler way to prove that a set S is countably infinite than finding an explicit one-to-one correspondence with $\mathbb{N}$: we just need to find an injection from S to any one of our growing collection of countably infinite sets. For example, we can now prove that

PROPOSITION 7.9. *The rational numbers, $\mathbb{Q}$, form a countably infinite set.*

PROOF. Every $q \in \mathbb{Q}$ can be uniquely expressed in lowest terms: $q = m/n$, where $m \in \mathbb{Z}$ and $n \in \mathbb{N}$ with $\text{GCD}(m,n) = 1$. We may use this fact to define a function $\tau : \mathbb{Q} \to \mathbb{Z} \times \mathbb{N}$ by $\tau(q) = (m,n)$, where $q = m/n$ in lowest terms.

Since distinct rational numbers can't have the same lowest term expression, τ is an injection. Since the set $\mathbb{Z} \times \mathbb{N}$ is countably infinite by Lemmas 7.4 and 7.6, Corollary 7.8 implies that $\mathbb{Q}$ is countably infinite. □

Exercise 7.12 Why is the function τ in the proof above not a surjection?

Finally, we present a result that is often stated informally as "a countable union of countable sets is countable."

LEMMA 7.10. *Let I be a countable index set, and for each $i \in I$ let A_i be a countable set. Then*

$$A = \bigcup_{i \in I} A_i$$

is countable.

PROOF. Since I is countable, it is not hard to show that there is an injective function $f : I \hookrightarrow \mathbb{N}$. Similarly, for each A_i there is an injection $f_i : A_i \hookrightarrow \mathbb{N}$. In the paragraph below we use these functions to construct an injection $F : A \hookrightarrow \mathbb{N} \times \mathbb{N}$.

Each $a \in A$ must come from at least one A_i, although it may well come from infinitely many A_i. The set $S_a = \{f(i) \mid a \in A_i\}$ is then a non-empty subset of $\mathbb{N}$, hence it has a least element by the Well-Ordering Principle (Theorem 4.3). Let i be the index such that $f(i)$ is this least element of S_a, and define

$$F(a) = (f(i), f_i(a)).$$

If $F(a) = F(a')$ then a and a' must both be in the set A_i, and $f_i(a) = f_i(a')$. But f_i is an injection, so $a = a'$. Thus the function F is an injection.

Since the set $\mathbb{N} \times \mathbb{N}$ is countably infinite, Corollary 7.8 implies that A is countably infinite. □

COROLLARY 7.11. *If A and B are countable sets, then so is $A \cup B$.*

REMARK 7.12. The argument for Lemma 7.10 illustrates an important style of proof involving injective functions and countability. However, as we discussed at the end of Chapter 2, discovering and presenting mathematics are distinct activities, so you might not be able to see the key insights that led to the proof just given.

Consider the special case when $I = \mathbb{N}$ and each A_i is countably infinite, which means that we can enumerate the elements of A_i as

$$A_i = \{a_{i1}, a_{i2}, a_{i3}, \ldots\}.$$

Figure 2 displays the elements of A in a grid so that row i contains all of the elements in A_i. The three bold elements in the grid correspond to a common element $a = a_{23} = a_{44} = a_{51}$, where of course there may be other occurrences of a in lower rows. The point is that we only need to account for one occurrence of a in the grid when considering the union A. The function F in the proof of Lemma 7.10 picks $\mathbf{a_{23}}$, the entry in the highest row of the grid, corresponding to 2 being the least element of $S_a = \{2, 4, 5, \ldots\}$. (We are assuming that the injection $f : I \hookrightarrow \mathbb{N}$ from the first line of the proof is simply the identity function.)

The display should now make it apparent that $F : A \hookrightarrow \mathbb{N} \times \mathbb{N}$ is an injection, because different values of a will show up in the grid as non-overlapping sets of bold elements,

$$\begin{array}{llllll}
A_1: & a_{11} & a_{12} & a_{13} & a_{14} & a_{15} & \cdots \\
A_2: & a_{21} & a_{22} & \mathbf{a_{23}} & a_{24} & a_{25} & \cdots \\
A_3: & a_{31} & a_{32} & a_{33} & a_{34} & a_{35} & \cdots \\
A_4: & a_{41} & a_{42} & a_{43} & \mathbf{a_{44}} & a_{45} & \cdots \\
A_5: & \mathbf{a_{51}} & a_{52} & a_{53} & a_{54} & a_{55} & \cdots \\
 & & & \vdots & & &
\end{array}$$

Figure 2. A visual aid for understanding the proof of Lemma 7.10.

so A corresponds to an infinite subset of the countable set $\mathbb{N} \times \mathbb{N}$. But the proof given for Lemma 7.10 is written to handle the most general situation, establishing the truth of the result with rigor that is unmatched by a single motivating figure.

7.5. Not Every Set is Countable

Might every infinite set be countable? And if there is an infinite set S that is not countably infinite, how could you show that there is no possible bijection $f : \mathbb{N} \to S$? We answer both questions with the following two examples.

7.5.1. Subsets of $\mathbb{N}$. Our first example of a set that is not countable is an instance of a more general result, sometimes referred to as *Cantor's Theorem*. It is also tangentially related to *Russell's Paradox*, which we briefly describe in Section 3.9.

THEOREM 7.13. *The set $\mathcal{P}(\mathbb{N})$ is not countable.*

In the proof below we assume to the contrary that $\mathcal{P}(\mathbb{N})$ is countable, so that there is a listing of all of the elements of $\mathcal{P}(\mathbb{N})$; but then we exhibit an element of $\mathcal{P}(\mathbb{N})$ not in the list. In fact, the proof essentially shows that there is no bijection between any set A and its power set $\mathcal{P}(A)$, which we ask you to prove in Exercise 7.14.

Before turning to those arguments, let's first consider the finite set $A = \{1, 2, 3\}$. Suppose there was a bijection $f : A \to \mathcal{P}(A)$. One out of many possible attempts to construct a bijection might be the function

$$\begin{array}{ccc}
 & f & \\
1 & \to & \{2,3\} \\
2 & \to & \emptyset \\
3 & \to & \{1,2,3\}
\end{array}$$

How do we know that this is not a bijection? Because not every element of $\mathcal{P}(A)$ is in Range(f): five sets are missed. One of them is

$$S_f = \{a \in A \mid a \notin f(a)\}.$$

As the notation suggests, the set S_f depends on the function f. For the given f, since $1 \notin \{2, 3\}$, $2 \notin \emptyset$, and $3 \in \{1, 2, 3\}$, we get $S_f = \{1, 2\}$, which we see is indeed an element in the codomain of f that was missed. In fact, for *any* function f this definition of S_f gives an element that's not in Range(f).

Exercise 7.13 Let $A = \{1, 2, 3, 4\}$, and create an injective map $f : A \to \mathcal{P}(A)$. What is the set $S_f = \{\ a \in A \mid a \notin f(a)\ \}$ for your function f? Verify that S_f is not in the range of f.

PROOF OF THEOREM 7.13. Suppose to the contrary that $\mathcal{P}(\mathbb{N})$ is countable. Then there exists a bijection $f : \mathbb{N} \to \mathcal{P}(\mathbb{N})$. We will show that the set

$$S_f = \{a \in \mathbb{N} \mid a \notin f(a)\}$$

is not in Range(f), contradicting our assumption that $\mathcal{P}(\mathbb{N})$ is countable.

Because we have assumed that $f : \mathbb{N} \to \mathcal{P}(\mathbb{N})$ is a bijection, there must be some $n \in \mathbb{N}$ where $f(n) = S_f$. However, ...

(a) ...if $n \in S_f$, then by the definition of S_f, $n \notin f(n) = S_f$,
(b) ...if $n \notin S_f$, then by the definition of S_f, $n \in f(n) = S_f$.

From this we conclude that in fact S_f is not equal to $f(n)$ for any $n \in \mathbb{N}$, which contradicts the fact that f is a bijection from $\mathbb{N}$ to $\mathcal{P}(\mathbb{N})$. □

DEFINITION 7.14. A set that is not countable is an *uncountable* set.

Exercise 7.14 Show that slight modifications of the proof of Theorem 7.13 allows you to prove that $A \not\asymp \mathcal{P}(A)$ for any non-empty set A. See Exercise 7.41 for further exploration of the relationship between a set and its power set.

7.5.2. The Real Numbers. We indicated that one of the questions in Exercise 5.42 is "REALLY hard." This is because it asks if there is a bijection from $\mathbb{N}$ to $\mathbb{R}$; using the terminology of this chapter, it is asking if the set of real numbers is countably infinite. In this section we prove that it is not.

Our argument makes frequent use of facts about the decimal expansions of real numbers. In particular, every $x \in (0, 1)$ can be expressed as a decimal expansion $x = 0.d_1d_2d_3\ldots$. This expression is unique as long as you add the caveat that the digits are not all equal to 9 after some point in the expansion. For example, $0.25000\ldots = 0.24999\ldots$, so we can exclude the expression with repeating 9s.

THEOREM 7.15. *The set of real numbers, $\mathbb{R}$, is an uncountable set.*

PROOF. Since $\mathbb{R} \asymp (0, 1)$ by Proposition 7.3, it suffices to show that $(0, 1)$ is uncountable. The argument we present for this is often referred to as *Cantor's Diagonal Argument*, which can be used in settings that don't involve decimal expansions.

Assume to the contrary that there is a bijection $f : \mathbb{N} \to (0, 1)$. Then we may list all of the real numbers between 0 and 1 as indicated:

$$\begin{array}{ccl} & f & \\ 1 & \to & 0.\mathbf{d_{11}}\, d_{12}\, d_{13}\, d_{14} \ldots \\ 2 & \to & 0.d_{21}\, \mathbf{d_{22}}\, d_{23}\, d_{24} \ldots \\ 3 & \to & 0.d_{31}\, d_{32}\, \mathbf{d_{33}}\, d_{34} \ldots \\ 4 & \to & 0.d_{41}\, d_{42}\, d_{43}\, \mathbf{d_{44}} \ldots \\ \vdots & \vdots & \qquad \vdots \end{array}$$

For each $n \in \mathbb{N}$ choose a digit $x_n \in \{1, 2, 3, \ldots, 8\} \setminus \{d_{nn}\}$. That is, x_n should be a digit that is not equal to 0, 9, or the nth digit in the decimal expansion of $f(n)$. Use these digits to define the real number

$$x = 0.x_1 x_2 x_3 x_4 \ldots$$

Since none of the x_i are 0 or 9, we know $x \in (0, 1)$. But $f(n) \neq x$ for any $n \in \mathbb{N}$: otherwise, we would have $x_n = d_{nn}$ for some $n \in \mathbb{N}$, which is not possible because of how we chose x_n. Thus f cannot be surjective, contradicting the claim that f is a bijection. □

Exercise 7.15 You may be uncomfortable with the step that begins "For each $n \in \mathbb{N}$ choose a digit ..." Verify that you can be more prescriptive than this by rewriting the proof with

$$x = 0.x_1 x_2 x_3 x_4 \ldots,$$

where

$$x_n = \begin{cases} 4 & \text{if } d_{nn} = 5, \\ 5 & \text{if } d_{nn} \neq 5. \end{cases}$$

Exercise 7.16 Prove that the irrational numbers, $\mathbb{Q}^c = \mathbb{R} - \mathbb{Q}$, form an uncountable set.

We know that $\mathbb{R}$ is uncountable, and by Exercise 7.16 we know that $\mathbb{Q}^c$ is uncountable, so $\mathbb{Q}^c$ is an uncountable subset of $\mathbb{R}$. By analogy with Corollary 7.8, you might think this sufficies to show that $\mathbb{Q}^c \asymp \mathbb{R}$. It's not that simple. The *Continuum Hypothesis* states that *any uncountable subset of* $\mathbb{R}$ *has the same cardinality as* $\mathbb{R}$. However, as its name suggests, this statement is not a theorem.[2] Thus the proof of the following proposition requires a different approach than the one we just attempted.

PROPOSITION 7.16. *The irrational numbers have the same cardinality as the full set of real numbers.*

A proof of this result is outlined in Exercise 7.47; the proof appeals to the important Schröder–Bernstein Theorem, introduced in the following section.

GOING BEYOND THIS BOOK. There is a well-known tale that as Georg Cantor was developing the concept of cardinality in the nineteenth century, and discovering the sort of counterintuitive results that appear in this section, he stated "Je le vois, mais je ne le crois pas" (roughly "I see it but I don't believe it"). A short history of Cantor's struggle with these ideas and the source for this tale can be found in [**Gou11**].

2 The Continuum Hypothesis is one of a set of 23 problems that David Hilbert presented in 1900. It has been shown that the Continuum Hypothesis can be neither proven nor disproven, starting from the most common axiomatic foundations for set theory. In 1940 Kurt Gödel proved that you cannot disprove the Continuum Hypothesis starting from the Zermelo–Fraenkel set theory axioms, even if you include the Axiom of Choice. In 1963 Paul Cohen proved that you cannot prove the Continuum Hypothesis in the same setting.

7.6. Using the Schröder–Bernstein Theorem

In the language of this chapter, Exercise 5.41 asked: *Does the half-open interval* $[0, 1)$ *have the same cardinality as the closed interval* $[0, 1]$*?* At this point you should not be fooled into thinking that the cardinality of $[0, 1]$ is larger because $[0, 1]$ has "one more point" than $[0, 1)$. On the other hand, it is not obvious how to construct a bijection between closed and half-open intervals. It is in situations like these that the Schröder–Bernstein Theorem is useful.

THEOREM 7.17 (The Schröder–Bernstein Theorem). *Let A and B be sets, and assume that there are injections $f : A \hookrightarrow B$ and $g : B \hookrightarrow A$. Then A and B have the same cardinality.*

As an example, inclusion gives us an injection from $[0, 1)$ to $[0, 1]$, and $g(x) = \frac{1}{2}x$ provides an injection from $[0, 1]$ to $[0, 1)$. It follows by the Schröder–Bernstein Theorem that these intervals have the same cardinality. Notice that we have made this claim even though we never constructed an explicit bijection between $[0, 1]$ and $[0, 1)$. That is the real power of the Schröder–Bernstein Theorem: it allows you to establish that two sets have the same cardinality even in situations where you cannot see an explicit bijection.

Exercise 7.17 Use the Schröder–Bernstein Theorem to prove that the intervals $(0, 1)$, $(0, 1]$, $[0, 1)$ and $[0, 1]$ all have the same cardinality.

Exercise 7.18 You can, in fact, construct an explicit bijection from $[0, 1]$ to $[0, 1)$. Define X to be the subset of $[0, 1]$ consisting of the non-negative powers of $\frac{1}{2}$:

$$X = \left\{ \ldots, \frac{1}{8}, \frac{1}{4}, \frac{1}{2}, 1 \right\}.$$

Define the function $h : [0, 1] \to [0, 1)$ by:

$$h(x) = \begin{cases} \frac{x}{2} & \text{if } x \in X, \\ x & \text{if } x \notin X. \end{cases}$$

This function is the identity on points of $[0, 1]$ that are not in X; the application of h to X is illustrated in Figure 3.

Prove that $h : [0, 1] \to [0, 1)$ is a bijection.

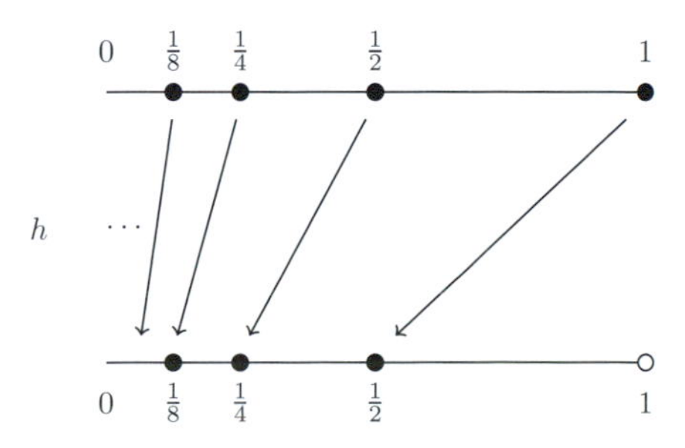

Figure 3. The bijection $h : [0, 1] \to [0, 1)$ restricted to $X \subset [0, 1]$. The function h is the identity on all other points.

We guide you through the hard work of proving the Schröder–Bernstein Theorem in Project 11.8. Our construction of the bijection $h : [0,1] \to [0,1)$ in Exercise 7.18 is derived directly from the proof outlined in that project. In the remainder of this chapter, we spend some time enjoying the consequences of the Schröder–Bernstein Theorem.

PROPOSITION 7.18. *The real numbers, $\mathbb{R}$, and the complement of the integers, $\mathbb{Z}^c = \mathbb{R} - \mathbb{Z}$, have the same cardinality.*

PROOF. One injection is straightfoward: $\mathbb{Z}^c \hookrightarrow \mathbb{R}$ by inclusion.

To construct the other injection, note that there is an injection from $\mathbb{R}$ into the open interval $(0,1)$ by Proposition 7.3. Composing this with the inclusion $(0,1) \subset \mathbb{Z}^c$ gives us an injection $\mathbb{R} \hookrightarrow \mathbb{Z}^c$.

Because we have injections between $\mathbb{R}$ and $\mathbb{Z}^c$, in both directions, the Schröder–Bernstein Theorem implies $\mathbb{R} \asymp \mathbb{Z}^c$. □

The next proposition is more challenging to prove, and we have outlined an argument that uses some lemmas, each interesting in its own right.

PROPOSITION 7.19. *The real numbers and the solid square,*

$$[0,1]^2 = \{(x,y) \in \mathbb{R}^2 \mid 0 \leq x \leq 1 \text{ and } 0 \leq y \leq 1\},$$

have the same cardinality.

LEMMA 7.20. *There is a bijection $f : (0,1) \times (0,1) \to (0,1)$.*

PROOF. By the Schröder–Bernstein Theorem we need injections between these two sets. One direction is easy. Let $g : (0,1) \hookrightarrow (0,1) \times (0,1)$ be the function defined by $g(x) = (x, 0.5)$.

Going in the other direction, define $f : (0,1) \times (0,1) \to (0,1)$ by

$$f(0.a_1a_2a_3\ldots, 0.b_1b_2b_3\ldots) = 0.a_1b_1a_2b_2a_3b_3\ldots,$$

where once again we do not allow for repeating 9s to end our decimal expansions. This map is an injection, as the decimal expansions in the coordinates of each element of $(0,1) \times (0,1)$ can be recovered from its image. (The function is not a surjection; for example, nothing is mapped to $0.\overline{01} = 0.010101\ldots$.) □

LEMMA 7.21. *The real line $\mathbb{R}$ and the plane $\mathbb{R}^2$ have the same cardinality.*

Exercise 7.19 Prove Lemma 7.21.

LEMMA 7.22. *The plane $\mathbb{R}^2$ has the same cardinality as the square $[0,1]^2$.*

PROOF. Inclusion gives us an injection $[0,1]^2 \hookrightarrow \mathbb{R}^2$.

For the other injection, Proposition 7.3 shows that there is an injection $f : \mathbb{R} \hookrightarrow (0,1)$. We can then define a function $F : \mathbb{R}^2 \to (0,1) \times (0,1)$ by

$$F(x,y) = (f(x), f(y)).$$

Because f is injective, so is the function F.[3] Since $(0,1)^2 \subset [0,1]^2$, this gives us an injection $\mathbb{R}^2 \hookrightarrow [0,1]^2$.

3 This follows from Exercise 5.38. If you did not do this exercise, then please verify this claim.

By the Schröder–Bernstein Theorem, we now know that $\mathbb{R}^2$ and $[0, 1]^2$ have the same cardinality. □

We leave the remainder of the proof of Proposition 7.19 as an exercise.

Exercise 7.20 Use Lemmas 7.20, 7.21, and 7.22 to prove Proposition 7.19.

GOING BEYOND THIS BOOK. Theorem 11.13 is sometimes referred to as the Cantor–Schröder–Bernstein Theorem and sometimes as the Cantor-Bernstein Theorem. This result's history is a bit complicated, as it was first stated by Cantor in 1887 but a correct proof – discovered nearly simultaneously and independently by Bernstein and Schröder – was not published until 1898. For more information on this history, see Arie Hinkis' book *Proofs of the Cantor–Bernstein Theorem* [**Hin13**].

7.7. End-of-Chapter Exercises

Exercises you can work on after Sections 7.1 and 7.2

7.21 Suppose that in his first turn Cookie Monster eats cookie #10; in his second turn he eats cookie #20; in his third turn he eats cookie #30; and so on. Which cookies will be left on his plate at noon? Compare your answer to the similar scenario that ends with Exercise 7.3.

7.22 Let $S \subseteq \mathbb{N}$ be any subset of the natural numbers. Prove that it is possible for Cookie Monster to eat his cookies in such a way that the cookies labelled by elements of S are the ones left on his plate at noon. (This may require Cookie Monster not to eat any cookies on some rounds.)

7.23 Let S be a finite subset of $\mathbb{N}$. Prove that $\mathbb{N} \asymp \mathbb{N} \setminus S$.

7.24 Construct a bijection from $\mathbb{Z}$ to the set of odd integers.

7.25 Construct a bijection from $\mathbb{R}$ to $\mathbb{R}$ that is not strictly increasing or strictly decreasing.

7.26 Use the functions given in the proof of Proposition 7.3 to create a function that is a bijection from $(0, 1)$ to $(-2, \infty)$.

7.27 Prove that the Euclidean plane and the interior of the first quadrant have the same cardinality. That is, prove $(\mathbb{R} \times \mathbb{R}) \asymp (\mathbb{R}^+ \times \mathbb{R}^+)$.

7.28 Prove that if $A \asymp B$, then $\mathcal{P}(A) \asymp \mathcal{P}(B)$.

Exercises you can work on after Sections 7.3 and 7.4

7.29 For any $n \in \mathbb{Z}$ define

$$\mathbb{Z}_{\geq n} = \{m \in \mathbb{Z} \mid m \geq n\}.$$

Prove that $\mathbb{Z}_{\geq n}$ is countably infinite for every n.

7.30 In Project 11.4 we explore the Gaussian integers, the complex numbers of the form $a + b\mathbf{i}$ where a and b are integers, and $\mathbf{i} = \sqrt{-1}$. Prove that the Gaussian integers are countably infinite.

7.31 Let $n \in \mathbb{N}$, and let $A_1, A_2, \ldots, A_n$ be countable sets. Generalize Lemma 7.6 by proving that

$$A_1 \times A_2 \times \cdots \times A_n$$

is countable.

7.32 Determine which of the following are true and which are false for non-empty sets A and B. Provide counterexamples for those that are false and proofs for those that are true.

(a) If $A \times B$ is countably infinite, then A is countably infinite.
(b) If $A \times A$ is countably infinite, then A is countably infinite.
(c) If $A \times B$ is countable, then A is countable.
(d) If $A \times A$ is countable, then A is countable.

7.33 Recall the definition of $\mathbb{Q}_n$ for $n \in \mathbb{N}$:

$$\mathbb{Q}_n = \left\{ q \in \mathbb{Q} \mid q = \frac{m}{n^k} \text{ for some } k, m \in \mathbb{Z} \right\}.$$

Prove that $\mathbb{Q}_n$ is countable for every $n \in \mathbb{N}$.

7.34 Find an example of two distinct sets A and B where

$$A \asymp B \asymp A \cup B \asymp A \cap B,$$

and prove that your example works.

7.35 In this question we return to the Hilbert Hotel. If you found Exercise 7.2 to be less than challenging, consider the following more difficult dilemma. What if infinitely many buses arrive at the Hilbert Hotel, each bus labelled with a natural number, and each of these buses containing a countably infinite number of passengers. Can the Hilbert Hotel accommodate all of the passengers?

7.36 Let S be a finite set and let $\mathcal{F}_{S \to \mathbb{N}}$ be the set of all functions from S to the natural numbers $\mathbb{N}$. Prove that $\mathcal{F}_{S \to \mathbb{N}}$ is countable.

Exercises you can work on after Section 7.5

7.37 Let A be a countable subset of $\mathbb{R}^2$. Prove there is a horizontal line that misses all the points in A.

7.38 Recall that the set of binary strings Σ^* consists of all finite strings of 0s and 1s, including the empty string ε.

(a) Prove that Σ^* is a countably infinite set.
(b) Define $\Sigma^*_{\mathbb{N}}$ to be the set of all infinite sequences of 0s and 1s. Use a diagonal argument to prove that $\Sigma^*_{\mathbb{N}}$ is uncountable.

7.39 This problem is about the power set of $\mathbb{N}$, $\mathcal{P}(\mathbb{N})$.

(a) Prove that the set of all finite subsets of $\mathbb{N}$ is a countably infinite set.
(b) Prove that the set of all infinite subsets of $\mathbb{N}$ is uncountable.

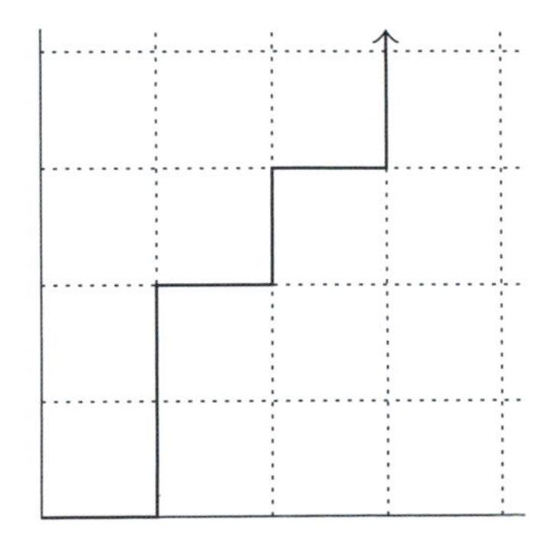
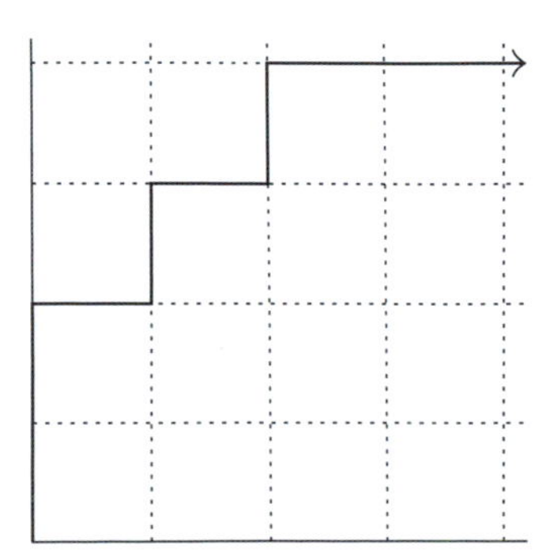

Figure 4. Two examples of staircases, as discussed in Exercise 7.40.

7.40 A *staircase* is a path that starts at the origin and continually goes up or to the right in the integer lattice inside of $\mathbb{R}^2$. The initial portions of two staircases are shown in Figure 4. Is the set of all possible staircases countable or uncountable?

7.41 In this exercise you will revisit the proof of Theorem 7.13, as modified in Exercise 7.14. For any non-empty set S, prove that ...

(a) ... there is an injective function from S to $P(S)$,
(b) ... there is a surjective function from $P(S)$ to S,
(c) ... there is no surjective function from S to $P(S)$,
(d) ... there is no injective function from $P(S)$ to S.

These results are usually interpreted as indicating that $P(S)$ is strictly larger than S. Use induction to show that there are infinitely many uncountably infinite sets, no two of which have the same cardinality. (Exercise 7.51 is similar to this one but perhaps a bit more complicated. Also look at Exercise 7.50.)

7.42 Let the universal set for this question be the real numbers, $\mathbb{R}$. Consider three subsets:

$\mathbb{Q}^+$ = the positive rationals numbers,
$[0, 1]$ = the closed interval from 0 to 1,
$\mathbb{Z}^c = \mathbb{R} - \mathbb{Z}$ = the complement of the integers as a subset of $\mathbb{R}$.

(a) A 3-set Venn diagram divides $\mathbb{R}$ into eight regions, as seen in Figure 5. Show that each of the eight regions is non-empty by providing explicit examples of real numbers in each region.
(b) Do any regions contain only finitely many real numbers?
(c) Are any countably infinite?
(d) Are any uncountable?

You should prove your claims for the final three questions!

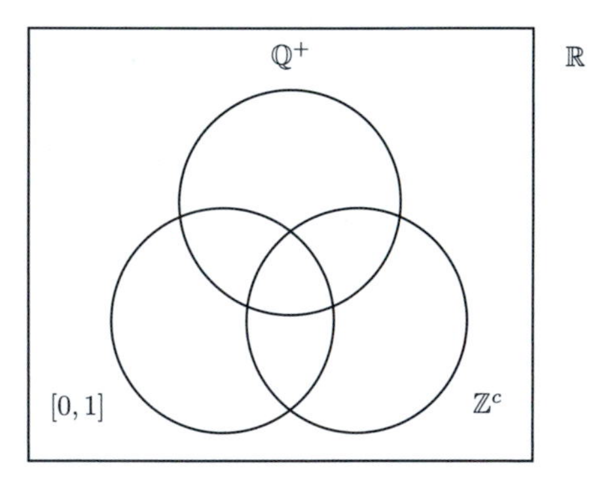

Figure 5. The Venn diagram for Exercise 7.42.

7.43 A real number r is *algebraic* if it is a root of a polynomial with integer coefficients. For example, $\sqrt{2}$ is an algebraic number, being a root of $x^2 - 2$. Denote the set of all algebraic numbers by $\mathbb{A}$. If $a \in \mathbb{A}$, then a has *degree* k if a is a root of a polynomial of degree k with integer coefficients, and k is as small as possible. For example, $\sqrt{2}$ has degree 2, and the algebraic numbers of degree 1 are the rational numbers.

(a) For each positive integer k let $\mathbb{Z}[x]_k$ be the set of polynomials of degree k with integer coefficients. Prove that $\mathbb{Z}[x]_k$ is countably infinite.
(b) Prove that the set of all polynomials with integer coefficients,

$$\mathbb{Z}[x] = \bigcup_{k \in \mathbb{N}} \mathbb{Z}[x]_k,$$

is countably infinite.
(c) Prove that $\mathbb{A}$ is countable.
(d) The *transcendental* numbers are the real numbers that are not algebraic. Prove that the set $\mathbb{R} - \mathbb{A}$ of transcendental numbers is uncountable.

Exercises you can work on after Section 7.6

7.44 Prove that if you have three sets A, B, and C and three injections $A \hookrightarrow B$, $B \hookrightarrow C$, and $C \hookrightarrow A$, then

$$A \asymp B \asymp C.$$

7.45 Prove that the set of points on the unit circle

$$\{(x, y) \in \mathbb{R}^2 \mid x^2 + y^2 = 1\}$$

has the same cardinality as $\mathbb{R}$.

7.46 Our first example of a set that is not countable was $\mathcal{P}(\mathbb{N})$, the power set of $\mathbb{N}$. Here you'll show that this set has the same cardinality as two other, seemingly unrelated, sets.

(a) Let $\Sigma_{\mathbb{N}}^*$ be the set of all infinite sequences of 0s and 1s as first defined in Exercise 7.38. Find an injection from $\mathcal{P}(\mathbb{N})$ to $\Sigma_{\mathbb{N}}^*$.
(b) Find an injection from $\Sigma_{\mathbb{N}}^*$ to $\mathbb{R}$.
(c) Find an injection from $\mathbb{R}$ to $\mathcal{P}(\mathbb{N})$. As a hint, start by pointing out that you only need to find an injection $(0, 1) \hookrightarrow \mathcal{P}(\mathbb{N})$. Then create such an injection by encoding the decimal (or binary) representation of $x \in (0, 1)$ as a subset of $\mathbb{N}$.
(d) Conclude via the Schröder–Bernstein Theorem that $\mathcal{P}(\mathbb{N}) \asymp \Sigma_{\mathbb{N}}^* \asymp \mathbb{R}$.

7.47 This exercise outlines a proof of Proposition 7.16; your first job is to fill in the details that prove each of the following claims. You will need to use the characterization of irrational numbers as those real numbers whose decimal expansions do not repeat.

(a) We know that $\sqrt{2}$ is irrational; its decimal expansion begins $\sqrt{2} = 1.4142\ldots$. Let s_i be the ith digit in the decimal expansion of $\sqrt{2}$, so that $s_1 = 4$, $s_2 = 1$, $s_3 = 4$, and so on. Prove that any real number of the form $0.s_1 a_1 s_2 a_2 s_3 a_3 \ldots$, where each $a_i = 0$ or 1, is also irrational.
(b) Prove that there is an injective function from $\mathcal{P}(\mathbb{N})$ to the irrational numbers.
(c) Use Exercise 7.46 to show there is an injection from $\mathbb{R}$ to $\mathbb{Q}^c = \mathbb{R} - \mathbb{Q}$.

Using the claims above, and the Schröder–Bernstein Theorem, you should now be able to construct a well-presented proof that $\mathbb{R} \asymp \mathbb{Q}^c$.

More Exercises!

7.48 Is $(\mathbb{R} \times \mathbb{N}) \asymp \mathbb{R}$?

7.49 Let $k \in \mathbb{N}$. We know by Exercise 7.31 that $\mathbb{N}^k = \underbrace{\mathbb{N} \times \cdots \times \mathbb{N}}_{k \text{ copies}}$ is countable.

(a) Define $\mathbb{N}^{\mathbb{N}} = \underbrace{\mathbb{N} \times \mathbb{N} \times \cdots \times \mathbb{N} \times \cdots}_{\text{countably many copies of } \mathbb{N}}$. Is $\mathbb{N}^{\mathbb{N}}$ countable or uncountable?

(b) Let $\underbrace{\mathbb{N} \oplus \mathbb{N} \oplus \cdots \oplus \mathbb{N} \oplus \cdots}_{\text{countably many copies of } \mathbb{N}}$ be the subset of $\mathbb{N}^{\mathbb{N}}$ consisting of those elements where all but finitely many of the entries are 0. Is this subset countable or uncountable?

7.50 Prove that "injects" behaves like a partial order. Specifically, prove or explain the following three statements.

(a) If A is any set then there is an injection $A \hookrightarrow A$.
(b) If there is an injection $A \hookrightarrow B$ and an injection $B \hookrightarrow C$, then there is an injection $A \hookrightarrow C$.
(c) Explain how the Schröder–Bernstein Theorem relates to the antisymmetric property of a partial order.

7.51 This exercise is a variation on Exercise 7.41. Let $\mathcal{F}_{S \to S}$ be the set of functions from a non-empty set S to itself. Prove ...

(a) ... there is an injective function from S to $\mathcal{F}_{S \to S}$,
(b) ... there is a surjective function from $\mathcal{F}_{S \to S}$ to S,
(c) ... there is no surjective function from S to $\mathcal{F}_{S \to S}$,
(d) ... there is no injective function from $\mathcal{F}_{S \to S}$ to S.

7.52 This exercise uses the staircases defined in Exercise 7.40. We recommend working on Exercise 7.46 before attempting this problem.

Define two staircases S_1 and S_2 to be *eventually equal* if their symmetric difference is bounded. That is, there is a real number r such that for all $(x, y) \in S_1 \triangle S_2$, $x^2 + y^2 \leq r$. Said another way, the set of all points that are on one staircase but not on the other can be contained inside a circle about the origin.

(a) Give an example of a pair of staircases that are eventually equal, and another example of a pair of staircases that intersect infinitely often, but are not eventually equal.
(b) Prove that this is an equivalence relation on the set of all staircases.
(c) Prove that each equivalence class is a countable set of staircases.
(d) Give a one-sentence proof that the number of equivalence classes is uncountable.
(e) Prove that the set of equivalence classes has the same cardinality as $\mathcal{P}(\mathbb{N})$.

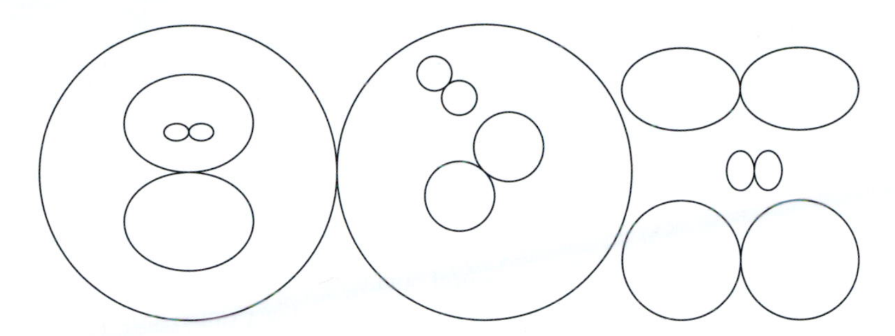

Figure 6. Some disjoint figure 8s in the plane, for Exercise 7.53.

7.53 In this problem you will prove a rather unexpected difference between circles in the plane and figure 8 diagrams.

(a) Prove that it is possible to embed an uncountable number of circles in the plane, no two intersecting at any point.

(b) Define a *figure 8* to be two circles or two ellipses that meet in a single point; several examples are shown in Figure 6. Prove that at most countably many disjoint figure 8s can be embedded in the plane.

8 The Real Numbers

In this chapter, we introduce several concepts that are crucial for developing an understanding of the real number system, including least upper bounds, completeness, and the Archimedean property. The chapter ends with proofs that the famous real numbers e and π are irrational. Key topics, such as limits of sequences and the comparison test for series, are developed here so that all sections except the final one do not require prior experience with calculus. However, our treatment of some of the topics is brief, so the full chapter is probably best appreciated by those who have previously studied infinite series in calculus.

8.1. Completeness

The rationals are properly contained in the real numbers, which immediately leads to the question of what properties distinguish $\mathbb{R}$ from $\mathbb{Q}$? In a course in analysis you would learn that the reals are the "completion" of the rationals, so in this section we give an indication of what this means.

Let $S \subseteq \mathbb{R}$ be a set of real numbers. An *upper bound* for S is a real number $m \in \mathbb{R}$ such that $s \leq m$ for all $s \in S$. Upper bounds are not unique; for example, consider the set

$$A = \left\{1 - \frac{1}{2^n} \mid n \in \mathbb{N}\right\} = \left\{\frac{1}{2}, \frac{3}{4}, \frac{7}{8}, \ldots\right\}.$$

The real number 6 is an upper bound for A, as is the number 1. In fact, every real number $x \geq 1$ is an upper bound for A.

Similarly, a *lower bound* for $S \subseteq \mathbb{R}$ is a number $\ell \in \mathbb{R}$ such that $\ell \leq s$ for all $s \in S$.

Exercise 8.1 Give examples of upper bounds and lower bounds for the half-open interval

$$(-1, 2] = \{x \in \mathbb{R} \mid -1 < x \leq 2\},$$

and draw a figure showing the set and your bounds.

A *least upper bound* for a set of numbers S is an upper bound $\mu \in \mathbb{R}$ with the property that if m is any other upper bound for S, then $\mu \leq m$. If A is the set defined above, then 1 is the least upper bound for A, a plausible claim that we will prove in Section 8.2. The number 4 is an upper bound for the open interval

$$(0, \pi) = \{x \in \mathbb{R} \mid 0 < x < \pi\},$$

while π is the least upper bound. To see this, note that the least upper bound is certainly positive, and no positive $m < \pi$ can be an upper bound because then $m < (m + \pi)/2$ with $(m + \pi)/2 \in (0, \pi)$.

Exercise 8.2 Does the empty set have any upper or lower bounds? Does it have a least upper bound?

DEFINITION 8.1. The *least upper bound property*, also referred to as the *completeness axiom* for the real numbers, states the following:

> (Completeness) If S is any non-empty subset of $\mathbb{R}$, and some $m \in \mathbb{R}$ is an upper bound for S, then there is a $\mu \in \mathbb{R}$ that is the least upper bound of S.

In many presentations of the development of the real numbers, this property is assumed as an axiom that is part of the description of $\mathbb{R}$, not as a property to be proven, and that is the approach we take as well.

The least upper bound property distinguishes $\mathbb{R}$ from $\mathbb{Q}$. Consider for example the set

$$S = \{q \in \mathbb{Q} \mid q > 0 \text{ and } q^2 < 2\}.$$

S is non-empty; for example $1 \in S$. Also, the set has upper bounds that are rational numbers; for example $m = 1.5$ is larger than every $s \in S$ since $1.5^2 = 2.25 > 2$. There is, however, no least upper bound $\mu \in \mathbb{Q}$: μ can't be less than $\sqrt{2} = 1.41421\ldots$ or else one of the rational numbers

$$1,\ 1.4,\ 1.41,\ 1.414,\ 1.4142,\ \ldots$$

would be greater than μ; and μ can't be greater than $\sqrt{2}$ or else one of the numbers

$$2,\ 1.5,\ 1.42,\ 1.415,\ 1.4143,\ \ldots$$

would be an upper bound less than μ. There is a least upper bound, namely $\sqrt{2}$, but this requires consideration of numbers outside of $\mathbb{Q}$. This hints at why $\mathbb{R}$ is the "completion" of $\mathbb{Q}$: if you start with the set $\mathbb{Q}$, augmenting it with the least upper bounds of all non-empty and bounded subsets leads to the set $\mathbb{R}$.

Analogous to the least upper bound, the *greatest lower bound* for a set $S \subseteq \mathbb{R}$ is a lower bound $\lambda \in \mathbb{R}$ with the property that if $\ell \in \mathbb{R}$ is any other lower bound for S, then $\ell \leq \lambda$. It is also true that every non-empty subset of the reals that has a lower bound has a greatest lower bound.

Exercise 8.3 Let

$$S = \{2.7, 2.71, 2.718, 2.7182, 2.71828, 2.718281, 2.7182818, \ldots\}$$

be the set of truncated decimal expansions of e.

(a) Find integers that are upper and lower bounds for S.
(b) What is the greatest lower bound for S?
(c) What is the least upper bound for S?

Exercise 8.4 Consider the set

$$A = (-2, 0) \cup \left\{\frac{1}{n} \mid n \in \mathbb{N}\right\}.$$

(a) Draw picture of A on the real number line.
(b) Describe the set of all lower bounds for A.

(c) Describe the set of all upper bounds for A.
(d) What is the greatest lower bound of A?
(e) What is the least upper bound of A?

GOING BEYOND THIS BOOK. There are other foundational properties that could be used in place of completeness to distinguish $\mathbb{R}$ from $\mathbb{Q}$; for example, the Monotone Convergence Theorem discussed in Section 8.5 is logically equivalent to completeness, as is the Nested Intervals Property described in Exercise 8.39. For more information on various axioms used to define the real numbers, some equivalent and some not, see "Real analysis in reverse," by James Propp [**Pro13**].

8.2. The Archimedean Property

The *Archimedean Property*, while not a distinguishing feature of the real numbers, is a fundamental property of $\mathbb{R}$. It was introduced by Archimedes as an axiom in his work on geometry.[1] In modern terminology it is the seemingly innocuous statement that the natural numbers do not have an upper bound.

Here, we prove the Archimedean Property as a consequence of completeness.

THEOREM 8.2 (Archimedean Property). *Given any $r \in \mathbb{R}$ there is an $n \in \mathbb{N}$ such that $r < n$.*

PROOF. Assume to the contrary that there is an upper bound for $\mathbb{N}$. By the least upper bound property there must be a least upper bound $\mu \in \mathbb{R}$ for $\mathbb{N}$. Note that if $n \in \mathbb{N}$, then $n + 1 \in \mathbb{N}$ as well. Thus we know that $n + 1 \leq \mu$ for all $n \in \mathbb{N}$, which implies that $n \leq \mu - 1$ for all $n \in \mathbb{N}$. So $\mu - 1$ is an upper bound for $\mathbb{N}$ that is smaller than the least upper bound μ. □

An alternate version of the Archimedean Property is given in Corollary 8.3, which follows easily from Theorem 8.2 by noting that the desired inequality is equivalent to saying there is an $n \in \mathbb{N}$ such that $s/r < n$.

COROLLARY 8.3. *Given any two positive real numbers r and s, there is an $n \in \mathbb{N}$ such that $s < nr$.*

Corollary 8.3 implies that the rational numbers $\{1/n \mid n \in \mathbb{N}\}$ get arbitrarily close to 0. To see this, just let $s = 1$ and divide the inequality by n.

COROLLARY 8.4. *Given any positive real number r, there is an $n \in \mathbb{N}$ such that $1/n < r$.*

Exercise 8.5 Show that, given any positive real number r, there is a $k \in \mathbb{N}$ such that $1/10^k < r$. Then revisit Proposition 3.9.

The least upper bound and Archimedean properties are powerful tools because they guarantee the existence of important numbers associated to subsets of $\mathbb{R}$. For example, the set $A = \left\{\frac{1}{2}, \frac{3}{4}, \frac{7}{8}, \ldots\right\}$ from Section 8.1 appears to have 1 as a least upper bound.

1 See Netz's translation *The Works of Archimedes, Vol. 1: The Two Books On the Sphere and the Cylinder* [**Arc04**].

We can now confirm this fact, for if we assume that $r < 1$ is an upper bound for A, then $0 < 1 - r$, and so by an easy variant of Exercise 8.5 there exists a $k \in \mathbb{N}$ such that $0 < 1/2^k < 1 - r$. A little algebra then shows that $r < 1 - 1/2^k < 1$, with $1 - 1/2^k \in A$, contradicting the assumption that r is an upper bound for A.

We conclude this section by examining upper and lower bounds for the sets

$$E(r) = \{r^n \mid n \in \mathbb{N}\}$$

consisting of powers of a single real number r. Examples of these sets include

$$E\left(\frac{1}{2}\right) = \left\{\frac{1}{2}, \frac{1}{4}, \frac{1}{8}, \dots\right\} \text{ and } E(-1) = \{\pm 1\} .$$

The existence and value of the least upper bound and greatest lower bound of the sets $E(r)$ depend on the value of r:

THEOREM 8.5. *Let r be a real number. Then the least upper bound and greatest lower bound for $E(r)$ are given in the table below.*

r is in	$(-\infty, -1)$	$\{-1\}$	$(-1, 0)$	$\{0\}$	$(0, 1)$	$\{1\}$	$(1, \infty)$
LUB of $E(r)$	None	1	r^2	0	r	1	None
GLB of $E(r)$	None	-1	r	0	0	1	r

The fact that we have presented this classification in the form of a table is an indication that the argument for it is best handled case by case. Some of the cases are immediate, for example the cases $r = \pm 1$; others are left to Exercise 8.22. Here we present proofs of two of the more interesting claims from Theorem 8.5.

PROOF THAT IF $r \in (0, 1)$, THE GREATEST LOWER BOUND OF $E(r)$ IS 0. Because the powers of r are all positive, 0 is a lower bound of $E(r)$. We need to show that there is no positive number that is a lower bound for $E(r)$, thereby establishing that 0 is the greatest lower bound. To this end, given any $\epsilon \in \mathbb{R}$ with $\epsilon > 0$, we produce an $n \in \mathbb{N}$ such that $r^n < \epsilon$.

Since $0 < r < 1$, we can define the positive real number

$$t = \frac{1 - r}{r}$$

so that

$$r = \frac{1}{1 + t}.$$

Bernoulli's inequality (Exercise 2.32) then implies that for any $n \in \mathbb{N}$,

$$r^n = \frac{1}{(1 + t)^n} \leq \frac{1}{1 + nt}.$$

A little algebra shows that

$$\frac{1}{1 + nt} < \epsilon$$

holds if $\epsilon t > 1/n$, and Corollary 8.4 insures that such an $n \in \mathbb{N}$ exists. We conclude that $r^n < \epsilon$. □

PROOF THAT IF $r \in (-\infty, -1)$, THEN $E(r)$ IS UNBOUNDED. We first show that $E(r)$ has no upper bound. Assume to the contrary that $m \in \mathbb{R}$ is an upper bound for $E(r)$. Since $r^2 > 1$, we know that

$$1 < r^{2n} \leq m$$

for all $n \in \mathbb{N}$. Taking logarithms[2] gives

$$0 < \ln(r^{2n}) = n \cdot \ln(r^2) \leq \ln(m)$$

for all $n \in \mathbb{N}$, where the middle equality follows from Proposition 2.16. Because $\ln(r^2)$ and $\ln(m)$ are positive, having the final inequality hold for all $n \in \mathbb{N}$ contradicts Corollary 8.3.

The argument that there is no lower bound is similar, but you need to be careful with negative numbers. The proof is left as the following exercise. □

Exercise 8.6 Show that $E(r)$ has no lower bound if $r \in (-\infty, -1)$.

REMARK 8.6. Our first proof used linearization in order to establish the necessary inequality, while our second proof used logarithms. We could have used one approach for both proofs, applying only linearization or using only logarithms. Sometimes it is helpful, though, to see that there are multiple techniques available.

8.3. Sequences of Real Numbers

A real-valued *sequence* is simply an ordered list of real numbers indexed by the natural numbers; recall part (c) of Example 5.5. It is not uncommon to have a sequence be indexed by non-negative integers, or at times by some other subset of the integers with a minimal element, but we rarely use this more general type of indexing until the final section of the chapter.

A sequence is displayed inside parentheses, to emphasize that this is an ordered list, not a set. Often a sequence can be described succinctly by a formula or other notation involving an index variable, as in the following examples:

(a) $(1/2^n) = \left(\frac{1}{2}, \frac{1}{4}, \frac{1}{8}, \frac{1}{16}, \ldots\right)$.

(b) $(0, 0, 0, 0, \ldots)$ is the constant 0 sequence.

(c) $(f_n) = (1, 1, 2, 3, 5, 8, \ldots)$ is the Fibonacci sequence.

(d) $\left(2, \frac{3}{2}, \frac{17}{12}, \frac{577}{408}, \ldots\right)$, where $a_1 = 2$ and $a_{n+1} = \frac{1}{2}\left(a_n + \frac{2}{a_n}\right)$.

Exercise 8.7 Before you tackle the precise definition of the limit of a real-valued sequence given below, try to guess the limits of each of the four example sequences given above: towards what value, if any, are the terms of each sequence ultimately heading?

DEFINITION 8.7. The real-valued sequence

$$(a_n) = (a_1, a_2, a_3, a_4, \ldots)$$

2 Using $\log_a$ for any base $a > 1$ works in this argument; here $\ln = \log_e$ is the natural log.

converges to a limit L if for every $\epsilon > 0$ there exists an $N \in \mathbb{N}$ such that

$$|L - a_n| < \epsilon$$

for all $n \geq N$. In this case, we write

$$\lim_{n\to\infty} a_n = L.$$

This definition has a universal quantifier ("*for every* $\epsilon > 0 \ldots$") followed by an existential quantifier ("*there exists* an $N \in \mathbb{N} \ldots$"), and even one additional universal quantifier ("*for all* $n \geq N$"). This makes the definition seem complicated, but it is necessary to describe the limit concept well.

Exercise 8.8 Convince yourself that the following sentence correctly rephrases Definition 8.7: "Regardless of how close you want the sequence terms to be to the limiting value, you will get your wish so long as you ignore an appropriate number of the initial terms of the sequence."

Exercise 8.8 implicitly defines the "tail" of a sequence. Given a sequence (a_n) and any $N \in \mathbb{N}$, the sequence that begins at N

$$(a_N, a_{N+1}, a_{N+2}, \ldots)$$

is a *tail* of (a_n). Thus another rephrasing of Definition 8.7 is "Given any $\epsilon > 0$, there is some tail of the sequence (a_n) whose terms are contained in $(L - \epsilon, L + \epsilon)$."

When presented with any definition, especially a complicated one, it is often helpful to explore the concept via a specific example. Consider Figure 1, which shows the first 50 terms of a sequence (a_n). The horizontal axis gives the index n and the vertical axis determines the value a_n, so the first few terms are $a_1 = 2.5$, $a_2 = 0.25$, and $a_3 = 0$. The values a_n *appear* to eventually cluster around a height of 1 as n gets larger, but is 1 the limit of the sequence? To answer this question, we need to know that in fact

$$a_n = 1 + \frac{3\cos(n\pi/3)}{n}$$

for all $n \in \mathbb{N}$.

As n gets larger, the influence of the term $3\cos(n\pi/3)/n$ decreases because cosine is bounded by -1 and 1, suggesting that the a_n are getting closer and closer to 1 and thus

$$\lim_{n\to\infty}\left(1 + \frac{3\cos(n\pi/3)}{n}\right) = 1\,. \tag{8.1}$$

To *prove* this, we need to show that given any $\epsilon > 0$ we can find an $N \in \mathbb{N}$ guaranteeing that $|1 - a_n| < \epsilon$ when $n \geq N$. In other words, we need to find a tail of (a_n) that is contained within ϵ of 1. For example, in Figure 1 the shaded band indicates the acceptable range for $\epsilon = 0.1$, namely the y-values between 0.9 and 1.1, corresponding to $|1 - y| < 0.1$. For this particular value of ϵ it appears that $N = 31$ might suffice. We can prove that $N = 31$ is sufficient for this particular ϵ by first noting the general result that

$$|1 - a_n| = \left|1 - \left(1 + \frac{3\cos(n\pi/3)}{n}\right)\right| = \left|\frac{3\cos(n\pi/3)}{n}\right| \leq \frac{3}{n} \tag{8.2}$$

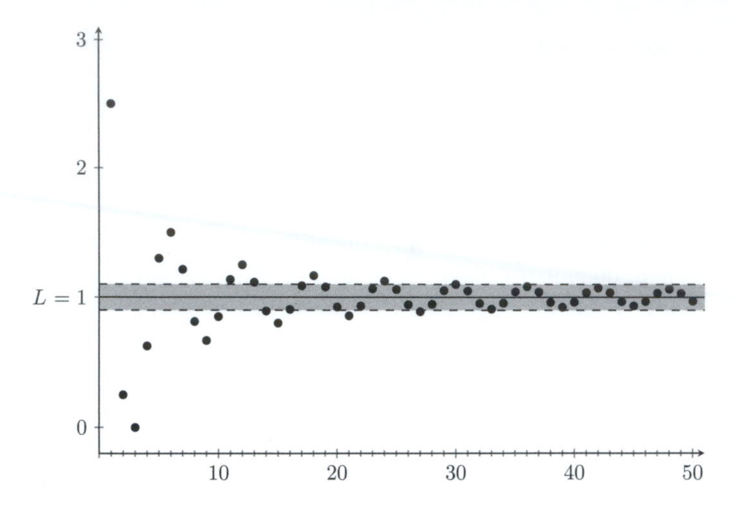

Figure 1. A sequence and its limit.

for all $n \in \mathbb{N}$, where the inequality holds because $|\cos(n\pi/3)|$ can never be larger than 1, and we are able to drop the absolute value signs because n is positive. Since

$$\frac{3}{n} < \frac{1}{10}$$

when $n \geq 31$, we know that $|1 - a_n| < 0.1$ for all $n \geq 31$.

Exercise 8.9 Show that $|1 - a_n| < 0.01$ for all $n \geq 301$. Comment on how the acceptable values of N depend on ϵ.

With these examples in hand, we are now well prepared to present a short proof of the limit claimed in Equation 8.1.

PROOF THAT THE LIMIT IS 1. By Inequality 8.2, we know that $|1 - a_n| \leq 3/n$ for all $n \in \mathbb{N}$. Thus we can establish the desired limit by showing that, given any $\epsilon > 0$, we can find an $N \in \mathbb{N}$ such that $3/N < \epsilon$, because $n \geq N$ implies $3/n \leq 3/N < \epsilon$. The inequality $3/N < \epsilon$ is equivalent to

$$\frac{1}{N} < \frac{\epsilon}{3}.$$

Corollary 8.4 guarantees that such an N exists. □

The result of the following exercise will be used in the next section on geometric series; for the proof, we encourage you to borrow ideas from Theorem 8.5.

Exercise 8.10 Let r be a real number in the open interval $(0, 1)$, and consider the sequence $(r^n) = (r, r^2, r^3, r^4, \ldots)$. Prove that

$$\lim_{n\to\infty} r^n = 0.$$

As an example of a sequence that does not converge, consider

$$\left((-1)^{n-1}\right) = (1, -1, 1, -1, \ldots).$$

To prove that this sequence does not converge, we need to find a troublesome value for ϵ so that the inequality

$$|L - a_n| < \epsilon$$

in the definition of convergence can never hold for sufficiently large $n \in \mathbb{N}$. For example, let $\epsilon = 1/2$, and assume to the contrary that the sequence converges to a limit L. Then there is an $N \in \mathbb{N}$ such that

$$|L - a_n| < 1/2$$

for all $n \geq N$. When the index n is odd, this implies that L must be a number where

$$|L - 1| < 1/2 \Rightarrow L \in (1/2, 3/2);$$

and, when the index n is even,

$$|L - (-1)| < 1/2 \Rightarrow L \in (-3/2, -1/2).$$

However, the interval $(-3/2, -1/2)$ does not intersect the interval $(1/2, 3/2)$, contradicting the existence of such an L.

REMARK 8.8. Another setting where we discuss the convergence of sequences, seemingly quite different than what is discussed in this section, is in Project 11.10. There we construct the Cantor Set, a subset of the unit interval with some striking and counterintuitive properties.

GOING BEYOND THIS BOOK. The intuitive notion of a "limit" of a sequence has a long history. It is implicit in the development of series expansions for sine and cosine in the Kerala school in fifteenth-century India, and later by Europeans developing the calculus. It is not, however, an elementary concept to define. The standard definition in use today was introduced by mathematicians in the 1800s, centuries after the concept was introduced. For a nice history of this fundamental topic, see Judith Grabiner's "Who gave you the epsilon? Cauchy and the origins of rigorous calculus" [**Gra83**].

8.4. Geometric Series

The definition of convergence for an infinite series builds directly upon the definition of convergence of a sequence. For an infinite series $\sum_{i=1}^{\infty} a_i$, you look at the associated sequence of *partial sums* (s_n), where $s_n = \sum_{i=1}^{n} a_i$. For example, the *harmonic series*

$$\sum_{i=1}^{\infty} \frac{1}{i} = 1 + \frac{1}{2} + \frac{1}{3} + \frac{1}{4} + \cdots$$

gives the sequence

$$(s_n) = \left(1,\ 1 + \frac{1}{2},\ 1 + \frac{1}{2} + \frac{1}{3},\ 1 + \frac{1}{2} + \frac{1}{3} + \frac{1}{4},\ \ldots\right) = \left(1, \frac{3}{2}, \frac{11}{6}, \frac{25}{12}, \ldots\right)$$

of partial sums. By focusing on partial sums, we never attempt to define the sum of an infinite number of terms; instead, we consider an infinite sequence of finite sums.

A series is said to converge to a limit L if and only if the sequence of partial sums converges to L. Written out fully, this means:

DEFINITION 8.9. The infinite series

$$\sum_{i=1}^{\infty} a_i = a_1 + a_2 + a_3 + \cdots$$

converges to a limit L if for every $\epsilon > 0$ there exists an $N \in \mathbb{N}$ such that

$$|L - s_n| < \epsilon$$

for all $n \geq N$, where $s_n = \sum_{i=1}^{n} a_i$. In this case, we write[3]

$$\sum_{i=1}^{\infty} a_i = L.$$

As we commented in the previous section, the definition of a convergent sequence is complicated. The definition of a convergent series has the same mix of quantifiers, with the added challenge of moving between the notion of an infinite series and its associated sequence of partial sums. In this section we examine the definition through a discussion of geometric series.

A *geometric series* is determined by two real numbers, the initial term a and the common ratio r. The ith term is $a_i = ar^{i-1}$, so the series is

$$\sum_{i=1}^{\infty} ar^{i-1} = a + ar + ar^2 + \cdots.$$

The most commonly encountered geometric series is perhaps

$$\sum_{i=1}^{\infty} \frac{1}{2^i} = \frac{1}{2} + \frac{1}{4} + \frac{1}{8} + \cdots,$$

with the initial term and common ratio both equal to $1/2$; the series happens to converge to 1.[4] Another geometric series with ancient roots arose in the context of Greek geometry. In computing the area of a region enclosed by a straight line crossing a parabola, Archimedes needed to sum the geometric series with initial term A – the area of a specific triangle embedded inside the region, as shown in Figure 2 – and common ratio $1/4$:

$$\sum_{i=1}^{\infty} A \frac{1}{4^{i-1}} = A + \frac{A}{4} + \frac{A}{16} + \frac{A}{64} + \cdots.$$

Archimedes proved that this series converges to $\frac{4}{3}A$. (See Sherman Stein's *Archimedes: What Did He Do Besides Cry Eureka?* [**Ste99**].)

3 While the definition involves a limit, the term "limit" doesn't appear here.

4 The Count and Cookie Monster are both aware of this result; recall subsection 7.1.2.

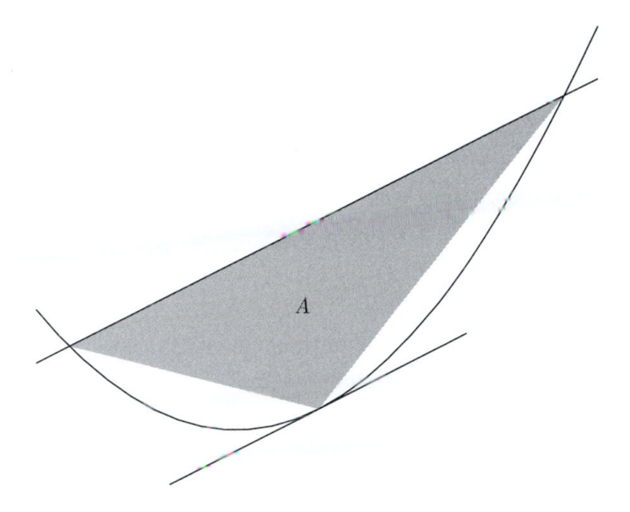

Figure 2. Archimedes showed that the area between the upper line segment and the parabola is $4A/3$.

What limit L might the general geometric series converge to? Suppose that you set

$$a + ar + ar^2 + \cdots = L.$$

Then you might try rescaling both sides by r to get

$$ar + ar^2 + ar^3 + \cdots = rL,$$

because the resulting expression has many terms in common with the original expression above it. If you then make the bold move of subtracting one of these equations from the other, you get

$$\{a + ar + ar^2 + \cdots\} - \{ar + ar^2 + ar^3 + \cdots\} = L - rL,$$

which leaves only $a = L - rL$ after canceling like terms, and thus

$$L = \frac{a}{1-r}.$$

We caution that this reasoning has made a number of completely unjustified assumptions about what one can do in terms of arithmetic operations on infinite series. On the other hand, this rather free-form approach has yielded a very valuable guess for the limiting value for a geometric series:

$$\sum_{i=1}^{\infty} ar^{i-1} = a + ar + ar^2 + \cdots = \frac{a}{1-r}.$$

Exercise 8.11 Verify that this guess is correct for the two geometric series examples given earlier in the section.

With this insight in hand, we can turn to establishing the general formula for geometric series using Definition 8.9.

THEOREM 8.10. *Let $a \in \mathbb{R}$, and let $r \in \mathbb{R}$ with $|r| < 1$. Then the geometric series*

$$\sum_{i=1}^{\infty} ar^{i-1} = a + ar + ar^2 + \cdots$$

with initial term a and common ratio r converges:

$$\sum_{i=1}^{\infty} ar^{i-1} = \frac{a}{1-r}.$$

PROOF. The result is immediate when $a = 0$. In order to prove this theorem we need to establish that for any $\epsilon > 0$ there is an $N \in \mathbb{N}$ such that

$$\left| \frac{a}{1-r} - \sum_{i=1}^{n} ar^{i-1} \right| < \epsilon$$

for all $n \geq N$. The key step involves arithmetic with cancellations based on the rescaling idea above, namely

$$(1-r)(a + ar + ar^2 + \cdots + ar^{n-1}) = a(1-r^n).$$

This implies that

$$\sum_{i=1}^{n} ar^{i-1} = a\frac{1-r^n}{1-r},$$

so we can prove the theorem by establishing that

$$\left| \frac{a}{1-r} - a\frac{1-r^n}{1-r} \right| = \left| \frac{ar^n}{1-r} \right| < \epsilon$$

for sufficiently large values of n.

Rewriting the above inequality shows that we need

$$|r|^n < \frac{\epsilon\,|1-r|}{|a|}.$$

This inequality is obviously true for all $n \in \mathbb{N}$ if $r = 0$. Otherwise, $0 < |r| < 1$, and Exercise 8.10 shows that $\lim_{n\to\infty} |r|^n = 0$. Since ϵ, $1 - r$, and $|a|$ are all positive real numbers, so is $\epsilon|1-r|/|a|$, and thus we know that the inequality holds for all n greater than some $N \in \mathbb{N}$, which establishes all of the prior displayed inequalities. □

Exercise 8.12 Explain why the condition $|r| < 1$ is necessary.

GOING BEYOND THIS BOOK. If you enjoyed the rather unjustified, free-form approach to guessing the limit of an infinite series, you should take a look at some of the things that the great mathematician Leonhard Euler did! William Dunham's book *Euler: The Master of Us All* [**Dun99**] has a chapter on Leonhard Euler's creative use of infinite series, and another interesting article is Morris Kline's "Euler and infinite series" [**Kli83**].

8.5. The Monotone Convergence Theorem

In this section, we use completeness to prove an important result about sequences commonly known as the Monotone Convergence Theorem. The name is a bit misleading, giving only half the truth: the theorem guarantees the convergence of a sequence if it is both monotone and bounded. Thus, we begin with two definitions:

DEFINITION 8.11. A sequence (a_n) is said to be *monotone increasing* (or simply *increasing*) if $a_n \leq a_{n+1}$ for all $n \in \mathbb{N}$. It is *monotone decreasing* (or *decreasing*) if $a_n \geq a_{n+1}$ for all $n \in \mathbb{N}$. Finally, it is *monotone* if either of these two conditions holds.

Exercise 8.13 What can be said of a sequence that is simultaneously monotone increasing and monotone decreasing?

DEFINITION 8.12. A sequence (a_n) is said to be *bounded* if there are real numbers ℓ and m such that $\ell \leq a_n \leq m$ for all $n \in \mathbb{N}$.

Exercise 8.14 Give examples of sequences that are ...

(a) ... monotone and bounded,
(b) ... bounded but not monotone,
(c) ... monotone but not bounded,
(d) ... neither monotone nor bounded.

We are now in position to prove the following.

THEOREM 8.13 (Monotone Convergence Theorem). *If (a_n) is a monotone and bounded sequence, then (a_n) converges. More specifically,*

(a) *If (a_n) is bounded and monotone increasing, and μ is the least upper bound of the set of values in the sequence, then $\lim_{n\to\infty} a_n = \mu$.*
(b) *If (a_n) is bounded and monotone decreasing, and λ is the greatest lower bound of the set of values in the sequence, then $\lim_{n\to\infty} a_n = \lambda$.*

PROOF OF (a). The set of sequence values is non-empty and bounded above, so by completeness it has a least upper bound μ. Thus, for any $\epsilon > 0$ there must be some a_N such that $\mu - \epsilon < a_N \leq \mu$; otherwise, $\mu - \epsilon$ would be an upper bound, contradicting the fact that μ is the least upper bound. Further, since (a_n) is monotone increasing we know that for any $n \geq N$ we must have

$$\mu - \epsilon < a_N \leq a_n \leq \mu \, .$$

Thus, for all $n \geq N$ we know $|\mu - a_n| < \epsilon$, which is what we needed to show. □

Exercise 8.15 Prove part (b) by minimally editing the proof given for part (a).

Notice that the Monotone Convergence Theorem does not provide a formula for the limit of the sequence, and in many instances the exact value of the limit may not be clear. Consider the sequence (a_n) where

$$a_n = \frac{1}{n+1} + \cdots + \frac{1}{2n}.$$

The first few terms are

$$a_1 = \frac{1}{2},$$
$$a_2 = \frac{1}{3} + \frac{1}{4} = \frac{7}{12},$$
$$a_3 = \frac{1}{4} + \frac{1}{5} + \frac{1}{6} = \frac{37}{60}.$$

All of the terms are positive, and for each $n \in \mathbb{N}$ we have

$$a_n = \frac{1}{n+1} + \cdots + \frac{1}{2n} \leq \underbrace{\frac{1}{n+1} + \cdots + \frac{1}{n+1}}_{n \text{ terms}} = \frac{n}{n+1} < 1.$$

Thus the terms of (a_n) are bounded below by 0 and above by 1. The sequence is also increasing, as

$$\begin{aligned} a_{n+1} - a_n &= \left(\frac{1}{n+2} + \cdots + \frac{1}{2n} + \frac{1}{2n+1} + \frac{1}{2n+2}\right) - \left(\frac{1}{n+1} + \frac{1}{n+2} \cdots + \frac{1}{2n}\right) \\ &= \frac{1}{2n+1} + \frac{1}{2n+2} - \frac{1}{n+1} = \frac{1}{2n+1} - \frac{1}{2n+2} \\ &= \frac{1}{(2n+1)(2n+2)} > 0. \end{aligned}$$

Since (a_n) is monotone and bounded, it must converge, but there is little in what we have presented that would illustrate the value it converges to (although if you've studied calculus, see Exercise 8.45). There are many similar situations in mathematics where one is able to prove that a quantity exists without necessarily being able to explicitly present it.

The Monotone Convergence Theorem has direct applications to series as well. In particular it can be used to prove the Comparison Test.

THEOREM 8.14 (Comparison Test). *Let $\sum_{n=1}^{\infty} a_n$ and $\sum_{n=1}^{\infty} b_n$ be series whose terms are positive, with $a_n \leq b_n$ for all $n \in \mathbb{N}$. If $\sum_{n=1}^{\infty} b_n$ converges, then $\sum_{n=1}^{\infty} a_n$ converges as well.*

PROOF. First, the sequence of partial sums for $\sum_{n=1}^{\infty} a_n$, whose nth term is given by $s_n = a_1 + \cdots + a_n$, is monotone increasing because each term a_i is positive. Thus, to prove that $\sum_{n=1}^{\infty} a_n$ converges, it suffices to establish that the sequence of partial sums is bounded.

Let $\sum_{n=1}^{\infty} b_n = B$. Because all of the b_i are positive we know that B is an upper bound for all the partial sums of this series:

$$b_1 + \cdots + b_n \leq B$$

for all $n \in \mathbb{N}$. Since each $a_n \leq b_n$, we know that the partial sums for $\sum_{n=1}^{\infty} a_n$ are bounded by B as well:

$$a_1 + \cdots + a_n \leq b_1 + \cdots + b_n \leq B.$$

Since the sequence of partial sums for $\sum_{n=1}^{\infty} a_n$ is a monotone and bounded sequence, it converges. □

We remark that the proof of the Comparison Test shows that

$$\sum_{n=1}^{\infty} a_n \leq \sum_{n=1}^{\infty} b_n.$$

Moreover, if $a_i < b_i$ for at least one index value i, then

$$\sum_{n=1}^{\infty} a_n < \sum_{n=1}^{\infty} b_n.$$

As an example of how you might apply Theorem 8.14, consider the series

$$\sum_{i=1}^{\infty} \frac{1}{i^i} = 1 + \frac{1}{4} + \frac{1}{27} + \frac{1}{256} + \cdots .$$

For $i \geq 2$ we know that $i^i \geq 2^i > 2^{i-1}$, so

$$\frac{1}{i^i} < \frac{1}{2^{i-1}}.$$

Thus, our original series is bounded above by the geometric series

$$\sum_{i=1}^{\infty} \frac{1}{2^{i-1}} = 1 + \frac{1}{2} + \frac{1}{4} + \frac{1}{8} + \cdots = 2.$$

So by Theorem 8.14, our original series must converge.

We note that the series $\sum_{i=1}^{\infty} \frac{1}{i^i}$ does not converge to 2, the limit of the bounding series. In fact, we do not know the value that our series converges to. Using techniques that we have not even hinted at, however, we do know that the decimal expansion for the limit of this series begins 1.29129, which is less than 2.

Exercise 8.16 What can you say about the convergence of

$$\sum_{i=1}^{\infty} \frac{1}{2^i + 3^i} = \frac{1}{5} + \frac{1}{13} + \frac{1}{35} + \cdots ?$$

8.6. Famous Irrationals

The irrational numbers – the gap between the rational numbers and the real numbers – are an accessible and intriguing part of mathematics. In this section we present or outline proofs that particular numbers are irrational. In addition to working your way through these arguments, we hope you spend some time thinking about the ways in which the arguments are distinct. You cannot memorize "the way to prove something is irrational" in the same way you might have memorized "the way to find local minima" in calculus.

8.6.1. $\sqrt{2}$ is Irrational. In Chapter 1, we presented a classic proof by contradiction showing that $\sqrt{2}$ is irrational (Proposition 1.14). You may have created similar proofs of irrationality at other points in your work, for numbers such as $\sqrt[3]{2}$ (Exercise 1.42), $\log_2(3)$ (Exercise 1.44), and the golden ratio (Exercise 4.36).

Exercise 4.37 asks you to determine the natural numbers k such that $\sqrt{k}$ is irrational. A key step in the proof that $\sqrt{2}$ is irrational is the claim that if $2|m^2$ then $2|m$. This claim is true if you replace 2 with any other prime number, but it is not true in general, as you might have shown in Exercise 4.33. For example, $\sqrt{12}$ is irrational, but it is not true that if $12|m^2$ then $12|m$. The value $m = 6$ provides a counterexample.

Thus even in situations that appear similar to the case you are thinking about – such as moving from proving $\sqrt{2}$ is irrational to proving that $\sqrt{12}$ is irrational – you are likely to need additional insights. Underlying many of the arguments that various roots and combinations of roots are irrational are the results proven in Chapter 4 on the structure of the integers. The next two examples are also proofs by contradiction, but they have little else in common with the proofs just described.

8.6.2. *e* is Irrational. There are various ways to define the number e, whose decimal expansion begins

$$e = 2.71828\ldots.$$

For this argument, we will use the definition of e as the limit of an infinite series:

$$e = \sum_{i=0}^{\infty} \frac{1}{i!} = 1 + 1 + \frac{1}{2!} + \frac{1}{3!} + \frac{1}{4!} + \cdots.$$

Exercise 8.17 Verify that the infinite series expression for e is a convergent series by comparing it to the series

$$1 + 1 + \frac{1}{2} + \frac{1}{4} + \frac{1}{8} + \cdots = 1 + \frac{1}{1 - \frac{1}{2}} = 3.$$

THEOREM 8.15. *The real number e is not rational.*

PROOF. This is a proof by contradiction. Assume to the contrary that $e = m/n$, where m and n are natural numbers and the fraction has been expressed in lowest terms. Unlike the argument for the irrationality of $\sqrt{2}$, taking powers of e does not appear to have any benefit, so mimicking the proof for $\sqrt{2}$ does not seem to be a useful tack to take.

We have

$$e = \frac{m}{n} = \sum_{i=0}^{\infty} \frac{1}{i!} = 1 + 1 + \frac{1}{2!} + \frac{1}{3!} + \cdots.$$

Multiplying all terms by $n!$ then gives

$$m(n-1)! = \sum_{i=0}^{\infty} \frac{n!}{i!} = n! + n! + \frac{n!}{2!} + \frac{n!}{3!} + \cdots + \frac{n!}{n!} + \frac{n!}{(n+1)!} + \frac{n!}{(n+2)!} + \cdots.$$

Since $m(n-1)!$ is an integer, and each of the terms in

$$n! + n! + \frac{n!}{2!} + \frac{n!}{3!} + \cdots + \frac{n!}{n!}$$

is an integer, it follows that the "tail" of the series,

$$\text{tail} = \frac{n!}{(n+1)!} + \frac{n!}{(n+2)!} + \cdots,$$

is an integer. We also know the tail is a positive integer, since all its terms are positive.

However, notice that $n!/(n+1)! = 1/(n+1)$, and

$$\frac{n!}{(n+2)!} = \frac{1}{(n+2)(n+1)} < \frac{1}{(n+1)^2},$$

and, in general,

$$\frac{n!}{(n+k)!} = \frac{1}{(n+k)\cdots(n+2)(n+1)} < \frac{1}{(n+1)^k}$$

for all $k \geq 2$. Thus we can use the Comparison Test to see that

$$\text{tail} = \frac{n!}{(n+1)!} + \frac{n!}{(n+2)!} + \cdots < \frac{1}{(n+1)} + \frac{1}{(n+1)^2} + \cdots.$$

The series on the right is a geometric series, where the first term and common ratio are $1/(n+1) < 1$, so Theorem 8.10 and a bit of arithmetic show that it equals $1/n$. Since n is a positive integer, we know $1/n \leq 1$. Thus

$$\text{tail} = \frac{n!}{(n+1)!} + \frac{n!}{(n+2)!} + \cdots < \frac{1}{n} \leq 1.$$

This is a contradiction, since we have just claimed there is a positive integer that is less than 1. □

8.6.3. π is Irrational. Perhaps the most famous irrational number is π. The known proofs of the irrationality of π are not as accessible as proofs of irrationality that we have already presented and, because of this, the argument we outline here does have some gaps. That is, we rely on some facts from calculus, which you may have learned but probably did not prove. We hope that these steps are sufficiently intuitive that you can believe (or show!) that the details can be filled in.[5]

THEOREM 8.16. *The real number π is not rational.*

This result is established via a proof by contradiction, beginning with the assumption that $\pi = m/n$, where m and n are natural numbers and the fraction is in lowest terms. The contradiction is similar to the one used above for e, in that an expression based on m and n must be both a positive integer and less than 1. However,

5 This argument was first presented by Ivan Niven in an article titled "A simple proof that π is irrational" [**Niv47**]. The reader who is interested in working through a complete presentation and motivation for this argument should consult the article "Discovering and proving that π is irrational" by Timothy Jones [**Jon10**].

the details could not differ more: in this case, the expression is a definite integral whose integrand is the product of sin(x) and a polynomial with a very special form and properties.

The polynomial has degree $2k$, with zeros at $x = 0$ and $x = m/n$:

$$f(x) = \frac{x^k(m - nx)^k}{k!}.$$

The constant k is a positive integer that will be directly discussed when we get to the end of the argument. Regardless of the specific value of k, we know

$$\int_0^{\pi} f(x)\sin(x)\,dx > 0$$

because $f(x)$ and $\sin(x)$, and thus their product, are positive for all values of x strictly between 0 and π.[6]

Define $F(x)$ to be the alternating sum of $f(x)$ and its first k even derivatives:

$$F(x) = f(x) - f''(x) + f^{(4)}(x) - \cdots + (-1)^k f^{(2k)}(x).$$

A careful, iterated integration by parts argument shows that

$$\int_0^{\pi} f(x)\sin(x)\,dx = F(\pi) + F(0). \tag{8.3}$$

Exercise 8.18 Use integration by parts to establish Equation 8.3 for the case when $k = 2$, which should convince you that result is plausible for any $k \in \mathbb{N}$. You might then also become convinced that the general result can be proved by inducting on k.

We are now in position to show that the integral in Equation 8.3 is a positive integer. As we have remarked before, it is often helpful in reading through a mathematical proof to be working an example on a separate piece of paper. The proof given for the following lemma is an excellent example of this sort of situation; on a first reading we advise you to verify the claim in the case of $k = 3$.

LEMMA 8.17. *$F(0)$ is an integer.*

PROOF. The polynomial $f(x)$ has no terms of degree $< k$, so we can express it as

$$f(x) = \frac{1}{k!}\left(c_{2k}x^{2k} + c_{2k-1}x^{2k-1} + \cdots + c_k x^k\right),$$

where the coefficients c_i are all integers. Since there is no constant term, $f(0) = 0$; and more generally, $f^{(i)}(0) = 0$ for each integer i where $i < k$. Moreover, we have $f^{(k)}(0) = \frac{1}{k!}(0 + 0 + \cdots + 0 + k!c_k) = c_k$; and in general, for each integer i between 0 and k, we have

$$f^{(k+i)}(0) = \frac{(k+i)!}{k!}c_{k+i} \in \mathbb{Z}.$$

Thus $F(0)$ is a finite sum of integers, and so it is an integer. □

6 This is one of the facts we are assuming from calculus, and from here on we won't always remark on the facts that are being assumed.

COROLLARY 8.18. *$\int_0^\pi f(x)\sin(x)\,dx$ is a positive integer.*

PROOF. Since we are assuming $\pi = m/n, f(\pi - x) = f(m/n - x)$. But

$$\begin{aligned} f(m/n - x) &= \frac{(m/n - x)^k(m - n(\frac{m}{n} - x))^k}{k!} = \frac{(m/n - x)^k(nx)^k}{k!} \\ &= \frac{\frac{1}{n^k}(m - nx)^k(n)^k(x)^k}{k!} = \frac{(m - nx)^k x^k}{k!} = f(x)\,. \end{aligned}$$

Since $\pi = m/n$, this means $f(\pi - x) = f(x)$. Hence $f'(x) = -f'(\pi - x)$, and in general $f^{(i)}(x) = (-1)^i f^{(i)}(\pi - x)$. We have already noted that $f^{(i)}(0)$ is an integer for each i, and so it is also the case that $f^{(i)}(\pi)$ must be an integer for each i as well. Thus $F(\pi)$ is an integer.

We now know that $F(0) + F(\pi)$ is an integer. But this is also the value of the integral $\int_0^\pi f(x)\sin(x)\,dx$, which we know is positive. □

In the concluding steps of this proof we finally discuss the number k.

PROOF OF THEOREM 8.16. The maximum value of $\sin(x)$ is 1. Because $x(m - nx)$ is a downward-pointing parabola, with roots at $x = 0$ and $x = m/n$, its maximum value occurs at $x = m/2n$, so

$$x(m - nx) \leq \frac{m}{2n}\frac{m}{2} = \frac{m^2}{4n} < \frac{m^2}{n}\,.$$

Thus for x in the interval $[0, \pi]$ we know that

$$0 \leq f(x)\sin(x) < \frac{m^{2k}}{n^k k!}. \tag{8.4}$$

We conclude that

$$\int_0^\pi f(x)\sin(x)\,dx < \pi \cdot \frac{m^{2k}}{n^k k!} = \frac{m^{2k+1}}{n^{k+1}k!} \tag{8.5}$$

for any natural number k.

However,

$$\lim_{k\to\infty} \frac{m^{2k+1}}{n^{k+1}k!} = 0.$$

In particular, there is some value of k where $m^{2k+1}/n^{k+1}k! < 1$. But then for this value of k it must be the case that $\int_0^\pi f(x)\sin(x)\,dx$ is a positive integer that is less than 1. □

Exercise 8.19 We have left some small gaps in the proof just given, including the justifications for Equations 8.4 and 8.5. Write up a more complete explanation, freely using results you know from your prior study of calculus.

We now have three arguments for irrationality that start along the same lines ("Assume to the contrary that the number is rational ...") but then exploit properties of integers, infinite series, or integration. This is then a good moment to return to a point we made in Chapter 2. The material you are learning requires a significant intellectual step beyond the processes that were successful in many if not all of your previous coursework in mathematics. And while seeing examples of proofs is useful, it is also of limited utility. In order to learn mathematics you cannot rely on studying earlier arguments. Rather, you need to spend time struggling with creating your own proofs, and then endure the process of having someone else read, critique, and perhaps even grade them.

8.7. End-of-Chapter Exercises

Exercises you can work on after Sections 8.1 and 8.2

8.20 Consider the set

$$A = \left\{ \frac{1}{2^i} \mid i \in \mathbb{N} \right\} \cup \left\{ 1 - \frac{1}{2^i} \mid i \in \mathbb{N} \right\}.$$

(a) What is the greatest lower bound of A?
(b) What is the least upper bound of A?

8.21 Let

$$S = \{q \in \mathbb{Q} \mid q^2 < 3\}.$$

(a) What is the greatest lower bound of S?
(b) What is the least upper bound of S?

8.22 Complete the proof of Theorem 8.5 by providing arguments that handle the cases not covered in Section 8.2.

8.23 Suppose that $A \subseteq \mathbb{R}$ has 1 as its greatest lower bound and 8 as an upper bound, and suppose that $B \subseteq \mathbb{R}$ has -2 as a lower bound and 10 as its least upper bound. Discuss the possible values for the greatest lower bound and least upper bound of $A \cup B$. Then do the same for $A \cap B$.

8.24 For each natural number n, define

$$A_n = \{x \in \mathbb{Z} \mid |x - \sqrt{2}| < 1/n\}.$$

What is

$$\bigcap_{n \in \mathbb{N}} A_n?$$

Answer the same question for

$$B_n = \{x \in \mathbb{Q} \mid |x - \sqrt{2}| < 1/n\}$$

and for

$$C_n = \{x \in \mathbb{R} \mid |x - \sqrt{2}| < 1/n\}.$$

Then change $\sqrt{2}$ to $7/5$ and do the three problems again.

Exercises you can work on after Sections 8.3 and 8.4

8.25 Prove that the sequence $(1, 1, \frac{1}{2}, \frac{1}{3}, \frac{1}{5}, \frac{1}{8}, \ldots)$ of reciprocals of the Fibonacci numbers converges to 0:

$$\lim_{n \to \infty} \frac{1}{f_n} = 0.$$

8.26 Prove that the sequence

$$\left(2 + \frac{1}{10^n}\right) = (2.1, 2.01, 2.001, \ldots)$$

converges.

8.27 We say that a sequence (a_n) *diverges to infinity* if, given any $x \in \mathbb{R}$, there is an $N \in \mathbb{N}$ such that, for all $n \geq N$, $a_n > x$. Prove that

(a) If $r > 1$, then the sequence $(r^n) = (r, r^2, r^3, \ldots)$ diverges to infinity.
(b) If $r < -1$, then not only does $(r^n) = (r, r^2, r^3, \ldots)$ not converge, it also does not diverge to infinity.

8.28 Recall the harmonic series is $1 + \frac{1}{2} + \frac{1}{3} + \frac{1}{4} + \cdots$.

(a) Prove that the harmonic series does not converge.
(b) Define what it would mean for a series to diverge to infinity (see Exercise 8.27).
(c) Prove that the harmonic series diverges to infinity.

As a hint, consider grouping terms like

$$1 + \frac{1}{2} + \underbrace{\frac{1}{3} + \frac{1}{4}}_{\geq 1/2} + \underbrace{\frac{1}{5} + \cdots + \frac{1}{8}}_{\geq 1/2} + \cdots$$

8.29 Prove that the infinite series $1 - 1 + 1 - 1 + \cdots$ does not converge. Does it diverge to infinity?

8.30 Find the limit of the series

$$\sum_{i=1}^{\infty} \frac{1 + (-1)^i}{2^i}.$$

8.31 J.J. Sylvester's sequence is defined by $a_1 = 2$ and

$$a_{n+1} = a_1 a_2 \cdots a_n + 1$$

for all $n \in \mathbb{N}$. The sequence begins $(2, 3, 7, 43, 1807, \ldots)$, which you may recall from Exercise 1.33. Sylvester was particularly interested in the series whose terms are the reciprocals of the a_i.

(a) Use an induction argument to prove that

$$\frac{1}{a_1} + \frac{1}{a_2} + \cdots + \frac{1}{a_n} + \frac{1}{a_1 a_2 \cdots a_n} = 1$$

for all $n \in \mathbb{N}$.

(b) Prove that

$$\sum_{i=1}^{\infty} \frac{1}{a_i} = 1.$$

As a hint for part (a), consider the case when $n = 3$:

$$\frac{1}{2} + \frac{1}{3} + \frac{1}{7} + \frac{1}{43} = \frac{3 \cdot 7 + 2 \cdot 7 + \overbrace{2 \cdot 3 + 1}^{7}}{2 \cdot 3 \cdot 7}.$$

8.32 Figure 3 exhibits a proof without words for the value of Archimedes' series:

$$A + \frac{A}{4} + \frac{A}{16} + \frac{A}{64} + \cdots = \frac{4}{3}A.$$

Use some words to explain the connection between the figure and the result.

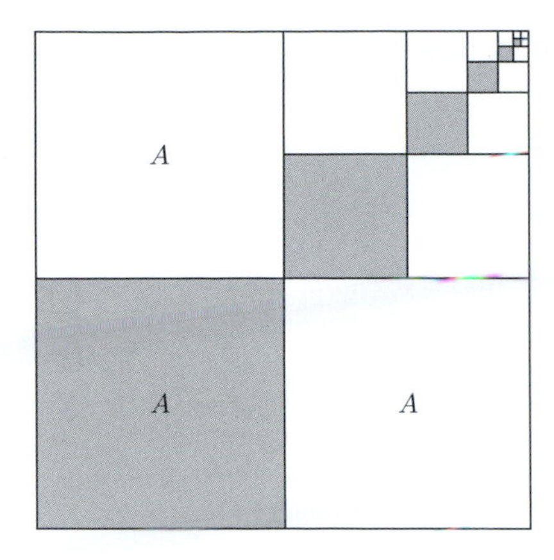

Figure 3. A proof without words for the value of Archimedes' series.

Exercises you can work on after Section 8.5

8.33 Prove that a convergent sequence is a bounded sequence. Show by example that a convergent sequence need not be monotone.

8.34 Define a sequence (a_n) by setting $a_1 = \sqrt{2}$ and defining

$$a_{n+1} = \sqrt{2 + a_n}$$

for all $n \in \mathbb{N}$.

(a) Show that (a_n) is monotone.
(b) Prove that (a_n) is bounded.
(c) Does (a_n) converge? If so, to what value?

8.35 The Babylonian square root approximation method produces a sequence that converges to the square root of a given positive number, S. The first step is to choose an initial approximation x_1, and then define

$$x_{n+1} = \frac{1}{2}\left(x_n + \frac{S}{x_n}\right)$$

for all $n \in \mathbb{N}$.

(a) The decimal expansion for the square root of 3 begins $\sqrt{3} = 1.7320\ldots$. Start the Babylonian approximation procedure with $x_1 = 2$, verify that $x_2 = 7/4 = 1.75$, and compute x_3.
(b) Prove that when $x_1 \geq \sqrt{S}$, the resulting sequence is monotone decreasing and bounded below by $\sqrt{S}$.
(c) Prove that when $x_1 \leq \sqrt{S}$, and $x_1 > 0$, the resulting sequence is monotone increasing and bounded above by $\sqrt{S}$.
(d) Prove that given a positive S and positive first guess x_1, the Babylonian approximation procedure always produces a sequence that converges to $\sqrt{S}$.

8.36 Prove that the series

$$\sum_{i=1}^{\infty} \frac{1+2^i}{3^{i-1}+4^{i-1}} = 3 + \frac{5}{7} + \frac{9}{25} + \cdots.$$

converges.

8.37 A sequence (a_n) is *eventually monotone increasing* if there is an $N \in \mathbb{N}$ such that $a_k \leq a_{k+1}$ for all $k \geq N$. A sequence (a_n) is *eventually monotone decreasing* if there is an $N \in \mathbb{N}$ such that $a_k \geq a_{k+1}$ for all $k \geq N$. The sequence (a_n) is *eventually monotone* if it is either eventually monotone increasing or eventually monotone decreasing.

(a) Rephrase these definitions in terms of tails of the sequence (a_n).
(b) Give an example of a sequence that is eventually monotone increasing, but is not monotone.
(c) Prove that a sequence which is eventually monotone increasing must have a lower bound; that is, there is an $\ell \in \mathbb{R}$ such that $\ell \leq a_n$ for every term a_n in the sequence.
(d) Give an example of a sequence that is eventually monotone decreasing, but is not monotone.
(e) Prove a statement similar to the one in part (c), but for eventually monotone decreasing sequences.

8.38 The notion of an eventually monotone sequence is defined in Exercise 8.37. Prove that every bounded, eventually monotone sequence must converge. If you are ambitious, state and prove a theorem similar to Theorem 8.13.

8.39 A sequence of closed and bounded intervals $[a_i, b_i]$ is *nested* if

$$[a_1, b_1] \supseteq [a_2, b_2] \supseteq [a_3, b_3] \supseteq \cdots.$$

The *Nested Intervals Property* states that a nested sequence of closed and bounded intervals has non-empty intersection:

$$\bigcap_{i=1}^{\infty} [a_i, b_i] \neq \emptyset.$$

(a) Show that the sequence of left endpoints (a_i) is monotone and bounded.
(b) Prove that $\mathbb{R}$ has the Nested Intervals Property.
(c) Is the result still true if we use open intervals (a_i, b_i) instead of closed intervals $[a_i, b_i]$?

With the Nested Intervals Property in hand, you are ready for Project 11.10 on the Cantor Set, which highlights some of the intricacies of the real number line.

Exercises you can work on after Section 8.6

8.40 Recall that the series expansion for sine, based at $x = 0$ and using radians, is

$$\sin(x) = \sum_{n=1}^{\infty} (-1)^{n-1} \frac{x^{2n-1}}{(2n-1)!} = x - \frac{x^3}{3!} + \frac{x^5}{5!} - \frac{x^7}{7!} + \cdots.$$

Prove that $\sin(1)$ is irrational.

More Exercises!

8.41 If (a_i) and (b_i) are sequences, then we say (a_i) and (b_i) are *eventually equal* if there is some $n \in \mathbb{N}$ such that $a_m = b_m$ for all $m \geq n$. We say (a_i) and (b_i) *have the same tail* if there is an $n \in \mathbb{N}$ and a $k \in \mathbb{Z}$ such that $a_{m+k} = b_m$ for all $m \geq n$.

(a) Find examples that illustrate that two sequences can be eventually equal without being identical.
(b) Find examples that illustrate that two sequences can have the same tail without being eventually equal.
(c) Prove that being eventually equal is an equivalence relation on the set of real-valued sequences.
(d) Prove that having the same tail is an equivalence relation on the set of real-valued sequences.

8.42 Consider the set of all real-valued sequences. Define a relation on this set by $(a_i) \sim (b_i)$ if the two sequences *almost always agree*:

$$(a_i) \sim (b_i) \Leftrightarrow \{n \in \mathbb{N} \mid a_n \neq b_n\} \text{ is a finite set.}$$

(a) Prove that this is an equivalence relation on the set of all real-valued sequences.
(b) Is the equivalence relation "almost always agree" different from the equivalence relation "eventually equal" described in Exercise 8.41? That is, are there real-valued sequences that are equivalent under one definition but not under the other?

8.43 Let $\mathbb{R}_{\geq 1} = \{x \in \mathbb{R} \mid x \geq 1\}$. For each $x \in \mathbb{R}_{\geq 1}$ let S_x be the 1×1 square

$$S_x = [1/x, 1 + 1/x] \times [1/x, 1 + 1/x] \subset \mathbb{R}^2 .$$

(a) Draw S_1 and S_2.
(b) If $I = \{1, 2, 3\}$, what points in the plane correspond to $\bigcup_{i \in I} S_i$?
(c) If $I = \{1, 2, 3\}$, what points in the place correspond to $\bigcap_{i \in I} S_i$?
(d) What is $\bigcap_{i \in \mathbb{N}} S_i$?
(e) What is $\bigcup_{n \in \mathbb{R}_{\geq 1}} S_n$? How does it differ from $\bigcup_{n \in \mathbb{N}} S_n$?

8.44 Define an ordering $\preceq$ on $\mathbb{R}^2 = \mathbb{R} \times \mathbb{R}$ by saying

$$(a, b) \preceq (c, d)$$

when $a < c$, or when $a = c$ and $b \leq d$. This ordering is an example of a dictionary ordering (more frequently called a lexicographic ordering) because this ordering is similar to an alphabetical ordering. We list the word BISCUIT before PANCAKE because B comes before P, and we list BACON before BISCUIT because A comes before I.

(a) Prove that $(\mathbb{R}^2, \preceq)$ is a partially ordered set.
(b) Prove that $(\mathbb{R}^2, \preceq)$ is a totally ordered set.

The notions of upper bound, lower bound, least upper bound, and greatest lower bound defined in Section 8.1 extend naturally to $(\mathbb{R}^2, \preceq)$.

(c) Let $\mathcal{D}$ be the unit disk in the plane:

$$\mathcal{D} = \{(x, y) \mid x^2 + y^2 \leq 1\} .$$

Explain why $(1, 0)$ is an upper bound for $\mathcal{D}$. That is, show that if $(x, y) \in \mathcal{D}$ then $(x, y) \preceq (1, 0)$.

(d) Explain why $(1, 0)$ is the least upper bound for $\mathcal{D}$.

(e) Describe the set of all upper bounds for $\mathcal{D}$.

(f) Let $\mathcal{H}$ be the set of points to the left of the y-axis,

$$\mathcal{H} = \{(x, y) \mid x < 0\}.$$

Explain why $(0, 0)$ is an upper bound for $\mathcal{H}$.

(g) Describe the set of all upper bounds for $\mathcal{H}$.

(h) Explain why $\mathcal{H}$ has no least upper bound.

If You Have Studied Calculus

8.45 Determine the limiting value of the sequence (a_n) in Section 8.5 given by

$$a_n = \frac{1}{n+1} + \cdots + \frac{1}{2n}$$

by viewing a_n as a Riemann sum approximation for $\int_0^1 \frac{1}{1+x} dx$.

9 Probability and Randomness

Probability is a beautiful and widely applicable branch of mathematics. Here we present a few of the basic concepts of probability as an application of set theory, functions, and other topics from the first several chapters. Probability functions, random variables, and expected value are introduced, motivated by the question of how you might guess if a sequence of coin flips is real or fake. Have some dice, coins, and cards on hand: many of our examples and exercises involve these and related games of chance.

9.1. A Class of Lyin' Weasels

One morning, Professor P passed out the following two-page homework assignment to a class of 40 students:

> *Page 1*: Flip a coin once. Write down H or T for the result: ____
>
> *Page 2*: If your result on Page 1 is H, flip that coin another 250 times, and record the sequence of results in the blanks below.
>
> If your result on Page 1 is T instead, don't flip the coin again. Instead, make up a sequence of Ts and Hs that you believe is a reasonable simulation of flipping a coin 250 times in a row, and record that sequence in the blanks below.
>
> ___, ___, ___, ___, ___, ___, ___, ___, ___, ___, ___, ___, ___, ___, ___, ___,
> ___, ___, ___, ___, ___, ___, ___, ___, ___, ___, ___, ___, ___, ___, ___, ___,
> ___, ___, ___, ___, . . .

At the next class meeting, Professor P told each student to keep Page 1 as a receipt and turn in Page 2. She then scanned the responses on Page 2 carefully and quickly, taking no more than 30 seconds to declare each submission either "flipped" or "faked." She wasn't always correct, but her success rate was quite high, greater than 80%. How is this possible? One of the submissions is presented in Figure 1. Can *you* tell if it was flipped or faked?

We will discuss her method of analysis later in this chapter, but for the moment let's informally introduce a few concepts that might assist with an initial approach. We will assume that the coin is fair, meaning that, on any single flip, it has the same likelihood of showing H or T. We will also assume that successive coin flips are independent, so that one flip doesn't have any bearing on another flip. These two assumptions imply, for example, that if you are about to flip a coin twice in succession, each of the 2-flip strings HH, HT, TH, and TT is equally likely to occur.

Exercise 9.1 If you are about to flip a coin three times in succession, how likely are you to get exactly two Hs among the three flips?

H, H, T, H, T, H, H, T, T, H, H, T, T, H, H, T, T, T, H, T, H, H, T, H, H,
H, H, H, T, T, T, H, T, H, T, H, H, T, H, H, H, H, T, T, T, T, T, H, H, H,
T, T, T, H, H, T, T, T, T, H, H, T, T, T, H, T, T, T, T, T, H, H, T, H, H,
H, H, H, T, H, T, T, H, T, T, H, T, H, H, T, H, H, H, H, T, T, H, H, T, T,
H, H, T, T, H, T, H, H, T, T, H, T, T, T, H, H, H, T, H, H, H, H, T, T, H,
T, T, H, T, T, H, H, T, T, H, T, H, T, H, T, T, T, T, H, T, T, T, T, T, H,
T, T, T, H, T, T, T, H, T, T, H, T, T, T, H, T, T, H, T, H, T, H, T, T, H,
H, H, T, T, H, H, T, T, T, H, H, H, T, H, H, T, H, T, H, H, T, T, T, T, H,
H, H, H, T, H, T, H, H, H, T, T, T, H, T, H, T, T, H, H, H, T, H, H, T, T,
H, H, H, T, T, H, H, H, T, T, H, H, T, H, T, H, H, H, H, H, T, H, T, T, H.

Figure 1. Was this sequence flipped or faked?

Exercise 9.2 If you are about to flip a coin four times in succession, how likely are you to get no more than two Hs among the four flips?

These ideas suggest an approach to analyzing the students' submissions. First, count the occurrences of Hs and Ts and see if they are roughly the same. Also, count the occurrences of the 2-flip strings and see if all four counts are roughly the same. Here, a question arises: should we only count the 2-flip strings that start in odd-numbered positions, to avoid issues of overlap? For example, if the first and second flips result in the 2-flip string HT, the second and third flips can only give TH or TT.

Exercise 9.3 Perform the counts just mentioned on the example in Figure 1. Do they provide evidence that the student actually flipped the coin? Or do you think the student faked the sequence of heads and tails?

After Professor P made all of her guesses and surprised the class with her accuracy, she noted that 32 of the 40 students had faked their flips. Given that a single fair coin flip determines which students should have faked their answers, it is reasonable to assume that approximately half the sequences would have been fake. How likely is it that as many as 32 were faked?

She playfully admonished the students as they left class that day. Although she couldn't point a finger at any particular student, she knew with great confidence that at least a handful of students had not followed the rules of the assignment.

9.2. Probability

In this chapter we explore randomness and some of the basic ideas and examples of the mathematical theory of probability. We begin by introducing common terms used in probability theory that merely rename concepts from Chapter 3.

DEFINITION 9.1. A *sample space* is a set S. A *sample point* is an element of S. An *event* is a subset of S.

In applications of probability, we usually think of these terms in the context of an experiment or study where the possible individual outcomes are the sample points and the set of all sample points is the sample space. For example, when flipping a coin twice in a row, the sample space is the 4-element set {HH, HT, TH, TT}, and the event described by "get at least one head" is the subset {HH, HT, TH} containing three sample points.

DEFINITION 9.2. Let S be a finite sample space. A function $p : S \to [0, 1]$ is a *probability function* on S if

$$\sum_{s \in S} p(s) = 1.$$

In our 2-flip example, the sample space is $S = \{\text{HH, HT, TH, TT}\}$. If we are considering the case of a fair coin and independent flips, then the probability p_1 assigned to each sample point should be the same:

$$p_1(\text{HH}) = p_1(\text{HT}) = p_1(\text{TH}) = p_1(\text{TT}) = 1/4.$$

But notice that our definition of a probability function is quite general and allows us to consider distributions of probability values that don't necessarily correspond to physical objects or prior experience. For example, defining p_2 by

$$p_2(\text{HH}) = p_2(\text{HT}) = 1/\pi, \quad p_2(\text{TH}) = 0, \quad p_2(\text{TT}) = 1 - 2/\pi$$

is totally fine from an abstract viewpoint because the individual probability values are non-negative and sum to 1.

DEFINITION 9.3. If $X \subseteq S$ is an event, then

$$p(X) = \sum_{s \in X} p(s).$$

Thus, the event X = "get at least one head" has probability $3/4$ using p_1 because

$$p_1(X) = p_1(\text{HH}) + p_1(\text{HT}) + p_1(\text{TH}) = \frac{1}{4} + \frac{1}{4} + \frac{1}{4} = \frac{3}{4},$$

while using p_2 the probability is

$$p_2(X) = p_2(\text{HH}) + p_2(\text{HT}) + p_2(\text{TH}) = \frac{1}{\pi} + \frac{1}{\pi} + 0 = \frac{2}{\pi}.$$

A probability function gives a *uniform distribution* when $p(s) = 1/|S|$ for all $s \in S$, and thus $p(X) = |X|/|S|$ for any event $X \subseteq S$.

EXAMPLE 9.4. In the game of American roulette, a ball comes to rest in one of 38 positions on a spinning wheel labelled

$$0, 00, 1, 2, 3, \ldots, 35, 36.$$

The cells 0 and 00 are usually green and give the casino the edge when players make bets on the outcome of a spin. Half of the other 36 cells are red and half of them are black. For a single spin of the wheel, the sample space S is the set containing all 38 of the labels listed above, and the gambler's assumption is that the probability distribution is uniform: the probability of landing in any particular cell is $1/38$. The event "lands on green" is $G = \{0, 00\}$, and the event "lands on red" is a subset $R \subset S$ with $|R| = 18$. Thus, $p(G) = 2/38$ and $p(R) = 18/38$.

Exercise 9.4 A European roulette wheel is similar, only it has no cell labeled 00. What are $p(G)$ and $p(R)$ in European roulette? Is a ball more likely to land on a red position on an American wheel or on a European wheel?

					(**6**, 1)					
				(**5**, 1)	(**5**, 2)	(**6**, 2)				
			(**4**, 1)	(**4**, 2)	(**4**, 3)	(**5**, 3)	(**6**, 3)			
		(**3**, 1)	(**3**, 2)	(**3**, 3)	(**3**, 4)	(**4**, 4)	(**5**, 4)	(**6**, 4)		
	(**2**, 1)	(**2**, 2)	(**2**, 3)	(**2**, 4)	(**2**, 5)	(**3**, 5)	(**4**, 5)	(**5**, 5)	(**6**, 5)	
(**1**, 1)	(**1**, 2)	(**1**, 3)	(**1**, 4)	(**1**, 5)	(**1**, 6)	(**2**, 6)	(**3**, 6)	(**4**, 6)	(**5**, 6)	(**6**, 6)
2	3	4	5	6	7	8	9	10	11	12

Figure 2. The 36 possible rolls of two dice, organized by their sum.

EXAMPLE 9.5. Roll two dice, one black and one white. Proposition 2.13 tells you that there are $6^2 = 36$ possible outcomes; the chart in Figure 2 has the outcomes grouped in columns by their sum, with the black roll in bold. The sample space S contains all of these ordered pairs, and under the assumption that the dice are fair and the rolls are independent, the probability distribution is uniform. The event "the sum is nine" is $N = \{(\mathbf{3}, 6), (\mathbf{4}, 5), (\mathbf{5}, 4), (\mathbf{6}, 3)\}$, and the event "the sum is less than seven" is a subset $L \subset S$ with $|L| = 15$. Thus, the probabilities of the events are $p(N) = 4/36$ and $p(L) = 15/36$.

We will refer to rolling a pair of dice several times in later sections of this chapter.

Exercise 9.5 Let E correspond to the event "the sum of two dice rolls is even," and let T correspond to "the sum of two dice rolls is 3, 7, or 11." What are $p(E)$ and $p(T)$?

Exercise 9.6 You are playing Monopoly, and in order to avoid landing on someone's property you need your roll of two dice to sum to a value other than 5, 6, or 8. What is the probability you will succeed?

EXAMPLE 9.6. There are many situations where probability functions are defined for infinite sample spaces, and here we present just one example. You decide to flip a fair coin until heads comes up for the first time. The process will most likely terminate after no more than a few flips, but there is no guarantee that it won't continue indefinitely. Thus, an appropriate sample space is the infinite set

$$S = \{\text{H, TH, TTH, TTTH, TTTTH}, \ldots\}.$$

Let $\ell(s)$ be the length of a flip sequence s. Our previous discussions of coin flips tells us that we should define $p(s) = 2^{-\ell(s)}$ for each $s \in S$, so

$$\sum_{s \in S} p(s) = \frac{1}{2} + \frac{1}{4} + \frac{1}{8} + \cdots$$

Quoting the geometric series formula from Section 8.4, this series converges to the value 1, so p is a probability function on S. The event $E =$ "the process ends by the third flip" is $E = \{\mathrm{H}, \mathrm{TH}, \mathrm{TTH}\}$, which has probability

$$P(E) = \frac{1}{2} + \frac{1}{4} + \frac{1}{8} = \frac{7}{8}.$$

Exercise 9.7 Get together with some friends and flip a coin until it lands heads up for the first time.[1] Repeat this activity at least 40 times, keeping a list how many flips were required. What proportion of the sequences are at least four flips in length?

We conclude this section by discussing just a few of the many ways that probability functions behave with respect to set-theoretic operations on events.

PROPOSITION 9.7. *Let A and B be events in a finite sample space S, with a probability function $p : S \to [0, 1]$.*

(a) *If $A \subseteq B$, then*

$$p(A) \leq p(B).$$

(b) *If A^c is the complement of A in S, then*

$$p(A^c) = 1 - p(A).$$

(c) *The probability of the union is given by*

$$p(A \cup B) = p(A) + p(B) - p(A \cap B).$$

PROOF. In each case we simply need to use the fact that

$$p(A) = \sum_{s \in A} p(s) \quad \text{and} \quad p(B) = \sum_{s \in B} p(s).$$

To establish (a), notice that when $A \subseteq B$,

$$p(B) - p(A) = \sum_{s \in B \setminus A} p(s),$$

which is non-negative because $p(s) \geq 0$ for all $s \in S$.

Since $A^c = S \setminus A$, we have

$$p(A^c) = \sum_{s \in S \setminus A} p(s) = \sum_{s \in S} p(s) - \sum_{s \in A} p(s) = 1 - p(A),$$

which proves (b).

Finally, if $s \in A \cap B$ then $p(s)$ is a term in both the sum for $p(A)$ and the sum for $p(B)$. Thus when $s \in A \cap B$ the probability $p(s)$ is double-counted in the sum $p(A) + p(B)$. So

$$\begin{aligned} p(A \cup B) &= \sum_{s \in A \cup B} p(s) = \sum_{s \in A} p(s) + \sum_{s \in B} p(s) - \sum_{s \in A \cap B} p(s) \\ &= p(A) + p(B) - p(A \cap B). \end{aligned}$$

□

1 Leaving the possibility that it never will!

You may have used an argument similar to the one just given for part (c) when you established the formula for $|A \cup B|$ in your proof for Exercise 3.18(a). We note that when A and B are *mutually exclusive* events, meaning A and B are disjoint sets, then the probability of their union is simply

$$p(A \cup B) = p(A) + p(B).$$

Proposition 9.7 holds for any probability function on S. Unless we state otherwise, for the rest of the chapter and exercises we will only consider probability distributions that are uniform, so the primary task is computing the sizes of sample spaces and events. To do this we often use Propostion 2.13, especially when each sample point corresponds to a sequence of outcomes like coin flips, and we also need combinations.

9.3. Revisiting Combinations

The enumeration ideas from Section 3.8 – where we introduced $\binom{n}{k}$ – make it easy to determine the probability of an event such as having a sequence of five coin flips contain exactly three heads. The values of $\binom{n}{k}$, which count k-combinations of an n-element set, are often presented in the form of Pascal's Triangle, as shown in Figure 3. There are many remarkable patterns in Pascal's Triangle, several of which are the focus of end-of-chapter exercises. The most important relationship occurs between two consecutive entries in a row and the entry below them.

THEOREM 9.8. *For* $0 \leq k < n$, *we have*

$$\binom{n+1}{k+1} = \binom{n}{k} + \binom{n}{k+1}.$$

PROOF. Let S be an $(n+1)$-element set, and let a be any element of S. There are $\binom{n+1}{k+1}$ subsets of S with $k+1$ elements, and any such subset X is in exactly one of the following two cases:

(1) $a \in X$,

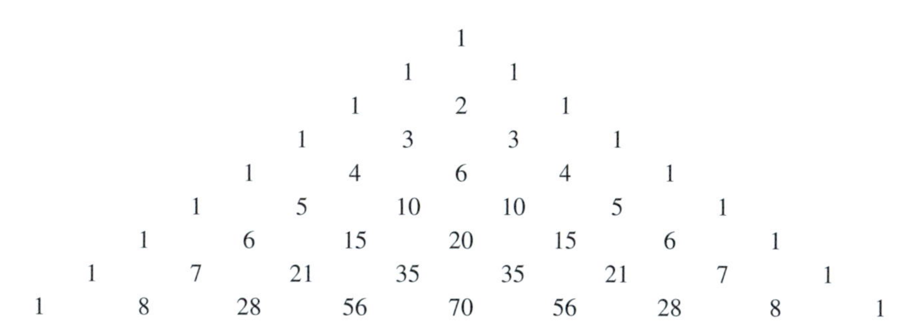

Figure 3. The first nine rows of Pascal's Triangle. If the numbering of both the rows and the entries within a row begin with 0, then the ith entry in row j is $\binom{j}{i}$. For example, the left-most 15 you see is $\binom{6}{2}$.

(2) $a \notin X$.

There are exactly $\binom{n}{k}$ different X that fall in case (1), because X contains a along with k of the remaining n elements in S. And there are exactly $\binom{n}{k+1}$ different X in case (2), because X contains $k+1$ of the n elements in $S \setminus \{a\}$. The formula follows from summing these two quantities together. □

Theorem 9.8 allows you to quickly generate the top rows of Pascal's triangle: after plotting the 1s on the left and right borders, every other entry is the sum of the two entries on either side of it in the previous row. For example, the first 28 in the bottom row of Figure 3 is the sum of the 7 and 21 that sit above it.

Exercise 9.8 You may notice that each row of Pascal's Triangle is both "symmetric" (it's the same reading from the left or from the right) and "unimodal" (the terms increase toward the middle and then decrease). Prove both of these properties using the factorial formula given in Theorem 3.18.

At the beginning of this section we posed the question: What is the probability that a sequence of five coin flips contains exactly three heads? The sample space consists of all sequences of heads and tails of length 5, for a total of $2^5 = 32$ sample points, so the probability of any given sequence is $1/32$. We can count the number of sample points in our event using a one-to-one correspondence between the sequences with exactly three heads and the 3-element subsets of $\{1, 2, 3, 4, 5\}$. The chosen elements correspond to the locations of the heads; for example, $\{1, 4, 5\}$ corresponds to HTTHH. Since there are $\binom{5}{3} = 10$ subsets of size 3 in $\{1, 2, 3, 4, 5\}$, there are 10 sequences with exactly three heads, so $p(\text{Exactly three heads}) = 10/32$.

Using this approach you can now return to Exercises 9.1 and 9.2 and quickly determine the answers using entries in the top rows of Pascal's triangle. We might also be interested in the value of $\binom{n}{k}$ when n and k are quite large. Generating the nth row of Pascal's Triangle might be impractical for large n, so we could attempt to use the formula involving factorials from Theorem 3.18. However, the value of $n!$ grows rapidly as n increases; for example, 100! is a 158-digit integer. For large values of n, there is an approximation for $n!$ called Stirling's Formula that is often very useful in applications:

$$n! \approx \sqrt{2\pi}\, n^{n+\frac{1}{2}}\, e^{-n}.$$

When $n = 100$, Stirling's Formula is within 0.1% of the true value of 100!.

Exercise 9.9 Use a computer to determine both 100! and its approximation via Stirling's Formula.

Exercise 9.10 The coefficients in the binomial expansion

$$(x + y)^4 = x^4 + 4x^3y + 6x^2y^2 + 4xy^3 + y^4$$

are the same numbers that appear in the 4th row in Pascal's triangle. Why is this the case? There is a general statement along these lines that explains why the numbers $\binom{n}{k}$ are often called binomial coefficients. State and prove this general result.

9.4. Events and Random Variables

The concept of a random variable will help us explain Professor P's concern that 32 out of 40 students had faked their flips.

DEFINITION 9.9. If S is sample space, a *random variable* X is a function $X : S \to \mathbb{R}$.

While the definition allows for random variables to be arbitrary functions, in practice random variables are usually employed for the purposes of counting or measuring. For example, let X be the sum of two dice rolls, with the sample space S displayed in Figure 2. Then $X((\mathbf{5}, 2)) = 7$ and $X((\mathbf{1}, 3)) = 4$. It is standard and natural notation in the study of probabilty to let $(X = k)$ denote the pre-image[2] of k under the map X, so that $(X = 4)$ is the event $X^{-1}(4) = \{(\mathbf{3}, 1), (\mathbf{2}, 2), (\mathbf{1}, 3)\}$. Thus we have

$$p(X = 4) = 3/36 = 1/12.$$

Exercise 9.11 Let X be the sum of two dice rolls. Compute $p(X = 6)$ and $p(X \leq 5)$.

Let S be the sample space of all length-40 sequences of heads and tails, and assign the probability $(1/2)^{40}$ to each sample point in S. Let X be the random variable that counts the number of heads in $s \in S$; the range of X is the set $\{0, 1, \ldots, 40\}$.

The students in the class were first asked to flip a coin once to determine if they should then flip a coin 250 times or simply fake their data. Listing the initial coin flips in alphabetical order by student name results in a sample point $s \in S$ with $X(s) = 32$. Since a fair coin is likely to produce a sequence with about 20 heads, only about half the students should have faked their sequence, which is why Professor P is surprised that the number of heads is as large as it is. She knows that $p(X \geq 32)$, the probability that the number of heads is 32 or larger, is rather small.

If m and n are distinct elements in $\{0, 1, \ldots, 40\}$, $(X = m)$ and $(X = n)$ are mutually exclusive events. Thus

$$p(X = m \text{ or } n) = p(X = m) + p(X = n),$$

which implies

$$p(X \geq 32) = p(X = 32) + p(X = 33) + \cdots + p(X = 40).$$

From the discussion of coin flip sequences in Section 9.3, we know that the number of sample points containing exactly k heads is $\binom{40}{k}$, so we have

$$p(X \geq 32) = \binom{40}{32}/2^{40} + \binom{40}{33}/2^{40} + \cdots + \binom{40}{40}/2^{40}.$$

Exercise 9.12 Use a computer to help you determine the values of the nine binomial coefficients displayed above and the resulting probability $p(X \geq 32)$. What conclusion do you draw about the 40 students?

2 We use the conventions described in Section 5.7, where you can talk of X^{-1} even when X is not a bijection.

9.5. Expected Value

The expected value of a random variable X, also known as its *mean* or *expectation*, is the average of the random variable values on the sample points, weighted by the probabilities:

DEFINITION 9.10. If X is a random variable on a finite sample space, then the *expected value* of X is

$$E(X) = \sum_{s \in S} X(s)p(s).$$

For example, let X be the sum of two fair dice rolls. Then

$$\begin{aligned} E(X) &= 2 \cdot \frac{1}{36} + 3 \cdot \frac{2}{36} + 4 \cdot \frac{3}{36} + \cdots + 11 \cdot \frac{2}{36} + 12 \cdot \frac{1}{36} \\ &= 252/36 = 7 \end{aligned}$$

is the expected value.

Exercise 9.13 Gather some friends and roll pairs of dice at least 50 times, recording the sum of each roll. Then compute the average of those sums. How close are you to the expected value of 7?

Exercise 9.14 Suppose that you are playing a game where you roll a single die. The points that you earn in a roll are double the normal value of an odd-valued face, while even-valued faces are worth nothing. What is the expected value for the points you earn on a roll?

Expected value helps to explain the long-term average behavior of repeated bets on certain games of chance.[3] Indeed, one of the primary sources for the modern theory of probability is the work of Blaise Pascal in the seventeenth century, who was asked by his friend the Chevalier de Mère to help explain the reason the Chevalier was ultimately losing money on a bet involving the repeated rolls of two fair dice. He would bet someone that in a sequence of 24 rolls of the pair of dice, double-sixes would show up at least once. If a double-sixes was seen, he would win the wagered amount; otherwise, he would lose as much.

Let S be the sample space of all possible sequences of 24 rolls of a pair of dice. Reminding ourselves of Example 9.5, we see that $|S| = 36^{24}$, and we assign the probability $(1/36)^{24}$ to each sample point. We are interested in the event A consisting of sequences with at least one double-six pair, but it is easier to count the number of sample points in A^c, the complementary event containing the sequences with *no* double-six pairs. By Proposition 2.13, $|A^c| = 35^{24}$, so

$$p(A^c) = 35^{24}/36^{24} \approx 0.5086,$$

and thus $p(A) = 1 - p(A^c) \approx 1 - 0.5086 = 0.4914$.

While it is now clear that getting no double-sixes is a bit more likely than getting at least one pair of them, let's focus on the wager. Let M be the random variable describing

3 Variance is another important concept, but we do not discuss it in this chapter.

the amount won by the Chevalier on a single bet: if x is wagered, then $M(s) = +x$ if $s \in A$ and $M(s) = -x$ if $s \in A^c$. We can compute the expected value of the bet from the Chevalier's perspective:

$$\begin{aligned} E(M) = \sum_{s \in S} M(s)p(s) &= \sum_{s \in A}(+x)(1/36)^{24} + \sum_{s \in A^c}(-x)(1/36)^{24} \\ &= x \cdot p(A) - x \cdot p(A^c) \\ &\approx (-0.0172)x. \end{aligned}$$

And here we see the predictive power of expected value: while there is no way to know with certainty the outcome of any particular wager, over a *very long* run of n individual wagers for the amount of x, the Chevalier can expect an average loss of $(0.0172)x$ on each one, for a total loss of approximately $(0.0172)xn$.

Exercise 9.15 The Chevalier also wagered on another dice game: if in the roll of four dice at least one 6 is seen, he would win the amount wagered, and otherwise he would lose the same. Why was he less concerned about this game?

EXAMPLE 9.11. The definition of expected value also makes sense for certain infinite sample spaces. Recall Example 9.6, where you flip a coin until you first get heads, and let L be the random variable measuring the length of a sequence. Then

$$E(L) = 1 \cdot \frac{1}{2} + 2 \cdot \frac{1}{4} + 3 \cdot \frac{1}{8} + \cdots,$$

which can be proven to converge to the value 2. How does the value $E(L) = 2$ compare with the arithmetic mean of the set of sequence lengths you recorded in Exercise 9.7?

GOING BEYOND THIS BOOK. If you are interested in the early history of probability, or find that you are prone to making errors in computing probabilities, then we recommend Gorroochurn's article "Errors of probability in historical context" [**Gor11**]. It contains a number of instances where prominent mathematicians made understandable errors in the early development of probability. Given Pascal's prominence in the development of probability, the story of one of his errors when working on a problem suggested by de Mère is particularly interesting.

9.6. Flipped or Faked?

We return to discuss Professor P's skill at determining which sequences of coin flips had been faked. When people try to simulate a random sequence of n heads or tails, many don't permit more than four or five results of the same type in a row, despite the fact that, as we show below, longer runs are to be expected when n is large enough.

Let X be the random variable measuring the length of the longest run of heads in a sequence of $n = 250$ flips of a fair coin. For example, if s is the sequence displayed in Figure 1, then $X(s) = 5$. The expected value of the longest run of heads is then

$$E(X) = 0 \cdot p(X = 0) + 1 \cdot p(X = 1) + 2 \cdot p(X = 2) + \cdots + 250 \cdot p(X = 250),$$

which we can approximate using a mixture of cleverness and computing.

Define $D_n(i)$ to be the number of sequences of length n whose longest sequence of heads has length exactly i. We are interested in these numbers because the probabilities in the equation for $E(X)$ are given by

$$p(X = i) = D_{250}(i)/2^{250}.$$

Define $A_n(i)$ to be the number of sequences of length n whose longest sequence of heads has length at most i. Thus

$$p(X \leq i) = A_{250}(i)/2^{250}.$$

The values of $D_n(i)$ and $A_n(i)$ are closely connected, as $D_n(0) = A_n(0) = 1$ (corresponding to the all tails sequence), and $D_n(i) = A_n(i) - A_n(i-1)$ for $i > 0$.

Any sequence contributing to $A_n(i)$ must begin with one of the following:

$$\text{T, HT, HHT, } \ldots, \underbrace{\text{HH}\cdots\text{H}}_{i \text{ terms}}\text{T.}$$

Since the remaining part of the sequence cannot contain a run of heads longer than i, we have the recursive formula

$$A_n(i) = A_{n-1}(i) + A_{n-2}(i) + A_{n-3}(i) + \cdots + A_{n-(i+1)}(i)$$

for $n > i$. This allows us to compute $A_n(i)$ and thereby to determine $D_n(i)$.

As an example, consider the case where $i = 1$. We know that $A_0(1) = 1$ because the empty sequence contains no heads. There are two strings of length 1, both of which have 1 or fewer heads, so $A_1(1) = 2$. Our recursion is

$$A_n(1) = A_{n-1}(1) + A_{n-2}(1),$$

so $A_2(1) = 3, A_3(1) = 5$, and we see that the values of $A_n(1)$ correspond to the Fibonacci sequence: $A_n(1) = f_{n+2}$. Thus $D_n(1) = A_n(1) - A_n(0) = f_{n+2} - 1$.

For $i \geq 2$ the pattern is similar but progressively more complicated. We have

$$A_0(i) = 1, \; A_1(i) = 2, \; A_2(i) = 4, \; \ldots, \; A_i(i) = 2^i,$$

which allows us to determine by hand $A_n(i)$ for moderate values of n and i, from which we can also derive the values of $D_n(i)$.

Exercise 9.16 Use the method described above to determine $D_5(2)$, the number of sequences of length 5 whose longest sequence of heads has length 2.

It is daunting to try to do all the necessary computations for $n = 250$ by hand, but it is not so difficult to implement the approach above on a computer. Approximate values of $p(X \leq i)$ and $p(X = i)$ are given for $i = 0, 1, \ldots, 12$ in Table 1.

Computing the values of $p(X = i)$ up to $i = 250$ and substituting those numbers into the formula

$$E(X) = 0 \cdot p(X = 0) + 1 \cdot p(X = 1) + 2 \cdot p(X = 2) + \cdots + 250 \cdot p(X = 250)$$

yields $E(X) \approx 7.3034$. Notice also that Table 1 shows there is less than a 14% chance that there will be at most five heads in a row, and the probability of seven or more heads in a row is

$$p(\text{at least 7 heads}) = 1 - p(\text{less than 7 heads}) \approx 1 - 0.3731 = 0.6269.$$

Table 1 *In a sequence of 250 coin flips, the probability $p(X \leq i)$ of having at most i heads in a row, and the probability $p(X = i)$ of having the longest run of heads be of length exactly i.*

i	$p(X \leq i)$	$p(X = i)$
0	< 0.0001	< 0.0001
1	< 0.0001	< 0.0001
2	< 0.0001	< 0.0001
3	0.0001	0.0001
4	0.0144	0.0143
5	0.1318	0.1174
6	0.3731	0.2413
7	0.6160	0.2429
8	0.7870	0.1710
9	0.8880	0.1010
10	0.9427	0.0547
11	0.9711	0.0284
12	0.9855	0.0144

Exercise 9.17 What is your conclusion about the sequence in Figure 1?

REMARK 9.12. Counting the longest run of heads in a sequence of 250 coin flips is something that can be done very efficiently on a computer. In fact, checking a complicated computation like the one we have just done against a simulation is a good way to avoid errors in logic. In Figure 4 we show the result of 10 000 trials, where each data point is the longest run of heads in a sequence of 250 coin flips. Our simulation yielded 2424 instances where the longest run was of length 7, which is quite close to what the probability predicts: $p(X = 7) = 0.2429\ldots$

How does Professor P guess which sequences were faked? Her basic strategy is to look for the length of the longest run of heads. If it's five or less, the sequence is labeled as a fake; if it's seven or more, the sequence is labeled as flipped. If the length is six, she is less confident and quickly looks for other features that might suggest its type. Of course, she always needs to take a few seconds to look out for obvious signs of fakery: among the responses she has received in the past, one consisted of a sequence of flips *exactly* alternating heads and tails, and another consisted of 25 heads, followed by 25 tails, followed by 25 heads, etc.

GOING BEYOND THIS BOOK. Our approach to the flip-or-fake question closely follows the presentation by Mark Schilling [**Sch90**], which also shows that if you are looking for the longest run of either heads or tails, the expected length is roughly 1

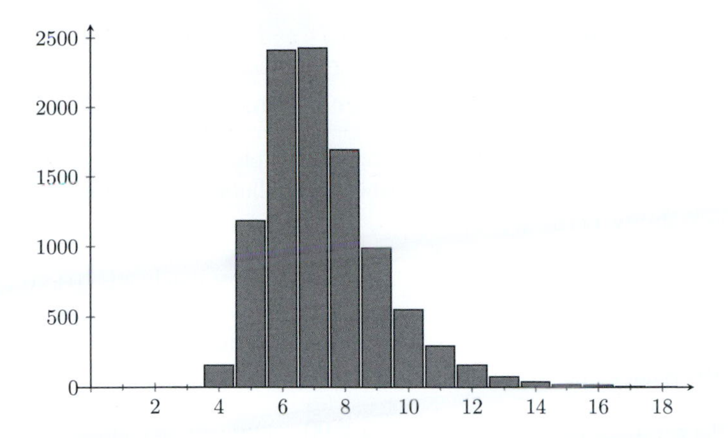

Figure 4. The results of 10 000 computer-generated trials. In each trial the computer "flipped a coin" 250 times and then determined the length of the longest run of heads. The distribution of those 10 000 longest runs is displayed, with the vertical axis giving the frequency.

greater than the expected length for the longest run of heads only. Finding longest runs and counting different subsequences of length k are just two of many approaches to investigating whether a sequence of outcomes has been randomly generated. Volume 2 of Donald Knuth's *The Art of Computer Progamming* is a classic reference on the subject [**Knu98**].

9.7. End-of-Chapter Exercises

Exercises you can work on after Sections 9.1 and 9.2

9.18 Let S be the sample space of all possible sequences of Hs and Ts resulting from four coin flips. Assume the probability distribution is uniform, so all sample points have probability $1/16$.

(a) What sample points are in the event "at most one head"?
(b) What is the probability that a four-flip sequence has at most one head?
(c) What sample points are in the event "half heads and half tails"?
(d) What is the probability that a four-flip sequence has half heads and half tails?

9.19 Professor P also teaches an 8:00 section of her class, with only five students enrolled. What is the probability that four or more students have a first flip of T, so that four or more students should fake their data?

9.20 American roulette is discussed in Example 9.4. There are many different bets that a gambler may make, based on the layout of the numbers on a large felt table, as indicated in Figure 5.

(a) The *top line* or *basket* bet wins if the roulette ball falls into a cell labelled 0, 00, 1, 2, or 3. What is the probability of winning a top line bet?
(b) There are three *column* bets, where you can bet on a set of twelve numbers such as $\{1, 4, 7, \ldots, 34\}$. (To see them as vertical columns, rotate Figure 5 a quarter-turn to the right.) What is the probability of winning a column bet?
(c) A *street* bet chooses any row of numbers on the layout. Examples would be $\{1, 2, 3\}$ and $\{31, 32, 33\}$. What is the probability of winning a street bet?

How do the probabilities for column bets and street bets change for European roulette, which has no "00"?

00	3	6	9	12	15	18	21	24	27	30	33	36
	2	5	8	11	14	17	20	23	26	29	32	35
0	1	4	7	10	13	16	19	22	25	28	31	34

Figure 5. "Place your bets! Lose some money!" ... over the long run.

9.21 You roll a pair of dice, one red and one blue.

(a) What is the probability that the two dice show the same number?
(b) What is the probability that the two dice show different numbers?
(c) What is the probability that the number showing on the red die is strictly less than the value showing on the blue die?
(d) What is the probability that the number showing on the red die is strictly greater than the value showing on the blue die?

9.22 You have carefully manufactured unfair dice where the faces labeled 1 and 6 are slightly more likely to arise than would be the case for fair dice. The probability function for a single die is given by

$$p(s) = \begin{cases} 0.18 & \text{for } s = 1 \text{ or } 6, \\ 0.16 & \text{for } s = 2, 3, 4, \text{ or } 5. \end{cases}$$

What is the probability of getting a total of 7 when you roll two of these unfair dice? How does this compare with rolling a 7 with two fair dice?

9.23 Flip a coin until you get heads for the first time. What is the probability that this takes an odd number of flips?

9.24 Let A, B, and C be events in a finite sample space S, with p a probability function on S. Express the probability of $A \cup B \cup C$ in terms of the probabilities of A, B, C, and their intersections.

9.25 We describe a standard deck of 52 cards in Exercise 3.82. Let S be the sample space of (unordered) hands of five cards, and let p be the probability function corresponding to the uniform distribution on S; thus, we are assuming the deck is well shuffled.

(a) If A is the event "Get 4-of-a-kind," what is $p(A)$?
(b) If B is the event "Get a full house," what is $p(B)$?
(c) Are A and B mutually exclusive?
(d) What is the probability that you are dealt four-of-a-kind or a full house?

9.26 Suppose that you randomly choose a sample point from the sample space $S = \mathcal{P}(\{a, b, c, d, e\})$. What is the probability that the element you choose is a subset of $\{a, b, d\}$?

9.27 Let D be the sample space of ordered pairs defined by

$$D = \{(m, n) \in \mathbb{N} \times \mathbb{N} \mid 1 \leq m < n \leq 10\}.$$

If you choose a sample point (m, n) at random from D, what is the probability that $m|n$?

9.28 If you have worked through subsection 7.1.2, recall that the Count counts and Cookie Monster eats cookies. On his turns, let's imagine that the Count counts out the next *two* cookies, not ten; and that on Cookie Monster's turns, he chooses one of his available cookies at random. For example, on Cookie Monster's first turn, he randomly selects a cookie to eat from the cookies numbered #1 and #2. On his second turn, he randomly selects a cookie to eat from the three cookies remaining on his plate (cookies #3 and #4, plus the one he didn't eat on his first turn). On his third turn, he randomly selects a cookie to eat from the four cookies remaining on his plate. And so on.

(a) What is the probability that cookie #1 gets eaten in one of Cookie Monster's first three turns? The best approach might be to consider the complementary event that cookie #1 isn't eaten.
(b) What is the probability that cookie #1 is eventually eaten?
(c) If n is any natural number, what is the probability that cookie #n is eventually eaten?

See Exercise 9.49 to consider the Count counting by 10.

Exercises you can work on after Section 9.3

9.29 In this problem you flip a fair coin an even number of times in succession, recording the sequence of Hs and Ts.

(a) If you flip the coin six times, what is the probability that exactly half of the flips are heads?
(b) If you flip the coin eight times, what is the probability that exactly half of the flips are heads?
(c) If you flip the coin ten times, what is the probability that exactly half of the flips are heads?
(d) Let P_{2n} be the probability that when you flip the coin $2n$ times, exactly half the flips are heads. Prove that $P_{2n} < P_{2m}$ for any natural numbers m and n such that $m < n$.

9.30 Let A and B be two finite sets, and let F be the sample space of all possible functions $f : A \to B$. Suppose that you randomly choose a function f from F.

(a) What is the probability that f is injective if $|A| = 2$ and $|B| = 3$?
(b) If n is a natural number, what is the probability that f is injective if $|A| = 3$ and $|B| = n$?
(c) What is the probability that f is surjective if $|A| = 4$ and $|B| = 3$?
(d) If n is a natural number, what is the probability that f is surjective if $|A| = n$ and $|B| = 2$?

9.31 Recall Exercise 3.34, where you proved that

$$\binom{n}{0} + \binom{n}{1} + \cdots + \binom{n}{n-1} + \binom{n}{n} = 2^n$$

by connecting the sum of the terms on the left to the size of the power set of $\{1, 2, \ldots, n\}$. Give another proof of this result by induction, using Theorem 9.8.

9.32 Prove that the alternating sum of the numbers in row n of Pascal's Triangle is 0, where n is any natural number. For example,

$$1 - 4 + 6 - 4 + 1 = 0\,.$$

9.33 Let S be an 8-element set. Suppose that you randomly choose a sample point from the sample space $T = \mathcal{P}(S)$.

(a) What is the probability that the element you choose contains an odd number of elements?
(b) Prove a generalization of your result in part (a) by letting S be an n-element set for any natural number n.

9.34 The shaded entries in Figure 6 are in a *shallow diagonal* of Pascal's Triangle, and their sum is $1+5+6+1 = 13$. The next shallow diagonal gives a sum of $1+6+10+4 = 21$.

You have seen these numbers in a sequence before. Make a conjecture, and prove it!

9.35 Copy the first nine rows of Pascal's triangle (Figure 3) onto a clean sheet of paper, and shade the odd entries. Describe any patterns you see, and add additional rows for more evidence that the patterns continue. Form conjectures, and then try to prove them.

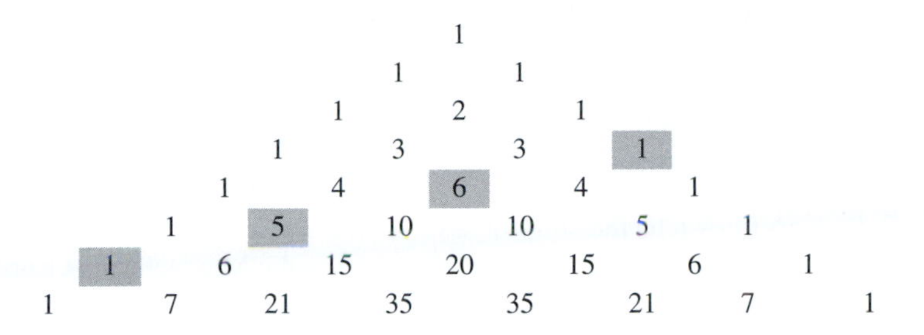

Figure 6. The four shaded numbers form a shallow diagonal in Pascal's Triangle.

Exercises you can work on after Sections 9.4–9.6

9.36 Let S be the sample space of possible rolls of two dice, and let $X : S \to \mathbb{N}$ be the random variable given by adding the values of the two dice. What is $p(X \geq 10)$?

9.37 Let S be the sample space of possible rolls of two dice, and let $M : S \to \mathbb{N}$ be the random variable given by taking the maximum value appearing on the dice.

(a) What is $p(M = 3)$?
(b) What is the expected value $E(M)$?

9.38 Following up on our discussion of roulette in Exercise 9.20, we now bring wagers into play.

(a) The *top line* or *basket* bet wins if the ball falls into a cell labelled 0, 00, 1, 2, or 3. The payout for a top line bet is 6 to 1, meaning that if you bet x and win, you will have your bet x returned to you and receive an additional $6x$; otherwise, you simply lose your bet x. What is the expected value of a top line bet?
(b) There are three *column* bets, where you can bet on a set of 12 numbers such as $\{1, 4, 7, \ldots, 34\}$. The payout for a column bet is 2 to 1. What is the expected value of a column bet?
(c) A *street* bet chooses any row of numbers on the layout. Examples would be $\{1, 2, 3\}$ and $\{31, 32, 33\}$. The payout for a street bet is 11 to 1. What is the expected value of a street bet?

There are many other types of bets in roulette. Their payouts are structured so that all bets except for the top line bet have a "house edge" of roughly 5.26%. This means that, over a very long run of bets, there is a good chance that you will lose more than 5% of all that you wager.

9.39 Repeat Exercise 9.38 for European roulette. What is the house edge for most bets?

9.40 After an evening in a casino, you have only two dollars in your pocket, but you realize that you need five dollars for the bus ride home. You head to the roulette table to try to solve your problem. If the minimum bet is one dollar, what is your best strategy?

9.41 Let $S = [64] = \{1, 2, \ldots, 64\}$, and let $H : S \to \mathbb{N}$ be the random variable that gives the highest power of 2 dividing an element of the sample space S. For example, $H(24) = 2^3 = 8$ and $H(17) = 2^0 = 1$.

(a) What is $p(H = 2)$?
(b) What is $p(H > 10)$?
(c) What is $E(H)$?
(d) Now let $S = [2^n] = \{1, 2, 3, \ldots, 2^n\}$ for any natural number n. Conjecture a value for $E(H)$, and then prove your conjecture.

9.42 Kari Lock tells the story of several enterprising math students from Williams College who took advantage of a poor decision by the Massachusetts Lottery [**Loc07**]. In a lottery game called 4 Spot Quick Draw, every five minutes a set S of 20 distinct numbers is randomly chosen from the set $\{1, 2, \ldots, 80\}$. Before S is selected, you can pay x to pick four numbers out of $\{1, 2, \ldots, 80\}$, and you are rewarded based on how many of your four numbers match the ones in S. If you match two numbers, you win x (which simply offsets your wager); if you match three numbers, you win $5x$; and if you match all four numbers, you win $55x$.

(a) It turns out that the approximate probabilities of matching exactly i numbers are given by

i	0	1	2	3	4
Probability	.308	.433	.213	.043	.003

If W represents your winnings on a single bet of x, what is the expected value $E(W)$?
(b) On certain Wednesdays, the lottery ran a promotion: the payoff would be doubled! For example, matching three numbers wins you $10x$, not $5x$ (so you end up with a total of $9x$ after accounting for the x you paid to play). Compute $E(W)$, and explain why the students were thrilled to have previously studied probability.
(c) Explain why the ith entry in the table of probabilities in part (a) is equal to

$$\frac{\binom{4}{i} \cdot \binom{76}{20-i}}{\binom{80}{20}}.$$

The value of $E(W)$ in part (a) for the regular 4 Spot Quick Draw game is typical for many US state lottery games.

9.43 Suppose that you toss a coin until you get two heads in a row. What is the probability that you stop on the eighth toss?

More Exercises!

9.44 As you approach a locked door, you pull out a keyring containing five keys that are essentially indistinguishable to you, but you know that only one of them opens the door.

(a) Assume that you are smart, so that you proceed through the keys in some order until you find the correct one. What is the expected number of keys that you will try before you find the correct one?
(b) But maybe you're not so smart, so that after every attempt with an incorrect key, you drop the ring on the ground, pick it up, and then try a key again, possibly one you've tried before. What is the expected number of keys you will try before you find the correct one? For this scenario, having dexterity with geometric series (if not keyrings) may prove helpful.

9.45 Assuming that birthdays are uniformly distributed over a year containing 365 days, what is the probability that at least two people in a group of 30 people have the same birthday?

9.46 A friend shuffles a deck of cards very well and slowly deals the cards one-at-a-time face up onto a table. At some moment of your choice before all of the cards are dealt, you must guess that the next card dealt will be red. What is your best strategy for the highest probability of success?

9.47 In volume 2 of *The Art of Computer Programming* [**Knu98**], Donald Knuth describes several methods for quickly generating arbitrarily long sequences of "pseudo-random" numbers. As an example, he presents John von Neumann's "middle square" method to generate a sequence $(a_1, a_2, a_3, \ldots)$ of n-digit integers.

Suppose that you start with any 4-digit integer a_1. For all $i \geq 1$, a_{i+1} will be the 4-digit integer created by the middle four of the eight digits of a_i^2. Thus an initial value like

$$a_1 = 2916$$

leads to

$$a_2 = 5030$$

because $a_1^2 = (2916)^2 = 08\underline{5030}56$; and this leads to

$$a_3 = 3009$$

because $a_2^2 = (5030)^2 = 25\underline{3009}00$; and so on.

(a) What are some of the desirable and undesirable features in this method of generating "random" 4-digit integers?
(b) Continue the sequence started above by computing a_4 and a_5, and then modify your answer to part (a) if necessary.
(c) Try starting a new 4-digit middle square sequence with $a_1 = 3792$.
(d) There are ways of generating pseudorandom sequences that are better, according to certain statistical tests, than the sequences produced by von Neumann's method. Try thinking of one yourself; but be warned that great care should be taken in the construction and analysis of pseudorandom sequences. Knuth puts it best: "Random numbers should not be generated with a method chosen at random."

If You Have Studied Calculus

9.48 Proving Stirling's approximation for $n!$ is beyond the scope of this chapter, but we can get fairly close. The idea is to bound

$$\ln(n!) = \ln(1) + \ln(2) + \cdots + \ln(n)$$

by thinking of the right-hand side as a Riemann sum for two integrals of the form $\int_a^b \ln(x)\,dx$.

(a) Show that $\int_0^n \ln(x)\,dx \leq \ln(n!)$.
(b) Show that $\ln(n!) \leq \int_1^{n+1} \ln(x)\,dx$.

(c) Use the fact that $x\ln(x) - x$ is an antiderivative for $\ln(x)$ to conclude that

$$n\ln(n) - n \leq \ln(n!) \leq (n+1)\ln(n+1) - n.$$

(d) Prove that

$$n^n e^{-n} \leq n! \leq (n+1)^{n+1} e^{-n},$$

and compare this with Stirling's Formula.

9.49 Reimagine the scenario in Exercise 9.28, with the Count on his turns counting out the next ten available cookies, not two. If Cookie Monster still randomly chooses one of his available cookies on his turns, what is the probability that cookie #1 is eventually eaten?

As a hint, one way to approach this problem is to remember that taking the logarithm of a product of terms gives you a sum of logarithms. And you may then find some assistance from the inequality $\ln(1+x) \leq x$, which holds for $x > -1$ and can be proven using calculus.

10 Algebra and Symmetry

Algebra and geometry are two of the oldest branches of mathematics. These two topics come together in the study of groups, algebraic objects that can be used to describe symmetry. In this chapter we introduce group theory, following a narrow and carefully chosen path toward the idea of isomorphism.

10.1. An Example from Modular Arithmetic

Modular arithmetic was introduced in Section 6.5 where we discussed addition, subtraction, and multiplication modulo n; in Section 6.6 we then determined that you can divide by m modulo n only when m and n are relatively prime. In this short introductory section, we examine addition in $\mathbb{Z}_4$ and multiplication in $\mathcal{U}_5$ (the invertible elements modulo 5). Each of these involve four equivalence classes, but little else looks similar. In particular, the equivalence classes are different:

$$\mathbb{Z}_4 = \{[0]_4, [1]_4, [2]_4, [3]_4\} \neq \{[1]_5, [2]_5, [3]_5, [4]_5\} = \mathcal{U}_5 .$$

To emphasize the differences, we note that

$$[1]_4 = \{\ldots, -7, -3, 1, 5, 9, \ldots\} \neq \{\ldots, -9, -4, 1, 6, 11, \ldots\} = [1]_5 .$$

In addition to the fact that the equivalence classes are different, a quick inspection of the addition table for $\mathbb{Z}_4$ and the multiplication table for $\mathcal{U}_5$ seems to indicate that addition modulo 4 and multiplication modulo 5 are not related, as seen in Figure 1.

+	0	1	2	3
0	0	1	2	3
1	1	2	3	0
2	2	3	0	1
3	3	0	1	2

·	1	2	3	4
1	1	2	3	4
2	2	4	1	3
3	3	1	4	2
4	4	3	2	1

Figure 1. Addition in $\mathbb{Z}_4$ and multiplication in $\mathcal{U}_5$. In both tables we have dropped the equivalence class notation to avoid clutter.

However, after reorganizing the order in which we list the elements of $\mathcal{U}_5$, we see that the resulting arithmetic tables look remarkably similar. As one example, look at the positions for **2** in the table for $\mathbb{Z}_4$ and **4** in the table for $\mathcal{U}_5$ in Figure 2. The locations for $0 \in \mathbb{Z}_4$ and $1 \in \mathcal{U}_5$ are also the exact same; the locations of $1 \in \mathbb{Z}_4$ and $2 \in \mathcal{U}_5$ are the same; and the locations of $3 \in \mathbb{Z}_4$ and $3 \in \mathcal{U}_5$ are the same. Overall, both tables follow the pattern illustrated by the card suit table on the right in Figure 2.

+	0	1	**2**	3
0	0	1	**2**	3
1	1	**2**	3	0
2	**2**	3	0	1
3	3	0	1	**2**

·	1	2	**4**	3
1	1	2	**4**	3
2	2	**4**	3	1
4	**4**	3	1	2
3	3	1	2	**4**

	♡	♣	♢	♠
♡	♡	♣	♢	♠
♣	♣	♢	♠	♡
♢	♢	♠	♡	♣
♠	♠	♡	♣	♢

Figure 2. The tables on the left and in the middle are the same as in Figure 1, but with the elements of $\mathcal{U}_5$ listed in a different order. The general pattern for both tables is shown by the card suits in the table on the right.

This small example hints at an idea worthy of further exploration, as it now appears that these two different mathematical objects may have some underlying structural similarity. Introducing the algebraic notion of groups and exploring how to describe this notion of similarity are the objectives of this chapter.

10.2. The Symmetries of a Square

We have constructed tables to display the structure of modular arithmetic. In this section we develop a similar table, but in a quite different context. Let $\mathcal{Q}$ denote a square, which in order to keep things concrete, we can think of as the 2×2 square

$$\mathcal{Q} = \{(x, y) \in \mathbb{R}^2 \mid |x| \leq 1 \text{ and } |y| \leq 1\}.$$

The *corners of* $\mathcal{Q}$ are the four points in

$$\{(x, y) \in \mathcal{Q} \mid x = \pm 1 \text{ and } y = \pm 1\}.$$

At an informal level, a symmetry of $\mathcal{Q}$ consists of any motion that does not change the square. In other words, if you look at the square, close your eyes while a symmetry is being performed, and then open them again, you would not see any difference. One example would be a $90°$ rotation of $\mathcal{Q}$, and another would be the reflection across the diagonal, both shown in Figure 3.

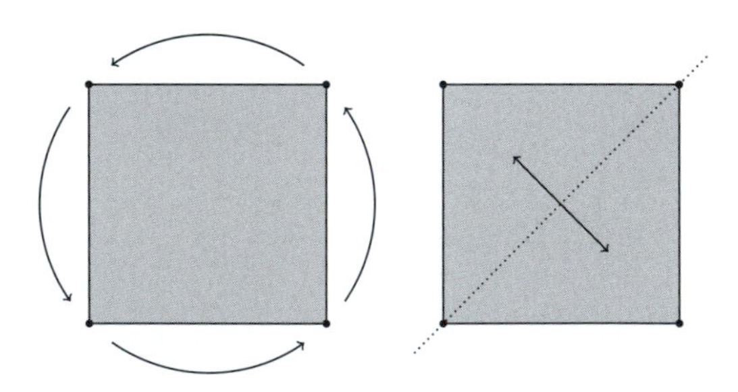

Figure 3. Two symmetries of $\mathcal{Q}$ are the counterclockwise rotation shown on the left and a reflection across the diagonal shown on the right.

While this characterization has some intuitive appeal, its lack of rigor makes it fairly useless for proving theorems. Therefore we define a *symmetry* of $\mathcal{Q}$ to be a function $f : \mathcal{Q} \to \mathcal{Q}$ that preserves distances: if the distance from (x_0, y_0) to (x_1, y_1) is d, then the distance from $f[(x_0, y_0)]$ to $f[(x_1, y_1)]$ is also d.

DEFINITION 10.1. The set of all symmetries of $\mathcal{Q}$ is denoted by $\text{SYM}(\mathcal{Q})$.

We would like to describe all the functions in $\text{SYM}(\mathcal{Q})$, and the following lemma is key to being able to do this.

LEMMA 10.2. *Any symmetry of $\mathcal{Q}$ is determined by what it does to the adjacent corners $(1, 1)$ and $(-1, 1)$. That is, if f and $g \in \text{SYM}(\mathcal{Q})$, where $f[(1, 1)] = g[(1, 1)]$ and $f[(-1, 1)] = g[(-1, 1)]$, then f and g are the same function.*

The idea behind our proof is that the location of any point in $\mathcal{Q}$ is determined by its distance from two adjacent corners; and by our definition, symmetries preserve distances. Thus if we know where the corners are moved by f or g, then we can determine the image of any point under f or g.

PROOF. A symmetry of $\mathcal{Q}$ has to take corners of $\mathcal{Q}$ to corners of $\mathcal{Q}$ because the distance between opposite corners is the maximum distance between any two points in $\mathcal{Q}$, and symmetries preserve distances. Adjacent corners, which are pairs of corners separated by a distance of 2, must then be mapped to adjacent corners.

Assume that $f[(1, 1)] = g[(1, 1)]$ and $f[(-1, 1)] = g[(-1, 1)]$, and let $(x, y) \in \mathcal{Q}$ be an arbitrary point. Our goal is to show that $f[(x, y)] = g[(x, y)]$.

Let d_- be the distance from $(-1, 1)$ to (x, y), and let d_+ be the distance from $(1, 1)$ to (x, y). Then (x, y) is on the circle of radius d_- centered at $(-1, 1)$ and is on the circle of radius d_+ centered at $(1, 1)$. Further, (x, y) is the only point in $\mathcal{Q}$ that sits on both circles; the circles may intersect twice, but only one of those intersections is contained in $\mathcal{Q}$, as seen in Figure 4. In general, every point in $\mathcal{Q}$ is determined by its distances to any two adjacent corners of $\mathcal{Q}$.

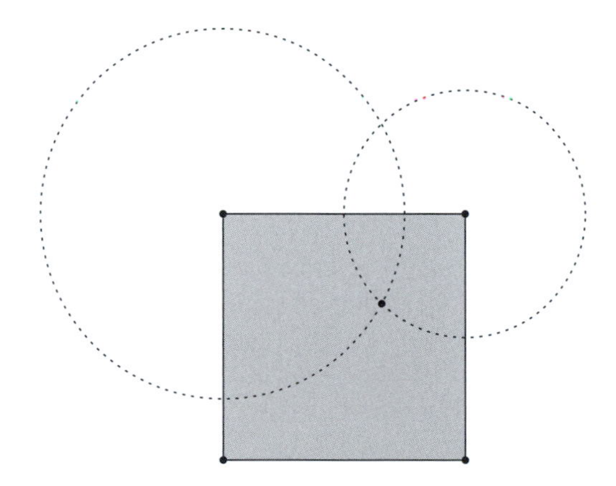

Figure 4. Each point in $\mathcal{Q}$ is determined by its distances to the corners $(-1, 1)$ and $(1, 1)$.

Since $f \in \text{SYM}(\mathcal{Q})$, the distance from $f[(-1, 1)]$ to $f[(x, y)]$ is d_-, and the distance from $f[(1, 1)]$ to $f[(x, y)]$ is d_+. Since $f[(-1, 1)]$ and $f[(1, 1)]$ are adjacent corners of $\mathcal{Q}$, it follows that there is only one possible location for $f[(x, y)]$. Since the same conditions hold for $g \in \text{SYM}(\mathcal{Q})$, we conclude that $f[(x, y)] = g[(x, y)]$. □

COROLLARY 10.3. *There are at most eight functions in* SYM($\mathcal{Q}$).

PROOF. In the proof of Lemma 10.2, we showed that the corner $(1, 1)$ has to go to one of the four corners of $\mathcal{Q}$. Thus if $f \in$ SYM($\mathcal{Q}$), there are four possibilities for $f[(1, 1)]$. The image of the corner $(-1, 1)$ must then be a corner of $\mathcal{Q}$ that is adjacent to $f[(1, 1)]$, so there are two possible locations for $f[(-1, 1)]$. By Lemma 10.2, once we know the location of $f[(1, 1)]$ and $f[(-1, 1)]$, we have determined $f \in$ SYM($\mathcal{Q}$). Thus there are at most $4 \cdot 2 = 8$ functions in the set SYM($\mathcal{Q}$). □

As was shown in Figure 3, a counterclockwise rotation through 90°, with the center of the rotation at $(0, 0)$, is one example of a symmetry of $\mathcal{Q}$. We denote this symmetry as ρ and note that there is a simple formula for this function:

$$\rho[(x, y)] = (-y, x).$$

Exercise 10.1 Prove that the formula for ρ is correct by showing that it is distance-preserving and sends two adjacent corners of $\mathcal{Q}$ to the correct locations.

Counterclockwise rotations through 180° and 270° are also symmetries of $\mathcal{Q}$. A counterclockwise rotation through an angle of 180° can be produced by applying ρ twice. Hence we may express this symmetry as $\rho^2 = \rho \circ \rho$. Similarly, a counterclockwise rotation through 270° can be expressed as $\rho^3 = \rho \circ \rho \circ \rho$.

Exercise 10.2 Show that the symmetry ρ^2 can be expressed by the formula

$$\rho^2[(x, y)] = (-x, -y),$$

and the symmetry ρ^3 can be expressed by the formula

$$\rho^3[(x, y)] = (y, -x).$$

We do not need to introduce separate notation for clockwise rotations, as each clockwise rotation is equivalent to a counterclockwise rotation. For example, let ϱ denote a 90° clockwise rotation of $\mathcal{Q}$. We then have $\varrho[(-1, 1)] = (1, 1)$ and $\varrho[(1, 1)] = (1, -1)$. But $\rho^3[(-1, 1)] = (1, 1)$ and $\rho^3[(1, 1)] = (1, -1)$, so by Lemma 10.2, $\varrho = \rho^3 \in$ SYM($\mathcal{Q}$). In the study of symmetry, it is the destination, not the journey, that matters!

In addition to the rotations described above, there are four reflections of $\mathcal{Q}$. In general, a *reflection* in the Euclidean plane is defined via a line ℓ in the plane, called the *axis* of the reflection. Each point $p \in \mathbb{R}^2 \setminus \ell$ has a corresponding point $p' \in \mathbb{R}^2 \setminus \ell$ such that the segment joining p to p' is perpendicular to ℓ, and the distances from p to ℓ and from p' to ℓ are the same. The reflection associated to the axis ℓ exchanges p and p' (for all such pairs), and it leaves the points on ℓ fixed.

The set SYM($\mathcal{Q}$) contains reflections across the x-axis, the y-axis, and across the two diagonals of $\mathcal{Q}$. Denote the main diagonal by

$$D = \{(x, y) \in \mathcal{Q} \mid y = x\}$$

and the off-diagonal by

$$O = \{(x, y) \in \mathcal{Q} \mid y = -x\}.$$

We name these four reflections using subscripts corresponding to their axes: the reflection across the y-axis is ϕ_Y, the reflection across the x-axis is ϕ_X; and the reflections across the diagonals are ϕ_D and ϕ_O.

Exercise 10.3 Find formulas for ϕ_D and ϕ_O, similar to the formulas you found in Exercise 10.2.

Finally, there is one more very important symmetry of $\mathcal{Q}$ that is easy to miss on first examination: the "do nothing" identity function $\mathrm{Id} : \mathcal{Q} \to \mathcal{Q}$, given by $\mathrm{Id}[(x, y)] = (x, y)$.

PROPOSITION 10.4. *The set* SYM($\mathcal{Q}$) *consists of the eight functions described above.*

PROOF. By Corollary 10.3, we know that there are at most eight functions in SYM($\mathcal{Q}$). The eight functions described above are all in SYM($\mathcal{Q}$), and they are all distinct, as can be verified by applying the functions to the corners $(-1, 1)$ and $(1, 1)$. □

Now, knowing that we have described all the symmetries of $\mathcal{Q}$, we can create a table that displays the result of function composition; see Table 1. For example, the second row is labeled by ρ and the fifth column by ϕ_X, so the entry in the table is then $\rho \circ \phi_X = \phi_D$. (If you think the answer should be ϕ_O then you composed your functions in the wrong order!)

Exercise 10.4 Verify that the entry in Table 1 in the row corresponding to ϕ_X and the column corresponding to ϕ_D should be ρ^3.

Table 1 *A table describing the composition of symmetries in* SYM($\mathcal{Q}$).

$\circ$	Id	ρ	ρ^2	ρ^3	ϕ_X	ϕ_D	ϕ_Y	ϕ_O
Id	Id	ρ	ρ^2	ρ^3	ϕ_X	ϕ_D	ϕ_Y	ϕ_O
ρ	ρ	ρ^2	ρ^3	Id	ϕ_D	ϕ_Y	ϕ_O	ϕ_X
ρ^2	ρ^2	ρ^3	Id	ρ	ϕ_Y	ϕ_O	ϕ_X	ϕ_D
ρ^3	ρ^3	Id	ρ	ρ^2	ϕ_O	ϕ_X	ϕ_D	ϕ_Y
ϕ_X	ϕ_X	ϕ_O	ϕ_Y	ϕ_D	Id	ρ^3	ρ^2	ρ
ϕ_D	ϕ_D	ϕ_X	ϕ_O	ϕ_Y	ρ	Id	ρ^3	ρ^2
ϕ_Y	ϕ_Y	ϕ_D	ϕ_X	ϕ_O	ρ^2	ρ	Id	ρ^3
ϕ_O	ϕ_O	ϕ_Y	ϕ_D	ϕ_X	ρ^3	ρ^2	ρ	Id

There are other 8×8 tables we already know. We could make the table describing addition modulo 8, and in Figure 4 (page 154) we displayed the table for multiplication of the eight invertible elements in $\mathcal{U}_{15}$. No two of these three tables – the one for SYM($\mathcal{Q}$), the one for $\mathbb{Z}_8$, and the one for $\mathcal{U}_{15}$ – appear to be the same. However, in the previous section we saw how simply permuting the elements can make two seemingly different tables become essentially identical. Checking that one table can or cannot be permuted to produce the other could well require all $8! = 40\,320$ permutations of the eight elements, which is one indication that blind permutation is not a good strategy.

GOING BEYOND THIS BOOK. Herman Weyl's classic book *Symmetrie*, translated into English as *Symmetry* [**Wey89**], provides an extended introduction to geometric symmetry by exploring the symmetry inherent in works of art and architecture, and then extending these ideas into the natural sciences.

10.3. Group Theory

The language and viewpoint of group theory is central in mathematics. Mathematical groups have proven to be a useful means of describing the symmetry of mathematical objects, from the sorts of geometric symmetries we have already seen, to notions of symmetry that are more subtle.

DEFINITION 10.5. A *group* $(G, \cdot)$ consists of a non-empty set G and a binary operation $\cdot : G \times G \to G$ such that:

(a) the binary operation is associative, that is, $a \cdot (b \cdot c) = (a \cdot b) \cdot c$ for all a, b and $c \in G$;
(b) there is an *identity* element $e \in G$ such that $e \cdot g = g \cdot e = g$ for all $g \in G$;
(c) for each $g \in G$ there is an *inverse* element $g^{-1} \in G$ such that $g \cdot g^{-1} = g^{-1} \cdot g = e$.

Many of the mathematical objects we have studied are groups, including those based on the associative operations from modular arithmetic.

EXAMPLE 10.6. For any $n \in \mathbb{N}$, $\mathbb{Z}_n$ and addition modulo n forms a group $(\mathbb{Z}_n, +)$. As we saw in Sections 6.5 and 6.6, the identity element is $[0] \in \mathbb{Z}_n$ and the inverse of $[k]$ is $[-k]$. The integers themselves, $\mathbb{Z}$, also form a group under addition.

EXAMPLE 10.7. For any $n \in \mathbb{N}$, the set $\mathcal{U}_n$ of units modulo n, with the binary operation of multiplication modulo n, forms a group $(\mathcal{U}_n, \cdot)$. You can see the Cayley table for $(\mathcal{U}_{15}, \cdot)$ in Section 6.6.

EXAMPLE 10.8. Finally, let's look at an example that is not based on a number system. Let S be a non-empty set, and let $\mathcal{P}(S)$ be its power set. Then the symmetric difference provides us with a binary operation $\mathcal{P}(S) \times \mathcal{P}(S) \to \mathcal{P}(S)$, since the symmetric difference of two subsets of S is another subset of S. You proved that the symmetric difference is associative in Exercise 3.17. The empty set is an identity element, as

$$\emptyset \triangle A = A \triangle \emptyset = A$$

for all $A \subseteq S$. Finally, because $A \triangle A = \emptyset$, each subset of S is its own inverse. Thus $(\mathcal{P}(S), \triangle)$ is a group.

REMARK 10.9. The notation $(\mathcal{P}(S), \triangle)$ represents a group, while the similar notation $(\mathcal{P}(S), \subseteq)$ from Section 6.2 represents a partially ordered set. The context makes things clear; in particular, $\subseteq$ is a relation, not a binary operation. To simplify notation, a group $(G, \cdot)$ is often called G when the operation is apparent.

Exercise 10.5 There are many binary operations on subsets of a non-empty set S; $\cup$ and $\cap$ are two examples. In this exercise you will show that $\mathcal{P}(S)$ with $\cup$ and $\mathcal{P}(S)$ with $\cap$ do not provide examples of groups.

(a) Show that the empty set $\emptyset$ is the only element of $\mathcal{P}(S)$ that can serve as the identity element for $\cup$.
(b) Show that there are elements of $\mathcal{P}(S)$ that do not have inverses with respect to $\cup$.
(c) Show that the full set S is the only element of $\mathcal{P}(S)$ that can serve as the identity element for $\cap$.
(d) Show that, as long as S is non-empty, there are elements of $\mathcal{P}(S)$ that do not have inverses with respect to $\cap$.

The symmetries of a square, $\text{SYM}(\mathcal{Q})$, form a group. The elements are symmetries of the square $\mathcal{Q}$ and the binary operation is function composition. In fact, much more is true.

DEFINITION 10.10. Given any subset $X \subseteq \mathbb{R}^n$, the set of all distance-preserving functions $f : X \to X$ is the *symmetry group* of X, denoted $\text{SYM}(X)$.[1]

THEOREM 10.11. *Let $X \subseteq \mathbb{R}^n$ and let $\text{SYM}(X)$ be the collection of all symmetries of X. Then $\text{SYM}(X)$ is closed under composition and the taking of inverses, and the set of symmetries along with function composition forms a group.*

Given our work developing $\text{SYM}(\mathcal{Q})$, Theorem 10.11 is not surprising, and we ask you to prove it in Exercise 10.27. You can find even more general results along these lines in [**Mei08**].

Given any two groups, we can use the Cartesian product to create a new group.

DEFINITION 10.12. In order to be very clear about the product structure, we let $(G, \cdot_G)$ be a group where "$\cdot_G$" is the operation and $(H, \cdot_H)$ be another group with "$\cdot_H$" as the operation. The *Cartesian product* of these two groups has as its underlying set the Cartesian product $G \times H$; the binary operation for $G \times H$ is defined component-by-component:

$$(g_1, h_1) \cdot (g_2, h_2) = (g_1 \cdot_G g_2, h_1 \cdot_H h_2).$$

For example, consider $\mathbb{Z}_4 \times \mathbb{Z}_2$, where the two operations are addition modulo 4 and addition modulo 2. In this group, we would have

$$([3]_4, [1]_2) \cdot ([2]_4, [0]_2) = ([3+2]_4, [1+0]_2) = ([1]_4, [1]_2),$$

where we have included all the notation for elements of $\mathbb{Z}_4$ and $\mathbb{Z}_2$.

LEMMA 10.13. *The Cartesian product of two groups is a group.*

Exercise 10.6 Prove Lemma 10.13.

You may have noticed that the definition of a group demands that there is "an identity" instead of "the identity," and that every element has "an inverse" and not necessarily a unique inverse.

1 Distance-preserving functions $f : X \to X$ are called *isometries* of X, and in this terminology we are defining the isometry group of X. That said, there is a lot of terminology to absorb in this chapter, so we will not use the term isometry.

PROPOSITION 10.14. *Let G be a group. Then there is only one identity element in G.*

PROOF. Assume to the contrary that both e and $f \in G$ are identities for G. Then $e \cdot f = f$, since e is an identity element. But it is also true that $e \cdot f = e$ since f is an identity element. Thus $e = e \cdot f = f$. □

PROPOSITION 10.15. *Let G be a group. Then each $g \in G$ has exactly one inverse.*

PROOF. Let g be any element of G, and assume to the contrary that both h and k are inverses of g. Thus $h \cdot g$ and $k \cdot g$ are both equal to the identity $e \in G$. Right multiplying both expressions by h then gives

$$h \cdot g \cdot h = k \cdot g \cdot h\,.$$

But $g \cdot h = e$, since h is an inverse of g. So we have $h \cdot e = k \cdot e$, hence $h = k$. □

10.4. Cayley Tables

Given a group G, we can make a table that displays all of the products of elements of G; the addition and multiplication tables we made for modular arithmetic are examples. The rows and columns of the table correspond to the elements of G, and the entry in the row corresponding to g and the column corresponding to h is the product $g \cdot h$. This is exactly what we did in constructing Table 1 for SYM($\mathcal{Q}$). A table displaying a group in this fashion is called a *Cayley table*.

Exercise 10.7 Construct the Cayley table for $\mathbb{Z}_4 \times \mathbb{Z}_2$.

Here we introduce another group and use it to construct an additional example of a Cayley table. Let $\mathfrak{R}$ be a rectangular box with three different side lengths. It may, for example, have length $L = 10$ cm, width $W = 12$ cm, and height $H = 15$ cm. As in the case of the square, let SYM($\mathfrak{R}$) be the collection of all functions from $\mathfrak{R}$ back to $\mathfrak{R}$ that preserve distances.

We can quickly identify a number of elements that are in SYM($\mathfrak{R}$). There is an identity element. There are three reflections, where instead of reflecting across a fixed line, we reflect across a fixed plane that divides $\mathfrak{R}$ into two congruent pieces; the box and one such plane are depicted in Figure 5. And there are three 180° rotations where the axis of rotation is a line that passes through the center of a rectangular face of $\mathfrak{R}$. We can denote the reflections by ϕ_L, ϕ_W, and ϕ_H, where we let the subscript denote the orientation of the reflecting plane. For example, the reflecting plane for ϕ_L is orthogonal to the four edges corresponding to the length $L = 10$ cm. Similarly we can denote the three rotations by ρ_L, ρ_W, and ρ_H, where for example the axis of rotation for ρ_L is parallel to the four edges corresponding to the length L.

Our census of symmetries might lead you to guess that SYM($\mathfrak{R}$) contains only these seven elements, and in order to verify this guess, you would like to have a result like Corollary 10.3. Because $\mathfrak{R}$ is irregular – having different length, width, and height – we get a slightly stronger result.

LEMMA 10.16. *Any symmetry of $\mathfrak{R}$ is determined by where it takes any single corner of $\mathfrak{R}$.*

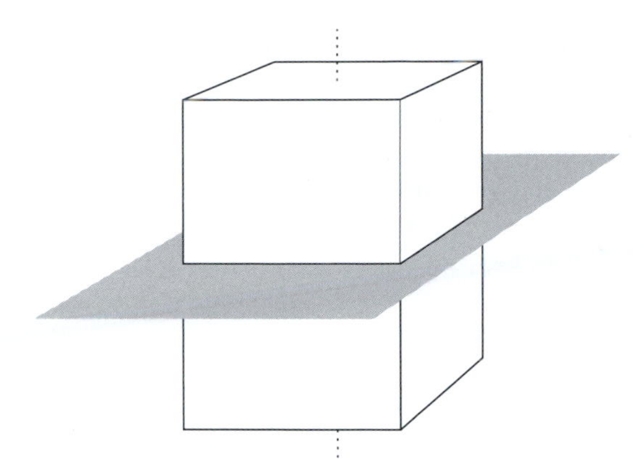

Figure 5. Two symmetries of an irregular box $\mathfrak{R}$. The 180° rotation about the "height" axis is denoted ρ_H; the reflection across the plane perpendicular to this axis is denoted ϕ_H. As shown in Table 2, the composition $\rho_H \circ \phi_H$ is the antipodal map we call α.

Exercise 10.8 Prove Lemma 10.16.

Since $\mathfrak{R}$ has eight corners, Lemma 10.16 immediately implies the following.

COROLLARY 10.17. *The symmetry group* $\text{SYM}(\mathfrak{R})$ *contains at most eight functions.*

There is a small gap between the maximum number of functions in $\text{SYM}(\mathfrak{R})$ and the number we have found. There is indeed one symmetry that we have not already described.

Exercise 10.9 Let $\alpha = \phi_L \circ \phi_D \circ \phi_H$. Prove that $\alpha \in \text{SYM}(\mathfrak{R})$ and that it is not one of the seven symmetries we previously identified.

We now know the eight elements in $\text{SYM}(\mathfrak{R})$:

$$\text{SYM}(\mathfrak{R}) = \{\text{Id}, \phi_L, \phi_W, \phi_H, \rho_L, \rho_W, \rho_H, \alpha\}.$$

The Cayley table for this group of symmetries is shown in Table 2.

Table 2 *Cayley table for the symmetries of the irregular box* $\mathfrak{R}$.

	Id	ϕ_L	ϕ_W	ϕ_H	ρ_L	ρ_W	ρ_H	α
Id	Id	ϕ_L	ϕ_W	ϕ_H	ρ_L	ρ_W	ρ_H	α
ϕ_L	ϕ_L	Id	ρ_H	ρ_W	α	ϕ_H	ϕ_W	ρ_L
ϕ_W	ϕ_W	ρ_H	Id	ρ_L	ϕ_H	α	ϕ_L	ρ_W
ϕ_H	ϕ_H	ρ_W	ρ_L	Id	ϕ_W	ϕ_L	α	ρ_H
ρ_L	ρ_L	α	ϕ_H	ϕ_W	Id	ρ_H	ρ_W	ϕ_L
ρ_W	ρ_W	ϕ_H	α	ϕ_L	ρ_H	Id	ρ_L	ϕ_W
ρ_H	ρ_H	ϕ_W	ϕ_L	α	ρ_W	ρ_L	Id	ϕ_H
α	α	ρ_L	ρ_W	ρ_H	ϕ_L	ϕ_W	ϕ_H	Id

Exercise 10.10 Verify three entries in the Cayley table for $\text{SYM}(\mathfrak{R})$, keeping in mind that using the identity element is a lazy strategy.

GOING BEYOND THIS BOOK. Arthur Cayley was an early advocate of studying groups like $\text{SYM}(\mathcal{Q})$; he frequently constructed composition tables like the one we have been exploring. For this reason such tables are often referred to as Cayley tables. *Groups and Their Graphs* by Grossman and Magnus [**GM64**] is an excellent text describing Cayley tables and geometric structures associated with groups, as is Nathan Carter's *Visual Group Theory* [**Car09**].

10.5. Group Properties

There are a handful of elementary properties which can be used to distinguish one group from another.

10.5.1. Abelian Groups. Not all groups have commutative operations; for example, $\rho \circ \phi_X \neq \phi_X \circ \rho$ in $\text{SYM}(\mathcal{Q})$.[2] However, many groups do have operations that are commutative.

DEFINITION 10.18. An *Abelian group* is a group where the binary operation is commutative. That is, for every g and $h \in G$ you have $g \cdot h = h \cdot g$.

REMARK 10.19. The adjective Abelian is used, instead of commutative, in honor of the pioneering mathematics of Niels Abel. Because the term is derived from his last name, it is often capitalized.

We have already seen a number of examples of Abelian groups. Addition in $\mathbb{Z}_n$ is a commutative operation, so $(\mathbb{Z}_n, +)$ is an Abelian group. Similarly multiplication modulo n is commutative, hence $(\mathcal{U}_n, \cdot)$ is an Abelian group. There are also infinite Abelian groups. The most accessible examples would be $\mathbb{Z}$, $\mathbb{Q}$, and $\mathbb{R}$, with $+$ as the binary operation. The sets $\mathbb{Q} - \{0\}$ and $\mathbb{R} - \{0\}$ with multiplication as the binary operation form Abelian groups as well.

Exercise 10.11 Verify some of the claims we have just made. At a minimum you should consider the following two items.

(a) Prove that $(\mathbb{Q} - \{0\}, \cdot)$ is an Abelian group.
(b) Why is $(\mathbb{Z} - \{0\}, \cdot)$ not an Abelian group?

Exercise 10.12 Is the symmetry group of the irregular box $\mathfrak{R}$ in Section 10.4 Abelian?

10.5.2. Cyclic Groups. In some groups, a single element can be used to express every element.

DEFINITION 10.20. If $g \in G$, then for $n \in \mathbb{N}$ define

$$g^n = \underbrace{g \cdot g \cdots g}_{n \text{ times}} \quad \text{and} \quad g^{-n} = \underbrace{g^{-1} \cdot g^{-1} \cdots g^{-1}}_{n \text{ times}}.$$

2 If you have studied matrix arithmetic, you have encountered non-commutative multiplication when you learned how to multiply square matrices (and you might be interested in Exercise 10.23 in the end-of-chapter exercises).

A group G is *cyclic* if there is some $g \in G$ such that $G = \{g^i \mid i \in \mathbb{Z}\}$. The element g is called a *cyclic generator* of G.

PROPOSITION 10.21. *For any $n \in \mathbb{N}$ the group $(\mathbb{Z}_n, +)$ is cyclic.*

PROOF. The element $[1] \in \mathbb{Z}_n$ is a cyclic generator of $\mathbb{Z}_n$ since for any $[k] \in \mathbb{Z}_n$

$$\underbrace{[1] + [1] + \cdots + [1]}_{k \text{ times}} = [k] .$$

Thus every element in $\mathbb{Z}_n$ can be written as a power[3] of [1]. □

Exercise 10.13 Show that cyclic generators may not be unique. That is, find a cyclic group G and distinct elements g and $h \in G$ where g and h are both cyclic generators of G.

While the groups $(\mathbb{Z}_n, +)$ are all cyclic groups, the situation for $(\mathcal{U}_n, \cdot)$ is a bit more complicated. As an example, the group $\mathcal{U}_7$ is cyclic, since [3] is a cyclic generator:

$$\mathcal{U}_7 = \{[3], [2], [6], [4], [5], [1]\} = \{[3]^1, [3]^2, [3]^3, [3]^4, [3]^5, [3]^6\} .$$

However, $\mathcal{U}_8 = \{[1], [3], [5], [7]\}$ is not cyclic. To verify this, we simply check the possibilities:

$$\{[1]^i \mid i \in \mathbb{Z}\} = \{[1]\} \neq \mathcal{U}_8,$$
$$\{[3]^i \mid i \in \mathbb{Z}\} = \{[1], [3]\} \neq \mathcal{U}_8,$$
$$\{[5]^i \mid i \in \mathbb{Z}\} = \{[1], [5]\} \neq \mathcal{U}_8,$$
$$\{[7]^i \mid i \in \mathbb{Z}\} = \{[1], [7]\} \neq \mathcal{U}_8.$$

Exercise 10.14 Find two more examples like those shown above, one where $\mathcal{U}_n$ is cyclic and the other where $\mathcal{U}_n$ is not cyclic.

10.5.3. Order of a Group and an Element. The word "order" has two meanings in group theory, which context makes clear.

DEFINITION 10.22. If a group G has a finite number of elements, then this number is the *order* of G. It is denoted $|G|$. If G is not finite, we say that G is an infinite group.

For example, $|\mathbb{Z}_n| = n$ and $|\mathcal{U}_{15}| = 8$, while $(\mathbb{Z}, +)$ is an infinite group.

DEFINITION 10.23. If there is an $n \in \mathbb{N}$ such that $g^n = e$, then we say that g has *finite order*. In this case, the *order* of g is

$$|g| = \text{the smallest } n \in \mathbb{N} \text{ such that } g^n = e .$$

For example, the order of 1 in $(\mathbb{Z}_8, +)$ is 8, while the order of 2 in $(\mathbb{Z}_8, +)$ is 4, and the order of 4 in $(\mathbb{Z}_8, +)$ is 2. Notice that stating "$g \in G$ has order 1" is the same thing as saying "$g \in G$ is the identity."

3 It seems odd to say "product" or "power" when the binary operation is addition, but this is the accepted way to phrase such statements.

Exercise 10.15 Consider the Abelian group $\mathbb{Z}_3 \times \mathcal{U}_{15}$, where addition modulo 3 is the operation in the first coordinate and multiplication modulo 15 is the operation in the second coordinate.

(a) What is the order of $\mathbb{Z}_3 \times \mathcal{U}_{15}$?
(b) What is the order of the element $(1, 1)$?
(c) What is the order of the element $(1, 7)$?

10.6. Isomorphism

In Section 10.1 we showed that the Cayley tables for $\mathbb{Z}_4$ and $\mathcal{U}_5$ have the same underlying form, even though the names of the elements and the binary operations are different. The differences, though, are simply cosmetic and are not part of the algebraic structure of either group.

DEFINITION 10.24. Two groups G and H are *isomorphic* if there exists a bijection $f : G \to H$ such that

$$f(a \cdot_G b) = f(a) \cdot_H f(b)$$

for each $a, b \in G$. We write $G \approx H$ when G and H are isomorphic.

Notice that the product on the left side of the equation above comes from G while the product on the right side comes from H; the subscripts on the operations are presented here for emphasis but are unnecessary in practice. Thus the equation is essentially saying that you can apply the binary operation from G and then f, and you'll get the same result as when you first apply f to the individual elements and then the operation from H. The phrase associated to this equation is "f respects the binary operations." A bijection $f : G \to H$ that respects the binary operations is an *isomorphism*.

An isomorphism between $\mathbb{Z}_4$ and $\mathcal{U}_5$ is given by the function

$$\begin{array}{ccc} & f & \\ [0]_4 & \to & [1]_5 \\ [1]_4 & \to & [2]_5 \\ [2]_4 & \to & [4]_5 \\ [3]_4 & \to & [3]_5 \end{array}$$

The fact that f is a bijection is clear from its description. What remains to be shown is that $f(a + b) = f(a) \cdot f(b)$ for any $a, b \in \mathbb{Z}_4$. As just one example, we can establish that

$$f([0]_4 + [2]_4) = f([0]_4) \cdot f([2]_4),$$

which holds because

$$f([0]_4 + [2]_4) = f([2]_4) = [4]_5 = [1]_5 \cdot [4]_5 = f([0]_4) \cdot f([2]_4).$$

Going through a similar computation for all sixteen possible choices of a and b is a numbing proposition. In most situations, however, you do not need to do this. For this example we can use the fact that $[1]_4$ is a cyclic generator of $\mathbb{Z}_4$ and $[2]_5$ is a cyclic generator of $\mathcal{U}_5$. In fact, the function f has been defined by beginning with $f([1]_4) = [2]_5$ and then extending this via

$$f([k]_4) = f(\underbrace{[1]_4 + [1]_4 + \cdots + [1]_4}_{k \text{ times}}) = [2]_5^k .$$

To show $f(a+b) = f(a) \cdot f(b)$, we note that

$$f([x]_4 + [y]_4) = f([x+y]_4) = [2]_5^{x+y}$$

and

$$f([x]_4) \cdot f([y]_4) = [2]_5^x \cdot [2]_5^y.$$

Since $[2]_5^{x+y} = [2]_5^x \cdot [2]_5^y$, the claim follows.

Exercise 10.16 Construct an isomorphism between $\mathbb{Z}_6$ and $\mathbb{Z}_2 \times \mathbb{Z}_3$.

Exercise 10.17 Let $S = \{a, b\}$, and let $G = (\mathcal{P}(S), \triangle)$ as in Example 10.8. The order of G is 4. Prove that $G \approx \mathbb{Z}_2 \times \mathbb{Z}_2$.

We close with a useful lemma.

LEMMA 10.25. *If $f : G \to H$ is an isomorphism, then its inverse $f^{-1} : H \to G$ is an isomorphism as well.*

PROOF. We know that f^{-1} is a bijection, so it remains only to prove the fundamental identity

$$f^{-1}(a \cdot b) = f^{-1}(a) \cdot f^{-1}(b)$$

for every $a, b \in H$. By applying the fact that f is an isomorphism and that f^{-1} is its inverse function, we see

$$a \cdot b = f(f^{-1}(a)) \cdot f(f^{-1}(b)) = f\left(f^{-1}(a) \cdot f^{-1}(b)\right),$$

and applying f^{-1} to the first and last terms then gives us $f^{-1}(a \cdot b) = f^{-1}(a) \cdot f^{-1}(b)$. □

10.7. Isomorphism and Group Properties

In this section we describe ways in which isomorphic groups share key properties. We begin with the following result, which follows immediately from the requirement that an isomorphism from G to H is a bijection.

PROPOSITION 10.26. *If G and H are isomorphic finite groups, then $|G| = |H|$.*

More interesting is that every isomorphism takes the identity element to the identity element.

LEMMA 10.27. *Let $f : G \to H$ be an isomorphism, and let e_G and e_H be the identity elements of G and H respectively. Then $f(e_G) = e_H$.*

PROOF. Since e_G is the identity in G, we know $e_G \cdot e_G = e_G$. Applying f to just the left side of this equation we get

$$f(e_G \cdot e_G) = f(e_G) \cdot f(e_G).$$

We also know that $f(e_G \cdot e_G) = f(e_G)$, so

$$f(e_G) \cdot f(e_G) = f(e_G).$$

Multiplying both sides of this equation on the left by $[f(e_G)]^{-1}$ gives us

$$[f(e_G)]^{-1} \cdot f(e_G) \cdot f(e_G) = [f(e_G)]^{-1} \cdot f(e_G). \tag{10.1}$$

These multiplications are occurring in the group H, so

$$[f(e_G)]^{-1} \cdot f(e_G) = e_H.$$

Applying this to both sides of (10.1) then gives

$$f(e_G) = e_H.$$

□

It is also the case that isomorphisms take inverses to inverses:

LEMMA 10.28. *If $f : G \to H$ is an isomorphism, then $f(g^{-1}) = [f(g)]^{-1}$ for all $g \in G$.*

PROOF. The proof is similar to the prior argument, in that we will start by applying f to an equation involving inverses. Namely, since $g \cdot g^{-1} = e_G$, we know:

$$f(g \cdot g^{-1}) = f(e_G) \Rightarrow f(g) \cdot f(g^{-1}) = e_H.$$

Thus $f(g^{-1})$ is the inverse[4] of $f(g)$. Said symbolically, we conclude that $f(g^{-1}) = [f(g)]^{-1}$. □

LEMMA 10.29. *If $f : G \to H$ is an isomorphism, then $f(g^k) = [f(g)]^k$ for any $k \in \mathbb{Z}$. Here we are using the convention that a^0 is the identity element for any group element a.*

Exercise 10.18 Prove Lemma 10.29. To do this, you may want to consider three cases.

(a) Prove that, for each $n \in \mathbb{N}, f(g^n) = [f(g)]^n$ using an induction argument.
(b) Prove that $f(g^0) = [f(g)]^0$ by quoting a previous result.
(c) Prove that, for each $n \in \mathbb{N}, f(g^{-n}) = [f(g)]^{-n}$ using Lemma 10.28 and an induction argument.

THEOREM 10.30. *Let $f : G \to H$ be an isomorphism. Then for any $g \in G$ the order of g equals the order of $f(g) \in H$.*

PROOF. In this argument we assume that both elements have finite order. We leave the case where they might be of infinite order as Exercise 10.41.

4 If you are concerned that we might have found only the "right inverse," congratulations! But see Exercise 10.29.

We first show that $|f(g)| \leq |g|$. By Lemma 10.29 we know that

$$f(g^k) = [f(g)]^k$$

for any $k \in \mathbb{N}$. If $k = |g|$ then $g^k = e_G$. It follows that $f(g^k) = f(e_G) = e_H$. Thus $[f(g)]^k = e_H$ and so $|f(g)| \leq |g|$.

It is also true that $|g| \leq |f(g)|$. To prove this, let $|f(g)| = k$, so that $[f(g)]^k = e_H$ and thus $f(g^k) = e_H$. Since f is a bijection and f takes the identity element of G to the identity element of H (by Lemma 10.27), $g^k = e_G$ and so $|g| \leq |f(g)|$.

The only way we can have $|f(g)| \leq |g|$ and $|g| \leq |f(g)|$ is if $|g| = |f(g)|$. □

Exercise 10.19 In the proof of Theorem 10.30 we appealed to the fact that if $g^k = e$, then $|g| \leq k$. Why can't we claim that if $g^k = e$ then $|g| = k$?

In Exercise 10.43 we ask you to prove the following corollary.

COROLLARY 10.31. *Let $f : G \to H$ be an isomorphism. Then, for any $n \in \mathbb{N}$, f induces a bijection between the elements of order n in G and the elements of order n in H.*

THEOREM 10.32. *If $f : G \to H$ is an isomorphism, then G is cyclic if and only if H is cyclic.*

PROOF. It suffices to prove that G being cyclic implies H is cyclic, as that argument can then be used with $f^{-1} : H \to G$ to show that when H is cyclic, G must be cyclic as well.

Let x be a cyclic generator of G. Given any $h \in H$ there is a $g \in G$ such that $f(g) = h$, and so there must be an n such that $f(x^n) = h$. But $f(x^n) = \left[f(x)\right]^n$ by Lemma 10.29. Thus given any $h \in H$ there is an n such that $\left[f(x)\right]^n = h$, and so $f(x)$ is a cyclic generator of H. □

THEOREM 10.33. *If $f : G \to H$ is an isomorphism, then G is Abelian if and only if H is Abelian.*

Exercise 10.20 Prove Theorem 10.33.

10.8. Examples of Isomorphic and Non-isomorphic Groups

Discovering that two groups are isomorphic usually entails understanding how those groups are connected to each other, with a benefit often being that additional insights are gained for at least one if not both of the groups. Proving that two groups are not isomorphic also involves discovery, this time of group-theoretic properties that are not the same for the two groups. In this section we provide some examples to help illustrate these ideas.

In Exercise 10.16 you showed that $\mathbb{Z}_6$ is isomorphic to $\mathbb{Z}_2 \times \mathbb{Z}_3$. The key is to realize that $\mathbb{Z}_2 \times \mathbb{Z}_3$ is also a cyclic group, with $([1]_2, [1]_3)$ being one example of a cyclic generator. From this it is easy to see that the map sending $[n]_6 \in \mathbb{Z}_6$ to

$$\underbrace{([1]_2,[1]_3)+([1]_2,[1]_3)+\cdots+([1]_2,[1]_3)}_{n \text{ times}} \in \mathbb{Z}_2 \times \mathbb{Z}_3$$

is an isomorphism.

The following result generalizes Exercise 10.17.

PROPOSITION 10.34. *Let $S = \{a_1, \ldots, a_n\}$ be a finite set with n elements, and let $G = (\mathcal{P}(S), \triangle)$. Then*

$$G \approx \underbrace{\mathbb{Z}_2 \times \mathbb{Z}_2 \times \cdots \times \mathbb{Z}_2}_{n \text{ copies}}.$$

PROOF. For convenience and following standard conventions, we denote

$$\underbrace{\mathbb{Z}_2 \times \mathbb{Z}_2 \times \cdots \times \mathbb{Z}_2}_{n \text{ copies}}$$

by $\mathbb{Z}_2^n$. An arbitrary element of $\mathbb{Z}_2^n$ can be written in coordinate notation as $(x_1, \ldots, x_n)$ with each $x_i \in \mathbb{Z}_2$. Our putative isomorphism from $\mathbb{Z}_2^n$ to G is then defined by

$$f[(x_1, \ldots, x_n)] = \{a_i \mid x_i = 1\} \subseteq S.$$

If $(x_1, \ldots, x_n)$ and $(y_1, \ldots, y_n)$ are distinct elements of $\mathbb{Z}_2^n$, then they must differ in at least one coordinate, hence $f[(x_1, \ldots, x_n)]$ and $f[(y_1, \ldots, y_n)]$ are distinct subsets of S. Thus f is injective. To show that f is surjective, note that, given any subset $A \subseteq S$, the element $(x_1, \ldots, x_n) \in \mathbb{Z}_2^n$ defined by

$$x_i = \begin{cases} 1 & \text{if } a_i \in A, \\ 0 & \text{if } a_i \notin A \end{cases}$$

is mapped to A by f.

Finally, we need to establish that f respects the binary operations. First note that if $(z_1, \ldots, z_n) = (x_1, \ldots, x_n) + (y_1, \ldots, y_n)$, then $z_i = 1$ if and only if exactly one of x_i and y_i is 1 (and the other is 0). So

$$f[(x_1, \ldots, x_n) + (y_1, \ldots, y_n)] = \{a_i \mid x_i = 1 \text{ or } y_i = 1 \text{ but not } x_i = y_i = 1\}.$$

Let the images of $(x_1, \ldots, x_n)$ and $(y_1, \ldots, y_n)$ be

$$A = f[(x_1, \ldots, x_n)] = \{a_i \mid x_i = 1\}$$

and

$$B = f[(y_1, \ldots, y_n)] = \{a_i \mid y_i = 1\}.$$

Then

$$\begin{aligned} f[(x_1, \ldots, x_n)] \cdot f[(y_1, \ldots, y_n)] &= A \triangle B \\ &= \{a_i \mid x_i = 1 \text{ or } y_i = 1 \text{ but not } x_i = y_i = 1\}. \end{aligned}$$

Since this is the exact same description given for $f[(x_1, \ldots, x_n) + (y_1, \ldots, y_n)]$, we see that f respects the binary operations. □

These two examples give some suggestions as to how you might create an isomorphism between two groups. Establishing that two groups are not isomorphic requires finding distinguishing properties of the groups. For example, $\mathbb{Z}_{12}$ and $\mathbb{Z}_6 \times \mathbb{Z}_2$ are both Abelian groups of order 12, but they are not isomorphic. Theorem 10.32 tells us that the property of being cyclic is preserved by isomorphisms, and $\mathbb{Z}_{12}$ is cyclic while $\mathbb{Z}_6 \times \mathbb{Z}_2$ is not.

Let $\mathcal{H}$ be a regular hexagon and let $\mathrm{SYM}(\mathcal{H})$ be its symmetry group. Like the symmetry group of a square discussed earlier in this chapter, any symmetry of $\mathcal{H}$ is determined by what it does to adjacent corners. An analysis almost exactly like that building up to Proposition 10.4 shows that $\mathrm{SYM}(\mathcal{H})$ has twelve elements: six reflections, five non-trivial rotations, and the identity element.

Exercise 10.21 Find two elements of $\mathrm{SYM}(\mathcal{H})$ that do not commute.

Like $\mathrm{SYM}(\mathcal{H})$, the groups $\mathbb{Z}_{12}$ and $\mathbb{Z}_6 \times \mathbb{Z}_2$ have order 12. But $\mathrm{SYM}(\mathcal{H})$ cannot be isomorphic to $\mathbb{Z}_{12}$ or $\mathbb{Z}_6 \times \mathbb{Z}_2$ or any other Abelian group of order 12, because you have just shown that $\mathrm{SYM}(\mathcal{H})$ is not Abelian, and in Exercise 10.20 you proved that isomorphic groups are either both Abelian or both non-Abelian.

Thus the properties of being cyclic and Abelian are sufficient to distinguish these three groups of order 12.

REMARK 10.35. If you enjoyed seeing how to distinguish these three groups of order 12, you should tackle Project 11.11, where you will identify and distinguish all of the groups of order 8.

GOING BEYOND THIS BOOK. Marcia Ascher's book *Ethnomathematics* [**Asc91**] contains two chapters that describe ways that groups arise in the context of anthropology. One is on "The logic of kin relations" and the other is "Symmetric strip patterns." These chapters, and the whole book in fact, are fascinating.

10.9. End-of-Chapter Exercises

Exercises you can work on after Sections 10.1 and 10.2

10.22 Lemma 10.2 states that symmetries in $\text{SYM}(\mathcal{Q})$ are determined by what they do to the points $(1, 1)$ and $(-1, 1)$. Prove or disprove the following variations on the statement of this lemma.

(a) Any symmetry of $\mathcal{Q}$ is determined by what it does to the points $(1, 1)$ and $(1, -1)$.
(b) Any symmetry of $\mathcal{Q}$ is determined by what it does to the points $(1, 1)$ and $(-1, -1)$.
(c) Any symmetry of $\mathcal{Q}$ is determined by what it does to the points $(0.1, 0.8)$ and $(0.7, 0.3)$.

10.23 In Section 10.2 we described the eight symmetries of a square. The rotation ρ was shown to be given by the formula $\rho[(x, y)] = (-y, x)$. If you are familiar with matrix multiplication you could also describe ρ as being given by the matrix

$$\rho = \begin{bmatrix} 0 & -1 \\ 1 & 0 \end{bmatrix}$$

and the function is then given by matrix multiplication

$$\begin{bmatrix} 0 & -1 \\ 1 & 0 \end{bmatrix}\begin{pmatrix} x \\ y \end{pmatrix} = \begin{pmatrix} -y \\ x \end{pmatrix}.$$

Determine the matrices associated with the other seven symmetries, and verify in at least two interesting cases that matrix multiplication yields the same results as function composition.

10.24 Let Δ be an equilateral triangle. The goal of this exercise is to show that $\text{SYM}(\Delta)$ contains six functions.

(a) Prove there are at most six symmetries of Δ.
(b) Find three reflections in $\text{SYM}(\Delta)$ and name them.
(c) Find two rotations in $\text{SYM}(\Delta)$ and name them.
(d) Show that you now have a collection of six distinct symmetries of Δ, meaning that you have identified all of the elements in $\text{SYM}(\Delta)$.

10.25 Let $\mathcal{O}$ be a regular octagon, and consider the group of symmetries $\text{SYM}(\mathcal{O})$.

(a) Describe all the elements of $\text{SYM}(\mathcal{O})$, and prove that your list is complete.
(b) Let ρ be a counterclockwise rotation through 45° and let f be any reflection in $\text{SYM}(\mathcal{O})$. Prove that $\rho^i \circ f = f \circ \rho^{8-i}$ for all i.

Exercises you can work on after Sections 10.3 and 10.4

10.26 In this problem you will describe the symmetries of a cube $\mathcal{C}$.

(a) Prove that a symmetry of $\mathcal{C}$ is determined by what it does to any face of $\mathcal{C}$.
(b) Prove that $\text{SYM}(\mathcal{C})$ consists of 48 functions.
(c) Describe the 48 symmetries of $\mathcal{C}$.

10.27 Prove Theorem 10.11.

10.28 There is a group with only one element, called the *trivial* group. Describe the structure of the trivial group, and verify that it is indeed a group.

10.29 Let $a \in G$ be an element of a group G, and let $e \in G$ be the identity element. Show that if there are elements g and h in G such that

$$a \cdot g = h \cdot a = e,$$

then $g = h$. Thus there is no reason to distinguish "left inverses" and "right inverses."

10.30 Create the Cayley table for $\mathbb{Z}_3 \times \mathbb{Z}_3$.

10.31 Create the Cayley table for $\text{SYM}(\Delta)$, described in Exercise 10.24.

10.32 Let G be any group. Prove that every element of G shows up once and only once in each row and each column of its Cayley table.

Exercises you can work on after Section 10.5

10.33 Let G and H be groups.

(a) Prove that if G and H are both Abelian, then their Cartesian product $G \times H$ is also Abelian.
(b) Prove the converse: If $G \times H$ is Abelian, then G and H are Abelian.

10.34 Prove that $\mathbb{Z}_8 \times \mathbb{Z}_3$ is a cyclic group.

10.35 Prove or disprove: $\mathbb{Z}_6 \times \mathbb{Z}_{10} \times \mathbb{Z}_{15}$ is a cyclic group.

10.36 Let $G = \text{SYM}(Q) \times \mathbb{Z}_5$.

(a) What is $|G|$?
(b) Find an element in G whose order is 10.
(c) Find an element in G whose order is 20.
(d) Prove that G is not cyclic.

10.37 Let G be an Abelian group of even order. Prove that some non-identity element must be its own inverse. That is, there is some $a \in G$ with $a \neq e$ and $a^2 = e$.

10.38 Here are some examples where $\text{SYM}(X)$ is infinite.

(a) Let X be the set of integer points inside of $\mathbb{R}$. Prove that $\text{SYM}(X)$ is a countably infinite group.
(b) Is $\text{SYM}(\mathbb{R})$ countable or uncountable?
(c) Let $\mathcal{L}$ be the integer grid in $\mathbb{R}^2$, that is,

$$\mathcal{L} = \{(x, y) \mid x \in \mathbb{Z} \text{ or } y \in \mathbb{Z}\}.$$

Is $\text{SYM}(\mathcal{L})$ countable or uncountable?

This problem continues with Exercise 10.50.

10.39 Let Δ be a regular tetrahedron – one where all the edges have the same length – with vertices $\{a, b, c, d\}$, as shown in Figure 6. The line segments joining vertices are referred to as the *edges* of Δ; the four triangles associated to triples of vertices are the *faces* of Δ.

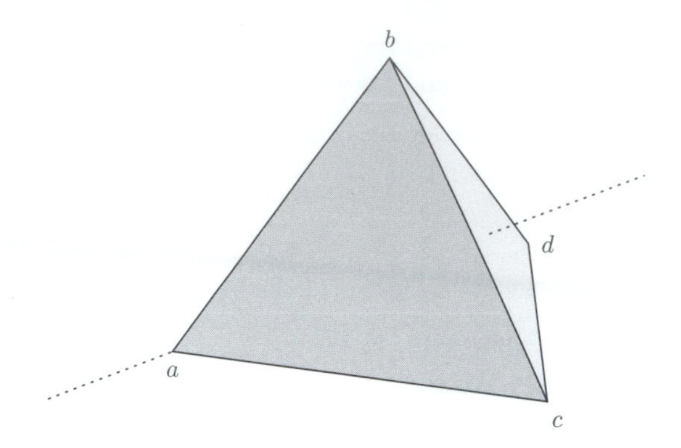

Figure 6. A 120° rotation around the axis passing through vertex a produces a symmetry of the regular tetrahedron.

Let $\mathrm{SO}(\Delta)$ be all of the rotational symmetries of Δ that you can perform on a solid, physical model of Δ.[5] For example, you can hold the vertex a fixed and rotate the face of Δ opposite of a by 120°, which has the effect of permuting the vertices b, c, and d.

(a) Prove that $\mathrm{SO}(\Delta)$ contains at most 12 elements.
(b) Find 12 distinct elements in $\mathrm{SO}(\Delta)$, thus proving that $|\mathrm{SO}(\Delta)| = 12$.
(c) Create the Cayley table for $\mathrm{SO}(\Delta)$, and highlight entries that demonstrate that $\mathrm{SO}(\Delta)$ is not Abelian.

Exercises you can work on after Sections 10.6–10.8

10.40 Prove

$$\mathbb{Z}_{30} \approx \mathbb{Z}_6 \times \mathbb{Z}_5 \approx \mathbb{Z}_3 \times \mathbb{Z}_{10} \approx \mathbb{Z}_2 \times \mathbb{Z}_3 \times \mathbb{Z}_5 .$$

10.41 Assume that $f : G \to H$ is an isomorphism. Prove that if $g \in G$ is an element of infinite order, then so is $f(g) \in H$.

10.42 Let Δ be a tetrahedron where two non-intersecting edges of Δ have length 2, while the remaining four edges have length $\sqrt{3}$. (See Figure 7.) If you would like a particular set of coordinates for the vertices, use $\{(1,0,0), (-1,0,0), (0,1,1), (0,-1,1)\}$. Since this tetrahedron was pointed out to us by Jon McCammond, we will refer to it as *McCammond's Tetrahedron*. Prove that the symmetry group of McCammond's Tetrahedron is isomorphic to the symmetry group of the square $\mathcal{Q}$.

10.43 Prove Corollary 10.31.

10.44 Use Corollary 10.31 to prove that $\mathrm{SYM}(\mathcal{H})$ and $\mathrm{SO}(\Delta)$ – the symmetries of a regular hexagon and the rotations of a regular tetrahedron – are not isomorphic.

10.45 Let $\mathcal{J}$ be the toy jack shown in Figure 8.

(a) Prove that the order of $\mathrm{SYM}(\mathcal{J})$ is 16.

5 The name we have chosen for this group, $\mathrm{SO}(\Delta)$, references a general class of groups including the "rotation group" that is commonly denoted $\mathrm{SO}(3)$. The group $\mathrm{SO}(3)$ is the group of all rotations in $\mathbb{R}^3$ that fix the origin.

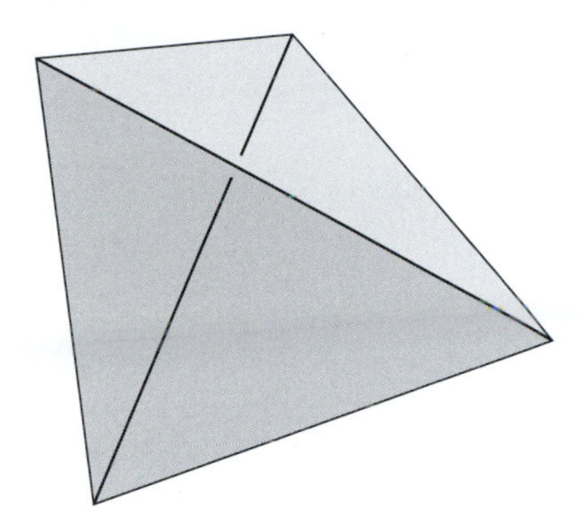

Figure 7. McCammond's Tetrahedron Δ with two sets of congruent edges.

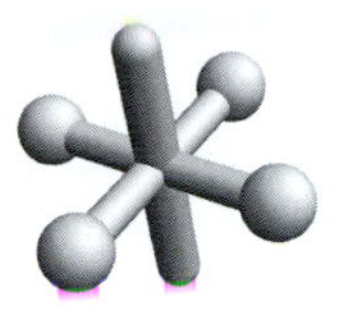

Figure 8. What is the symmetry group of a toy jack?

(b) The symmetry group of a toy jack and the symmetry group of an octagon are both non-Abelian groups of order 16. Prove that these two groups are not isomorphic.

More Exercises!

10.46 Let $\mathcal{C}$ be the circle $\{(x, y) \mid x^2 + y^2 = 1\} \subset \mathbb{R}^2$. Prove that $\text{Sym}(\mathcal{C})$ is an uncountable, non-Abelian group.

10.47 Prove that "isomorphic" is an equivalence relation on the set of all groups.

10.48 There are five Abelian groups of order 16. List them, and prove that no two groups on your list are isomorphic.

10.49 In this problem we focus on the collection of all finite groups.

(a) For a fixed $n \in \mathbb{N}$, prove there can be only finitely many groups of order n.
(b) Prove that the collection of all finite groups is countably infinite.

10.50 In this problem we explore elements of the groups introduced in Exercise 10.38.

(a) Let X be the set of integer points inside of $\mathbb{R}$. Prove that if $g \in \text{Sym}(X)$, then $|g| = 1$ or 2, or g has infinite order.
(b) Let $\mathcal{L}$ be the integer grid in $\mathbb{R}^2$, that is,

$$\mathcal{L} = \{(x, y) \mid x \in \mathbb{Z} \text{ or } y \in \mathbb{Z}\}.$$

What can you say about the orders of the elements in $\text{Sym}(\mathcal{L})$?

11 Projects

The projects on the following pages are intended to guide you in the exploration of various topics in mathematics. While each project provides steps that guide you toward proofs of the theorems in each section, these projects can be quite demanding. At the start of each project we indicate the background that is needed in order to tackle the topic being presented.

11.1. The Pythagorean Theorem

The Pythagorean Theorem is often referred to as the "first great theorem," so it is very suitable to examine it, even if you have only read Chapter 1. We begin by presenting one proof of the theorem, and then asking you to write up additional proofs and consider ways to extend the result.

THEOREM 11.1. *Let a and b be the lengths of the two shortest sides of a triangle $\mathcal{T}$, and let c be the length of the longest side. Then $\mathcal{T}$ is a right triangle if and only if*

$$a^2 + b^2 = c^2.$$

The history of this theorem is long and convoluted. It may or may not have been proven by Pythagoras of Samos (circa 500 BCE); it is a highlight of the first book of Euclid's *Elements*; and it certainly was known in ancient China. In fact, Figure 1 appears in *Arithmetic Classic of the Gnomon and the Circular Paths of Heaven* (China, circa 600 BCE). The accompanying argument is written using the specific values of 3, 4, and 5, but it is easy to use modern notation and convert this ancient argument-by-example into what would be accepted as a modern mathematical proof.

A PROOF OF THE PYTHAGOREAN THEOREM. Let $\mathcal{T}$ be a right triangle, with sides of length a, b, and c, where for convenience[1] we have $a \leq b \leq c$. Then you can arrange four copies of $\mathcal{T}$ as in Figure 1, where the right angles are in the interior of the figure, and the four hypotenuses form a bounding quadrilateral. Call this quadrilateral $\mathcal{F}$, since it has $\mathcal{F}$our sides.

Because the sum of the interior angles in any triangle is 180°, the sum of the two non-right angles of $\mathcal{T}$ is 90°. At each corner of the boundary quadrilateral $\mathcal{F}$ there is a copy of each of these non-right interior angles, hence at each corner of $\mathcal{F}$ there is a right angle. Thus $\mathcal{F}$ is a rectangle. Furthermore, each side of $\mathcal{F}$ comes from the hypotenuse of a copy of $\mathcal{T}$, so $\mathcal{F}$ is a rectangle with four congruent sides, hence $\mathcal{F}$ is a square. The area of $\mathcal{F}$ is then c^2. Thinking of $\mathcal{F}$ as being composed of four triangles and a smaller

1 Why is it important for us to say "for convenience" when making this assumption? Why is it also a legitimate step?

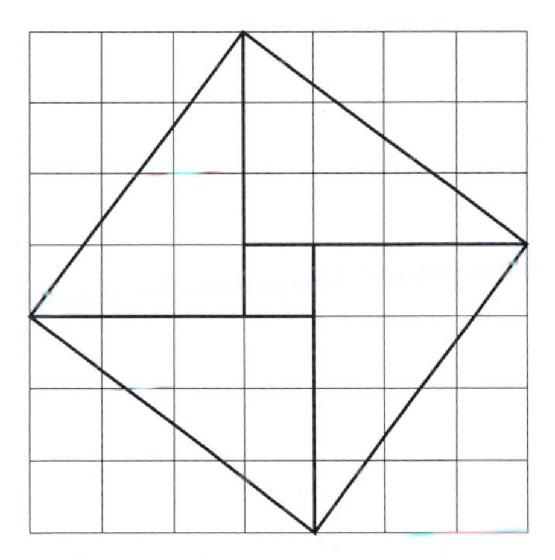

Figure 1. The illustration accompanying an ancient Chinese proof of the Pythagorean Theorem.

interior square, we know the area of $\mathcal{F}$ can also be expressed as $4\left(\frac{1}{2}ab\right) + (b-a)^2$. This implies

$$\begin{aligned} c^2 &= 4\left(\frac{1}{2}ab\right) + (b-a)^2 \\ &= 2ab + b^2 - 2ab + a^2 \\ &= a^2 + b^2. \end{aligned}$$

Notice, however, that we have not yet established the full Pythagorean Theorem. This argument shows only half of it: that the lengths of the three sides of a right triangle have a certain arithmetic property. The other fact claimed in the Pythagorean Theorem, the other half of the "if and only if" statement, is equally important: triangles with this arithmetic property are right triangles. This fact was used by the ancient Egyptians in surveying and construction, and it is still used by carpenters to construct right angles.

To prove the other direction, assume we have a triangle $\mathcal{T}$ where $a^2 + b^2 = c^2$. Let $\mathcal{T}^{\perp}$ be a right triangle with sides a and b and hypotenuse h. By the argument above we know $h^2 = a^2 + b^2$, so $h = c$. Thus $\mathcal{T}$ and $\mathcal{T}^{\perp}$ are congruent, and therefore $\mathcal{T}$ is a right triangle. □

You do not need to write multiple proofs to establish the validity of a mathematical theorem. But multiple proofs can provide additional insights, and there are many fun and insightful proofs of the Pythagorean Theorem. In the following example and exercise, we are proving only that the sides of right triangles satisfy the arithmetic condition.

President James Garfield constructed a proof of the Pythagorean Theorem, long before he became President of the United States. His argument begins by forming a trapezoid using two copies of a right triangle, arranged so that side a of one copy and side b of the other copy lie on a line, as in Figure 2.

Exercise 11.1 Verify that Garfield's trapezoid is composed of three right triangles. Then establish the arithmetic identity $a^2 + b^2 = c^2$ by comparing the area formula

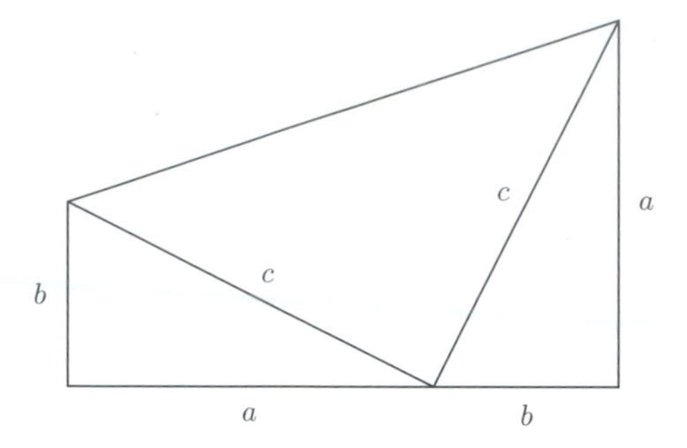

Figure 2. The trapezoid used in President Garfield's proof of the Pythagorean Theorem.

for a trapezoid with the sum of the areas of the three triangles that comprise Garfield's trapezoid.

Exercise 11.2 Figure 3 is commonly used in proving the Pythagorean Theorem. Use this figure as inspiration to construct yet another proof of the Pythagorean Theorem.

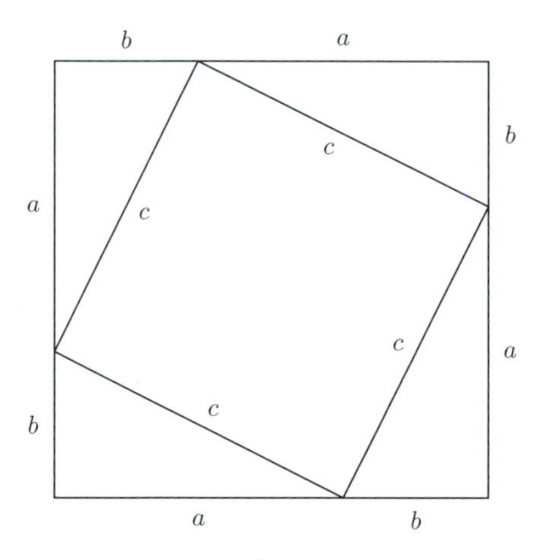

Figure 3. Another arrangement that leads to a proof of the Pythagorean Theorem.

Exercise 11.3 Find one additional proof of the Pythagorean Theorem, so that you will have seen at least four proofs of this result. You can try to find this proof on your own, but we are really only asking you to consult available resources and read a fourth proof. Which of these proofs do you think give the most insight into why the theorem is true? Explain!

Exercise 11.4 Let A, B, and C be the three vertices of a triangle Δ, where the side opposite to A is a, the side opposite to B is b, and the side opposite to C is c. If the angle at C is less than $90°$, what can you conclude, if anything, about the two values $a^2 + b^2$ and c^2? Why? What if the angle at C is greater than $90°$?

Exercise 11.5 Let x, y, and z be three positive real numbers, and let Δ be the tetrahedron in $\mathbb{R}^3$ formed by the points

$$O = (0,0,0),$$
$$X = (x,0,0),$$
$$Y = (0,y,0),$$
$$Z = (0,0,z).$$

Because one corner of Δ corresponds to a corner of an octant in $\mathbb{R}^3$, Δ is a three-dimensional analogue of a right triangle. In fact, a Pythagorean-style formula holds, connecting the areas of the four faces of Δ. Let A, B, and C be the areas of the faces that touch the origin, and let D be the area of the triangle formed by the points X, Y, and Z. Prove that

$$A^2 + B^2 + C^2 = D^2 .$$

11.2. Chomp and the Divisor Game

Chomp is a game that is easy to describe and play, and which is amenable to mathematical analysis. You should be able to work on this project after having worked through the first two chapters of this book. One of the questions makes use of a result from Chapter 4, but it is not an essential question that is necessary for the successful completion of this project. Chomp appears in Exercise 3.86, which is a nice problem, but not one that is necessary for this project.

Chomp is a fun game for two players, who we'll call Alice and Bob. They start with an $m \times n$ rectangular "cake," such as the 4×6 cake shown in Figure 4.

The players alternate turns, with Alice going first. On a turn, the player takes a right-angled "bite" from the northeast: a square is removed along with all other squares to the north and east and anywhere in between these directions. In Figure 4 the darkest shading shows the cake remaining after Alice first takes a bite (no shading), followed by a bite by Bob (lighter shading).

The loser of this game is the person who is forced to take the last bite. In the mathematical study of two player strategy games, here's an important question: If both players use optimal strategy, who wins?

Exercise 11.6 Play Chomp with a friend on boards of varying sizes, such as 2×3, 2×6, 3×3, and 5×5. Play several games of each size, and let each player be Alice half of the time. Make conjectures about which board sizes favor Alice, and which favor Bob. Are there certain boards that you think are guaranteed wins for Alice if she plays optimally? Guaranteed wins for Bob?

Exercise 11.7 Prove that Alice has a winning strategy for any $1 \times n$ or $n \times 1$ board, when $n \geq 2$.

Exercise 11.8 For this exercise you should focus on the case of a 2×3 game of Chomp. Label each of the six squares with an "A" or a "B" depending on who has a winning strategy if Alice's first bite is based at that square. We have begun this process

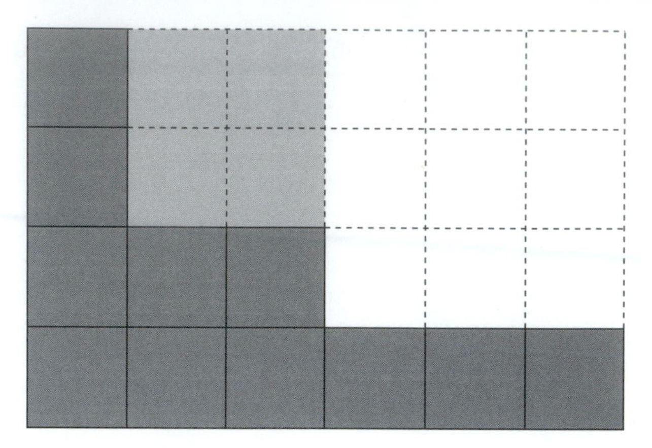

Figure 4. Two plays into the game Chomp. The nine unshaded squares were eaten first; the four lightly shaded squares were eaten second.

in Figure 5. Verify that our two labels are correct, and then label the remaining four squares.

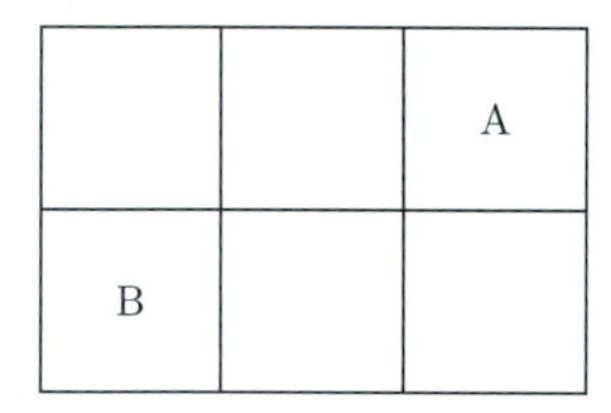

Figure 5. Beginning to label a 2×3 game of Chomp. The "A" indicates that if Alice bites this position, then she has a winning strategy. The "B" indicates that if Alice bites in this position, then Bob has a winning strategy.

Exercise 11.9 Prove that, for any $m \times n$ board, Alice has a winning strategy so long as $mn > 1$. One way to do this is via a proof by contradiction, starting with the assumption that Bob has a winning strategy, and then showing that Alice could always make use of that strategy herself.

Exercise 11.10 Here's a related game called the Divisor Game, and you will need the Fundamental Theorem of Arithmetic (Theorem 4.11) in order to fully analyze this game. Alice and Bob start with a natural number n. They alternate turns in which a player calls out a natural number that divides n, with the following restriction: each number called out must divide n, but it *can't* be a multiple of any number that has already been called out. The loser is the person forced to say 1.

Here's an example game for $n = 30$:

> Alice starts by saying 10; then Bob says 2; then Alice says 15; then Bob says 3; then Alice says 5; and then Bob is forced to say 1. Alice wins!

Who wins for $n = 45$? For $n = 60$? For any n?

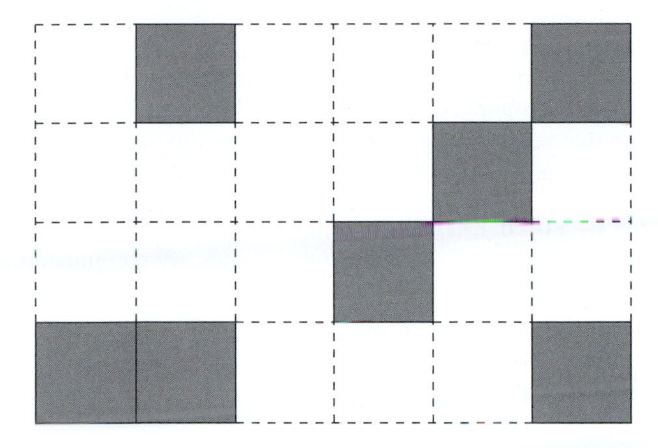

Figure 6. A "cake" made of seven squares.

The game of Chomp can be generalized in many ways. One generalization is to remove the assumption that you start with a rectangular cake, so that you allow Alice and Bob to play Chomp on a cake *of any shape* made up of n 1×1 squares in the square lattice. An example of the sort of cake we are allowing is shown in Figure 6. As in the original version, a bite removes some remaining square and all other squares in the northeast quadrant from it, and the player forced to remove the last square loses. In Figure 6, Alice can win by eating the square adjacent to the southwest corner square on her first turn.

Exercise 11.11 Consider the $n = 2$ case and the $n = 3$ case.

(a) When $n = 2$, Alice has two choices of which square to eat, and she does not want to eat a square that would remove both squares, as then Bob would win. By examining the possible configurations of two squares, show that Alice always has a choice that will leave one remaining square, meaning that Alice has a winning strategy.

(b) Create a configuration for $n = 3$ where Alice has a winning strategy and a different configuration for $n = 3$ where Bob has a winning strategy.

In this more general situation we cannot give a simple statement of which player has a winning strategy. We do, however, know that one of them does have a winning strategy.

THEOREM 11.2. *For any $n \in \mathbb{N}$ and any cake made up of n 1×1 squares in the square lattice, one of the two players has a winning strategy for the game of Chomp beginning with that configuration.*

This is an example of a more general result about certain types of two-player games: you may not know *who* has the winning strategy, but you know that someone does!

Exercise 11.12 Here we outline a proof by induction for Theorem 11.2. Start by establishing the base case when $n = 1$.

For the inductive step, assume that for any value of k such that $1 \leq k \leq n$, someone has a winning strategy for each possible cake shape made up of k little 1×1 squares.

Now, consider Alice eating first from a cake made up of $n+1$ little 1×1 squares. Since after any bite she makes there will be no more than n little 1×1 squares left, by the inductive assumption we know that either Alice or Bob has a winning strategy for the cake that remains after her first bite. Use this insight to prove that either Alice or Bob must have a winning strategy for the original cake.

GOING BEYOND THIS BOOK. If you would like to know more about Chomp, or the mathematics underlying many other games, we recommend *Winning Ways for Your Mathematical Plays* by Berlekamp, Conway, and Guy [**BCG01**]. Over the years there have been different editions and volumes of this work – we have listed some in the bibliography – and they are all excellent.

11.3. Arithmetic–Geometric Mean Inequality

The arithmetic–geometric mean inequality is a fundamental result in analysis. The two-variable version was stated in Exercise 1.1: *If a and b are two non-negative real numbers, then*

$$\frac{a+b}{2} \geq \sqrt{ab}\,.$$

In this project we will guide you through a proof of this result as well as its natural generalization to any finite number of non-negative real numbers. The argument is a somewhat complicated application of mathematical induction, and so we recommend having worked though Chapters 1 and 2 before beginning this project.

There are a number of ways to compute an average for a given a set of positive numbers, $\{a_1, \ldots, a_n\}$. The most common method is to add them up and divide:

$$\mu = \frac{a_1 + a_2 + \cdots + a_n}{n}\,.$$

This is the *arithmetic mean* of $\{a_1, \ldots, a_n\}$. Another approach is to multiply and then take a root:

$$\gamma = \sqrt[n]{a_1 a_2 \cdots a_n}\,.$$

This is the *geometric mean*. We note that the arithmetic mean can easily be applied to a mix of positive and negative numbers, while the geometric mean cannot. Hence for the remainder of this project we will be limiting ourselves to non-negative real numbers.

Exercise 11.13 Gather five or more friends. Give each of them n randomly chosen, non-negative real numbers, where n is some integer greater than 2. Your friends' job is to compute the arithmetic and geometric means of the numbers you gave them. We will wager that in all cases $\frac{a_1+\cdots+a_n}{n} > \sqrt[n]{a_1 \cdots a_n}$. If you can't find five friends willing to help you with these computations, perhaps you could have a computer assist you in collecting evidence.

You should have found that the arithmetic means were always at least as large as the geometric means of your random sets of real numbers chosen for Exercise 11.13, and that the arithmetic mean was usually strictly larger than the geometric mean. The *arithmetic–geometric mean inequality* says that the results of Exercise 11.13 were not random. If you didn't get such results, some of your friends probably need a refresher course on arithmetic.

THEOREM 11.3. *For any n non-negative real numbers* $\{a_1, a_2, \ldots, a_n\}$,

$$\frac{a_1 + a_2 + \cdots + a_n}{n} \geq \sqrt[n]{a_1 \cdot a_2 \cdots a_n}\,. \tag{11.1}$$

Further, the inequality is strict unless $a_1 = a_2 = \cdots = a_n$.

The second statement, that the inequality is strict, means that

$$\frac{a_1 + a_2 + \cdots + a_n}{n} > \sqrt[n]{a_1 \cdot a_2 \cdots a_n}$$

whenever there is an a_i that is not equal to some a_j in $\{a_1, a_2, \ldots, a_n\}$.

You will establish the arithmetic–geometric mean inequality via two lemmas.

LEMMA 11.4. *Inequality 11.1 holds for all* $n = 2^k$, $k \in \mathbb{N}$. *Further, when* $n = 2^k$ *the inequality is strict whenever* $a_i \neq a_j$ *for some* $i \neq j$.

The proof of Lemma 11.4 is by induction on the exponent k.

Exercise 11.14 Begin by establishing the base case, when $k = 1$, of the inequality. Note that in this case $a_1 = \alpha^2$ and $a_2 = \beta^2$ for some $\alpha, \beta \in \mathbb{R}$. Explain why it suffices to establish

$$\frac{\alpha^2 + \beta^2}{2} \geq \alpha\beta\,,$$

for all $\alpha, \beta \in \mathbb{R}$. And then use algebra to show that this is equivalent to showing that

$$(\alpha - \beta)^2 \geq 0\,.$$

Since squares of real numbers are non-negative, the result follows.

Exercise 11.15 Explain how the second claim, that the inequality is usually strict, follows from the fact that when $\alpha \neq \beta$, $(\alpha - \beta)^2 > 0$.

For the inductive step, assume the inequality holds for $n = 2^k$ up to some $k \geq 1$, and let $m = 2n = 2^{k+1}$. We need to establish for any list of $2n$ non-negative real numbers, the inequality

$$\frac{a_1 + a_2 + \cdots + a_{2n}}{2n} \geq (a_1 \cdot a_2 \cdots a_{2n})^{1/2n}$$

holds.

Exercise 11.16 Define $b_1 = \frac{a_1 + a_2}{2}$ and more generally, $b_i = \frac{a_{2i-1} + a_{2i}}{2}$. Since each $b_i \geq 0$, our induction hypothesis says

$$\frac{b_1 + b_2 + \cdots + b_n}{n} \geq (b_1 \cdot b_2 \cdots b_n)^{1/n}\,.$$

Use the induction hypothesis and the fact that $b_i = \frac{1}{2}(a_{2i-1} + a_{2i})$ to establish the inequality.

Exercise 11.17 Prove that the inequality will be strict should $a_i \neq a_j$ for some $i \neq j$.

The second lemma in the proof of Theorem 11.3 works backwards, and can be used to fill in the gaps between 2^k and 2^{k+1}.

LEMMA 11.5. *If Inequality 11.1 holds for a natural number n, then it also holds for* $n-1$.

Exercise 11.18 We want to show that the inequality holds for a list of non-negative numbers a_1 to a_{n-1}, assuming it holds for lists with one additional number. Define

$$\mu = \frac{a_1 + \cdots + a_{n-1}}{n-1},$$

the arithmetic mean of the original list of numbers. Then start from the assumption that

$$\frac{a_1 + a_2 + \cdots + a_{n-1} + \mu}{n} \geq (a_1 \cdot a_2 \cdots a_{n-1} \cdot \mu)^{1/n}$$

to establish Inequality 11.1 for the original list of $n-1$ non-negative real numbers.

Exercise 11.19 Verify that you continue to get a strict inequality as long as some $a_i \neq a_j$.

We can now establish the arithmetic–geometric mean inequality.

PROOF OF THEOREM 11.3. Given any $n \in \mathbb{N}$ there is a $k \in \mathbb{N}$ such that $n \leq 2^k$. Lemma 11.4 shows that Inequality 11.1 holds for lists of length 2^k; Lemma 11.5 then shows that the inequality holds for lists of length n. See Figure 7. □

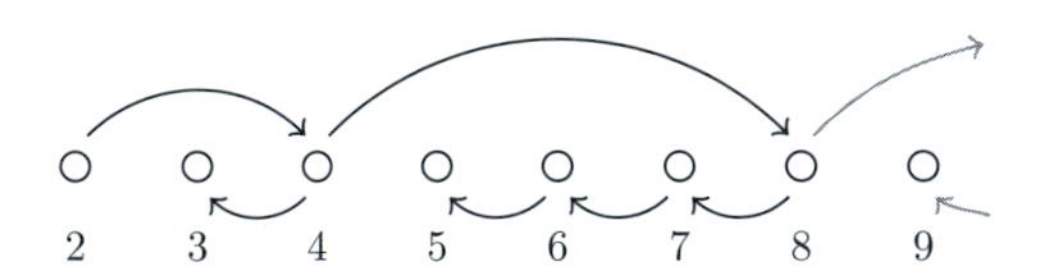

Figure 7. The structure of the proof of the arithmetic–geometric mean inequality. The arrows pointing right are applications of Lemma 11.4, while the ones pointing left are Lemma 11.5.

GOING BEYOND THIS BOOK. This proof of the arithmetic–geometric mean inequality is essentially due to Augustin Cauchy. Our presentation in this project is heavily inspired by *An Introduction to Inequalities* by Beckenbach and Bellman [**BB61**]. If this section strikes you as interesting, we strongly recommend you take a look at that book.

11.4. Complex Numbers and the Gaussian Integers

This project introduces the complex numbers, $\mathbb{C}$, and focuses on the Gaussian integers, $\mathbb{Z}[\mathbf{i}]$. Anyone who has worked through the first four chapters should be able to work on this project. We do use some of the language of functions, which is covered in Chapter 5, but our use of this language is minimal and we expect that it is not necessary to have completed Chapter 5 before taking up this project.

The complex numbers $\mathbb{C}$ are formed by adding $\mathbf{i} = \sqrt{-1}$ to the real numbers. Every element of $\mathbb{C}$ can be expressed as $a + b\mathbf{i}$, where $a, b \in \mathbb{R}$. Addition, subtraction, and multiplication in $\mathbb{C}$ are defined keeping in mind that $\mathbf{i} \notin \mathbb{R}$ and $\mathbf{i}^2 = -1$. Thus

$$(a + b\mathbf{i}) \pm (c + d\mathbf{i}) = (a \pm c) + (b \pm d)\mathbf{i}, \text{ and}$$
$$(a + b\mathbf{i}) \cdot (c + d\mathbf{i}) = (ac - bd) + (ad + bc)\mathbf{i}\,.$$

The conjugate of a complex number changes the sign of the coefficient of $\mathbf{i}$. That is, the conjugate of $a + b\mathbf{i}$ is $a - b\mathbf{i}$.[2] Conjugation helps explain division by non-zero complex numbers. In general, as long as $c + d\mathbf{i} \neq 0 + 0\mathbf{i}$,

$$\frac{a + b\mathbf{i}}{c + d\mathbf{i}} = \frac{a + b\mathbf{i}}{c + d\mathbf{i}} \cdot \frac{c - d\mathbf{i}}{c - d\mathbf{i}} = \frac{(a + b\mathbf{i})(c - d\mathbf{i})}{c^2 + d^2}\,.$$

For example,

$$\frac{1 - 2\mathbf{i}}{3 + 4\mathbf{i}} = \frac{1 - 2\mathbf{i}}{3 + 4\mathbf{i}} \cdot \frac{3 - 4\mathbf{i}}{3 - 4\mathbf{i}} = \frac{-5 - 10\mathbf{i}}{25} = -\frac{1}{5} - \frac{2}{5}\mathbf{i}\,.$$

The complex numbers are usually represented by points in the plane, where $a + b\mathbf{i}$ corresponds to the point $(a, b) \in \mathbb{R}^2$. The distance from the origin to (a, b) is $\sqrt{a^2 + b^2}$, and so the product of $a + b\mathbf{i}$ with its conjugate $a - b\mathbf{i}$ is the square of the distance to the origin. As we shall see, this norm

$$N(a + b\mathbf{i}) = a^2 + b^2$$

is a very useful function from $\mathbb{C}$ to $\mathbb{R}$.

Exercise 11.20 The *Gaussian integers* are the subset of the complex numbers consisting of all integer combinations of 1 and $\mathbf{i} = \sqrt{-1}$. That is,

$$\mathbb{Z}[\mathbf{i}] = \{a + b\mathbf{i} \mid a, b \in \mathbb{Z}\} \subset \mathbb{C}\,.$$

Show that the Gaussian integers are closed under addition, subtraction, and multiplication.

Exercise 11.21 If $\alpha, \beta \in \mathbb{Z}[\mathbf{i}]$, we say α *divides* β if there is a $\gamma \in \mathbb{Z}[\mathbf{i}] - \{0 + 0\mathbf{i}\}$ with $\alpha \cdot \gamma = \beta$. As you might expect, we denote this $\alpha | \beta$, and when α does not divide β we write $\alpha \nmid \beta$.

(a) Show that $(2 + \mathbf{i}) | (-6 + 7\mathbf{i})$.
(b) Show that the only divisors of $2 + \mathbf{i}$ are $\pm 1, \pm\mathbf{i}$, and $\pm 2 \pm \mathbf{i}$.
(c) Show that $(2 + \mathbf{i}) | 5$, and conclude that $5 \in \mathbb{Z}[\mathbf{i}]$ does not deserve to be called a prime number (in the context of Gaussian integers).

Exercise 11.22 Let $N : \mathbb{C} \to \mathbb{R}$ be the norm described above.

(a) Prove that, for every α and $\beta \in \mathbb{C}$, $N(\alpha \cdot \beta) = N(\alpha)N(\beta)$.
(b) Show that the restriction of the norm to $\mathbb{Z}[\mathbf{i}]$ produces a function from $\mathbb{Z}[\mathbf{i}]$ to $\mathbb{Z}$.

2 The conjugate of a complex number α is usually denoted by placing a bar over α, as in $\overline{\alpha}$.

(c) Is there any $\alpha \in \mathbb{Z}[\mathbf{i}]$ where $N(\alpha) = 3$?
(d) There are 12 Gaussian integers whose norm is 25. Find them, and plot them in the complex plane.

The Gaussian integers seem like a reasonable analog of the integers, just with the addition of the element **i**. It is reasonable to see if Gaussian integers have similar divisibility properties, including unique factorization.

As a cautionary example, we note that

$$8 + 10\mathbf{i} = (1 + \mathbf{i})^2(5 - 4\mathbf{i}) = (1 + \mathbf{i})(1 - \mathbf{i})(4 + 5\mathbf{i}),$$

which might temper your hopes for unique factorization. The situation might seem less severe if one were to notice:

$$(1 + \mathbf{i}) = (\mathbf{i})(1 - \mathbf{i}),$$
$$(5 - 4\mathbf{i}) = (-\mathbf{i})(4 + 5\mathbf{i}).$$

Thus the two expressions given above for $8 + 10\mathbf{i}$ are connected by

$$\begin{aligned}(1 + \mathbf{i})(1 - \mathbf{i})(4 + 5\mathbf{i}) &= (\mathbf{i})(-\mathbf{i})(1 + \mathbf{i})(1 - \mathbf{i})(4 + 5\mathbf{i}) \\ &= (1 + \mathbf{i}) \cdot (\mathbf{i})(1 - \mathbf{i}) \cdot (-\mathbf{i})(4 + 5\mathbf{i}) = (1 + \mathbf{i})^2(5 - 4\mathbf{i}).\end{aligned}$$

Define the *units* of $\mathbb{Z}[\mathbf{i}]$ to be those elements $\alpha \in \mathbb{Z}[\mathbf{i}]$ where $\alpha \cdot \beta = 1$ for some $\beta \in \mathbb{Z}[\mathbf{i}]$.

Exercise 11.23 Use the norm to show that if α is a unit in $\mathbb{Z}[\mathbf{i}]$, then $N(\alpha) = 1$. After establishing this lemma, verify that the units of $\mathbb{Z}[\mathbf{i}]$ are $\{\pm 1, \pm \mathbf{i}\}$.

The norm can help in finding factorizations. For example, $N(7 + 3\mathbf{i}) = 49 + 9 = 58$. Since $58 = 2 \cdot 29$, there is a chance that $7 + 3\mathbf{i} = \alpha \cdot \beta$ where $N(\alpha) = 2$ and $N(\beta) = 29$. In fact,

$$7 + 3\mathbf{i} = (1 + \mathbf{i})(5 - 2\mathbf{i}).$$

Exercise 11.24 Use the norm to assist you with the following questions.

(a) Find a factorization of $3 + 5\mathbf{i}$ where neither factor is a unit.
(b) Show that if $5 - 4\mathbf{i} = \alpha \cdot \beta$, then either α or β must be a unit.
(c) Just to keep your enthusiasm in check, verify that $N(2 + \mathbf{i})$ divides $N(4 + 3\mathbf{i})$ but $(2 + \mathbf{i}) \nmid (4 + 3\mathbf{i})$ in $\mathbb{Z}[\mathbf{i}]$.

We are now ready to define the notion of a prime number for the Gaussian integers.

DEFINITION 11.6. An element $\pi \in \mathbb{Z}[\mathbf{i}]$ is *prime* if whenever $\pi = \alpha \cdot \beta$, then exactly one of α or β is a unit of $\mathbb{Z}[\mathbf{i}]$.

Exercise 11.25 Determine if the following Gaussian integers are prime or not. They are arranged in order of increasing norm. While there are quite a few of them, you will gain some critical familiarity with $\mathbb{Z}[\mathbf{i}]$ by checking them all.

(a) $1 + \mathbf{i}$
(b) 2

(c) $2 - \mathbf{i}$
(d) $2 + 2\mathbf{i}$
(e) 3
(f) $3 + \mathbf{i}$
(g) $2 + 3\mathbf{i}$

And, given your work above, you should now be able to do the following quite quickly.

(h) Prove that if $N(\alpha)$ is a prime number in $\mathbb{Z}$, then α is a prime in $\mathbb{Z}[\mathbf{i}]$.
(i) Show by example that there are integers that are prime in $\mathbb{Z}$ which are not prime in $\mathbb{Z}[\mathbf{i}]$.

Exercise 11.26 Prove that every Gaussian integer is a Gaussian prime or can be expressed as a product of Gaussian primes. We recommend a proof by strong induction where you induct on the norm of the Gaussian integer.

Exercise 11.27 At this point we have finished guiding you! Review the proof of unique factorization of the integers, and following that general approach, you should now be able to state and prove an analogous result for the Gaussian integers. We aren't saying this will be easy; we are merely claiming that we have great confidence in your mathematical abilities.

GOING BEYOND THIS BOOK. If you would like to know more about the Gaussian integers and similar number systems, you should read *On Quaternions and Octonions* [**CS03**].

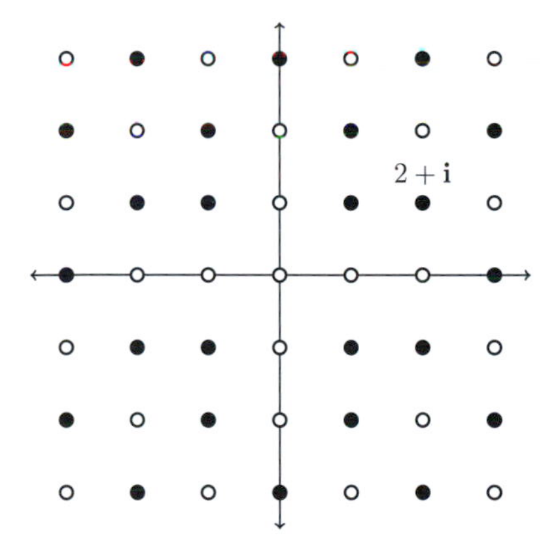

Figure 8. The Gaussian integers inside of the complex plane. The filled circles correspond to the Gaussian primes.

Exercise 1.36 asks you to prove that there are arbitrarily large gaps between consecutive primes in $\mathbb{Z}$, so it is impossible for someone to walk from 0 to infinity along the real line, where the steps are of bounded length and the person is only allowed to step on prime integers. A currently unsolved problem asks the same question in the context of the Gaussian integers, thought of as lattice points in the plane. Namely, is it possible for someone to walk from $(0, 0)$ to infinity when the steps are of bounded length and the person must only step on points (a, b) where $a + b\mathbf{i}$ is a Gaussian prime? Figure 8 displays Gaussian primes near the origin. You can read more about this problem in [**GWW98**], [**Loh07**], and the references cited by these papers.

11.5. Pigeons!

In this project we explore ...

> THE PIGEONHOLE PRINCIPLE (Lemma 5.21). *Let $f : A \to B$, where A and B are finite, non-empty sets. If $|A| > |B|$, then f is not injective.*

We proved this result at the end of Section 5.5. Here we take up some of its consequences and ask you to state and prove a generalized Pigeonhole Principle. We expect that to be successful in this project you will need to have finished the first three chapters and Chapter 5, through the proof of the Pigeonhole Principle in Section 5.5. It will be even easier if in fact you have worked through all of Chapter 5.

11.5.1. Lots of Pigeons! The Pigeonhole Principle can be generalized. For example, if you have 25 pigeons fly into 12 coops, it is certainly the case that at least one coop has more than one pigeon. In fact, it must be the case that at least one coop has more than two pigeons.

Exercise 11.28 In this exercise you will discover and prove a generalization of the Pigeonhole Principle.

(a) Let A and B be two finite sets, with $|B| = n$. How large must A be to ensure that given any $f : A \to B$ there is some $b \in B$ where at least three elements of A are mapped to b?
(b) Let A and B be two finite sets, with $|B| = n$. How large must A be to ensure that given any $f : A \to B$ there is some $b \in B$ where at least k elements of A are mapped to b?
(c) State a conjectured generalization of the Pigeonhole Principle.
(d) Prove your conjecture.

11.5.2. Pigeonhole Projects. The exercises in this subsection are independent of each other, so you may pick and choose those you find most interesting. Their diversity illustrates the broad applicability of the Pigeonhole Principle, and each is written in a manner that requires you to do some creative exploration of the application. Our recommendation is that you write up a presentation and proof of your generalized Pigeonhole Principle, and illustrate this result by presenting your work on one of the following exercises.

Exercise 11.29 Define a *rectangular game board* to be an $m \times n$ rectangle, with $m, n \in \mathbb{N}$ and both at least 2, which is divided into colored unit squares. A standard chessboard is one example; another example is shown in Figure 9.

(a) A 3×7 board is divided into 21 unit squares, each colored black or white. Prove that there must be a rectangular game board contained in this board whose four corner squares all have the same color.

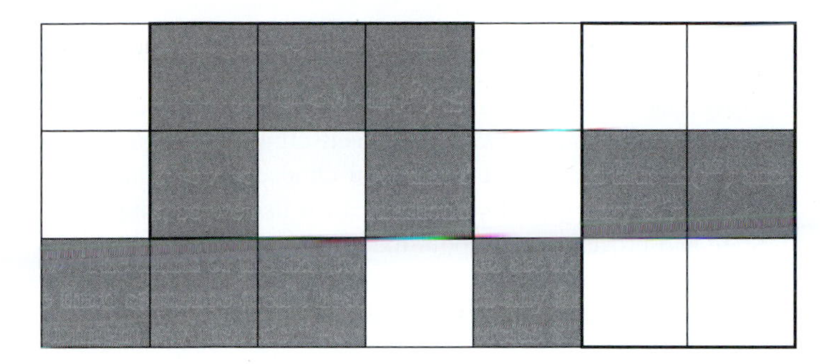

Figure 9. A 3×7 game board colored by two colors. The thick boundaries highlight a 2×3 game board, and a 3×2 game board, where in each case the four corners have the same coloring.

(b) Find examples that illustrate that the claim in part (a) is not true on a $2 \times n$ game board for any $n \geq 2$.
(c) Find another example that illustrates that the claim in part (a) is not always true for 3×6 game boards.
(d) Consider $m \times n$ game boards where the squares can be colored by three different colors. Can you find values of m and n that guarantee there is a game board contained in the $m \times n$ board where all four corners have the same color?
(e) Can you state and prove a result that works for c colors, where c is an integer that is greater than or equal to 2?
(f) How might you frame questions like these for rectangular bricks which have been divided into unit cubes? What sort of results can you find in this context?

Exercise 11.30 The *positive difference* of a pair of distinct integers $\{a, b\}$ is $|a - b|$. In this exercise you will consider subsets $S \subseteq \{1, 2, \ldots, n\}$ and prove results about the positive differences of pairs chosen from S. You may assume that the number of pairs of elements contained in a set with k elements is $k(k-1)/2$. More information about this and related formulas is presented in Section 3.8.

(a) Consider subsets $S \subseteq \{1, 2, \ldots, 10\}$. Prove that if $|S| \geq 5$, then there are at least two pairs of elements in S with the same positive difference.
(b) Give an example that shows the claim in (a) is false if $|S| = 4$.
(c) Consider subsets $S \subseteq \{1, 2, \ldots, 16\}$. Find a value of k such that when $|S| \geq k$, then there are at least two pairs of elements in S with the same positive difference.
(d) Conjecture a result along the lines of

Let $S \subseteq \{1, 2, \ldots, n\}$. When $|S| \geq f(n)$, then there are at least two pairs of elements in S with the same positive difference.

Here you should give an explicit formula for f in terms of the variable n. And of course you should prove that your conjecture is correct.
(e) How might you extend your ideas to the situation where S is sufficiently large as to guarantee that there are at least three pairs of elements in S with the same positive difference?

Exercise 11.31 A *lattice point* in $\mathbb{R}^2$ is any point (x, y) where x and y are both integers.

(a) Prove that if five lattice points are chosen in the plane, then there must be a pair where the midpoint between them is itself a lattice point.
(b) How many lattice points must you choose to guarantee that there are two pairs where the midpoint between each pair is itself a lattice point.
(c) State and prove a result about the number of lattice points that will guarantee that there are n pairs where the midpoint between each pair is a lattice point.
(d) Extend your ideas to lattice points in $\mathbb{R}^3$.

Exercise 11.32 In this problem you will explore the clustering of points chosen inside of a unit square.

(a) Prove that given any 9 points in the unit square, there must be three that are contained in a circle of radius $2/5$.
(b) Prove that given any 19 points in the unit square, there must be three that are contained in a circle of radius $1/4$.
(c) Prove that given any 28 points in the unit square, there must be four that are contained in a circle of radius $1/4$.
(d) Conjecture a result along the lines of the previous parts of this exercise, and prove that your conjecture is correct.
(e) Conjecture a result along the lines of

Given $f(n)$ points in the unit square, there must be n that are contained in a circle of radius _.

Here you should give an explicit formula for f in terms of the variable n. And of course you should prove that your conjecture is correct.

GOING BEYOND THIS BOOK. Like any fundamental result in mathematics, the history of the Pigeonhole Principle is fascinating and complicated. If you are interested in this history, we recommend the article by Rittaud and Heeffer [**RH14**] as a good entry point to the literature.

11.6. Mirsky's Theorem

Mirsky's Theorem is a general result about partially ordered sets, so you should have completed Section 6.2 before working on this project, including the definition and exercises on chains and maximal chains in the end-of-chapter exercises for Chapter 6.

Consider the following sequence of nine integers:

$$30, 20, 10, 31, 21, 11, 32, 22, 12\,.$$

By focusing on particular entries, you can find subsequences that are increasing like the bold entries in

$$30, 20, \mathbf{10}, 31, 21, \mathbf{11}, 32, \mathbf{22}, 12$$

and subsequences that are decreasing like the bold entries in

$$\mathbf{30}, \mathbf{20}, 10, 31, 21, \mathbf{11}, 32, 22, 12\,.$$

Call such a subsequence a *monotone* subsequence. We designed our sequence of nine numbers so that the longest monotone subsequences have three terms.

Exercise 11.33 Verify that our sequence contains no monotone subsequences with four terms.

Something interesting happens when a tenth integer is added into this sequence. We randomly selected a new integer and inserted it into a random location in our original sequence to produce

$$30, 20, 10, \underline{23}, 31, 21, 11, 32, 22, 12\,.$$

This sequence now contains monotone subsequences with four terms, for example,

$$\mathbf{30}, 20, 10, \mathbf{23}, 31, 21, 11, 32, \mathbf{22}, \mathbf{12}$$

and

$$30, \mathbf{20}, 10, \mathbf{23}, \mathbf{31}, 21, 11, \mathbf{32}, 22, 12\,.$$

Exercise 11.34 Do some experimenting to see the emergence of a monotone subsequence with four terms was an accident or if it might be an example of a larger pattern.

(a) Find your own additional integer and insert it into the sequence

$$30, 20, 10, 31, 21, 11, 32, 22, 12$$

at a random location. Can you find a monotone subsequence with four terms contained in your new sequence?

(b) Repeat part (a).

(c) Repeat part (a) again.

It is a fact that any sequence of ten distinct integers will have a monotone subsequence with four (or possibly more) terms. This is a special case, an example really, of a result first proved by Erdös and Szekeres in the 1930s. After we explore Mirsky's Theorem, we will return to the Erdös–Szekeres Theorem and use Mirsky's Theorem to prove it.

Mirsky's Theorem connects the height of a poset to its antichains. Let $(S, \preceq)$ be a partially ordered set. An *antichain* is a subset of S where no two elements are comparable. For example, if $S = \mathcal{P}(\{1, 2, 3, 4\})$ is ordered by containment, then

$$\{\{1, 2\}, \{1, 4\}, \{2, 3, 4\}\}$$

is an antichain.

Exercise 11.35 In this problem we will continue to work with the poset of subsets of $\{1, 2, 3, 4\}$ ordered by containment.

(a) The poset of subsets of $\{1, 2, 3, 4\}$, ordered by containment, contains 168 antichains. Find five of them.

(b) The largest antichains in the poset of subsets of $\{1, 2, 3, 4\}$ contain six elements. Find such an antichain.

(c) Find another antichain with six elements, or prove that the antichain with six elements is unique.

A *maximal* antichain is one that is not properly contained in any other antichain.

Exercise 11.36 For $n \in \mathbb{N}$, let $\mathcal{Z}_n$ be a zigzag poset with n elements. Find a formula for the size of the largest antichain in $\mathcal{Z}_n$, and prove that your formula is correct. (Zigzag posets are defined in Exercise 6.49.)

DEFINITION 11.7. An *antichain cover* of a poset is a collection of antichains, whose union contains all the elements of the partially ordered set. If the antichain cover consists of the antichains $\{A_1, A_2, \ldots, A_k\}$, then the cover has *size* k.

Exercise 11.37 The following questions provide a bit of practice with antichain covers, in the context of the poset of subsets of $[n] = \{1, 2, \ldots n\}$ ordered by containment.

(a) Prove that for any fixed integer k, $0 \leq k \leq n$, the collection of k-element subsets is a maximal antichain.
(b) Prove that the poset $([n], \subseteq)$ can be covered by $n + 1$ antichains.

THEOREM 11.8 (Mirsky's Theorem). *Let S be a finite set, partially ordered by $\preceq$. Then the height of the poset $(S, \preceq)$ equals the size of a minimal antichain cover of S.*

Below we outline two approaches to proving Mirsky's Theorem. The first proof connects the height directly to antichains, and the second is a proof by induction.

Exercise 11.38 Let $(S, \preceq)$ be a finite poset.

(a) Prove that the intersection of any antichain with a chain can contain at most one element.
(b) Prove that if the height of $(S, \preceq)$ is H, then any antichain cover must contain at least H antichains.

Exercise 11.39 Let $(S, \preceq)$ be a finite poset, and define the *height of an element $s \in S$* to be the maximum number of terms that occur in any chain with s as its largest element. These heights give us a function $h : S \to \mathbb{N}$, where $h(s)$ is the height of the element s.

(a) If H is the height of the poset $(S, \preceq)$, prove that h gives a surjection onto $[H] = \{1, 2, \ldots, H\}$.
(b) Prove that if $i \in [H]$, then $h^{-1}(i)$ is an antichain.
(c) Prove there is an antichain cover of $(S, \preceq)$ consisting of H antichains.

Exercises 11.38 and 11.39 immediately imply Mirsky's Theorem:

PROOF OF MIRSKY'S THEOREM. Let $(S, \preceq)$ be a finite poset of height H, where a minimal antichain cover of $(S, \preceq)$ contains A antichains. Then Exercise 11.38 shows that $H \leq A$, and Exercise 11.39 shows that $A \leq H$. Thus $A = H$. □

Mirsky's Theorem can also be proven by induction on the height of the poset.

Exercise 11.40 When you complete the following steps you will have outlined an inductive proof for Mirsky's Theorem.

(a) Base Case: Prove that Mirsky's Theorem holds for finite posets of height 1.
(b) Lemma: Prove that the set of all maximal elements of a finite poset forms an antichain.
(c) Lemma: Let M be the set of all maximal elements of the poset $(S, \preceq)$. Prove that $(S \setminus M, \preceq)$ is a poset whose height is smaller than the height of $(S, \preceq)$.
(d) Use the two lemmas above to establish the inductive step.

Having proven Mirsky's Theorem (twice!), we are now able to head toward a proof of the Erdös–Szekeres Theorem.

COROLLARY 11.9. *Let m and n be natural numbers. Then in any partially ordered set $(S, \preceq)$ with $mn + 1$ or more elements, there is a chain with $m + 1$ elements or an antichain with $n + 1$ elements.*

PROOF. Using the argument from Exercise 11.39 we see that if the height of the poset is less than $m + 1$ then there is an antichain cover consisting of fewer than $m + 1$ antichains, where no two antichains overlap. Denote these antichains by $A_1, A_2, \ldots, A_k$ $(k \leq m)$. Thus $|S| = |A_1| + |A_2| + \cdots + |A_k|$, and so

$$mn + 1 \leq m \cdot \max\{|A_i| \mid 1 \leq i \leq k\}.$$

Thus $n + 1 \leq \max\{|A_i| \mid 1 \leq i \leq k\}$, and so there must be an antichain with at least $n + 1$ elements. □

Exercise 11.41 Use Corollary 11.9 to prove the Erdös–Szekeres Theorem: *Each sequence of $m^2 + 1$ distinct real numbers possesses a monotone subsequence with $m+1$ terms.* As a hint, define an ordering on the terms of the sequence by saying $a_i \preceq a_j$ if $i \leq j$ and $a_i \leq a_j$.

For a graphical interpretation of this result, see Figure 10.

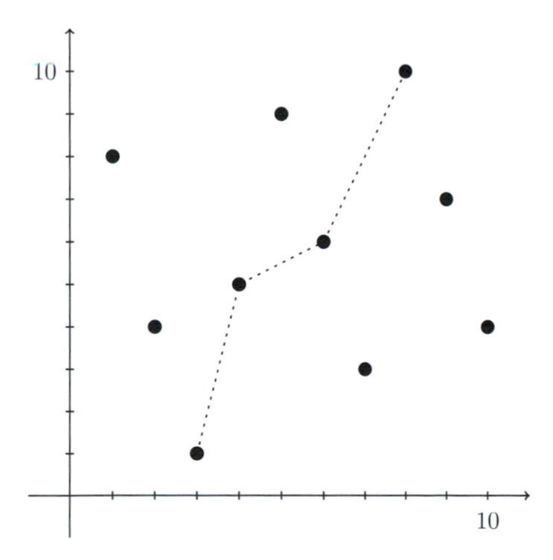

Figure 10. Mirsky's Theorem guarantees that you can find an increasing or decreasing constellation of length $m + 1$ in the graph of a sequence with $m^2 + 1$ distinct terms.

Exercise 11.42 In Exercise 6.8 we described a partial ordering on closed and bounded intervals $[a, b] \subset \mathbb{R}$. Use Mirsky's Theorem to prove that for any collection of $m^2 + 1$ closed and bounded intervals there must be either $m + 1$ that are disjoint or $m + 1$ that share a common point.

GOING BEYOND THIS BOOK. Mirsky first published his now eponymous theorem in [**Mir71**]. That paper is quite readable and is recommended, particularly for the way it discusses the connection between the result presented in this project and a related result, Dilworth's Theorem. The original paper by Erdös and Szekeres [**ES35**] is also worth reading, particularly with an eye toward how these ideas have evolved over the decades since their first presentation.

11.7. Euler's Totient Function

We have seen that knowing the order of an Abelian group can be useful. It is immediate that $|\mathbb{Z}_m| = m$, but the order of $\mathcal{U}_m$ is less clear. Since the elements of $\mathcal{U}_m$ correspond to the integers k where $1 \leq k \leq m$ and $\text{GCD}(k, m) = 1$, we are simply wondering how many natural numbers less than or equal to m are relatively prime to m. The function

$$\phi(m) = \text{the number of } k \in \mathbb{N} \text{ such that } k \leq m \text{ and } \text{GCD}(k, m) = 1$$

is known as *Euler's totient function*. Our goal in this section is to find a nice formula for ϕ. As part of the motivation for studying ϕ comes from the units arising in modular arithmetic, this project is appropriate for students who have worked through Section 6.6.

Exercise 11.43 Explain why $\phi(p) = p - 1$ for all primes p.

Here's one that is a bit more difficult.

Exercise 11.44 Prove that if p is prime, then $\phi(p^k) = p^k - p^{k-1}$.

DEFINITION 11.10. Let $m, n \in \mathbb{N}$, where both are at least 2. Define the *remainder map* $\rho : \mathbb{Z}_{mn} \to \mathbb{Z}_m \times \mathbb{Z}_n$ by

$$\rho([a]_{mn}) = ([a]_m, [a]_n).$$

For example, if you are considering $\rho : \mathbb{Z}_6 \to \mathbb{Z}_2 \times \mathbb{Z}_3$ then

$$\rho([5]_6) = ([5]_2, [5]_3) = ([1]_2, [2]_3).$$

We do need to verify that this remainder map is well defined.[3] To do this, consider any integer b where $a \equiv b$ (MOD mn). Thus $b = a + k \cdot mn$. So

$$\rho([b]_{mn}) = \rho([a + k \cdot mn]_{mn}) = ([a + kn \cdot m]_m, [a + km \cdot n]_n) = ([a]_m, [a]_n).$$

Thus our definition does not depend on the representative used in computing the answer.

3 Remember 'A Cautionary Tale' in Section 6.5?

We now get to play with this remainder map. Let's first look at $\rho : \mathbb{Z}_6 \to \mathbb{Z}_2 \times \mathbb{Z}_3$.

$$\rho = \begin{cases} [0]_6 \mapsto ([0]_2, [0]_3) \\ [1]_6 \mapsto ([1]_2, [1]_3) \\ [2]_6 \mapsto ([0]_2, [2]_3) \\ [3]_6 \mapsto ([1]_2, [0]_3) \\ [4]_6 \mapsto ([0]_2, [1]_3) \\ [5]_6 \mapsto ([1]_2, [2]_3) \end{cases}$$

The remainder map has given us a bijection between the six elements in $\mathbb{Z}_6$ and the six elements in $\mathbb{Z}_2 \times \mathbb{Z}_3$. Before we make any hasty conjectures, let's look at another example, this time considering $\rho : \mathbb{Z}_{12} \to \mathbb{Z}_2 \times \mathbb{Z}_6$:

$$\rho = \begin{cases} [0]_{12} \mapsto ([0]_2, [0]_6) \\ [1]_{12} \mapsto ([1]_2, [1]_6) \\ [2]_{12} \mapsto ([0]_2, [2]_6) \\ [3]_{12} \mapsto ([1]_2, [3]_6) \\ [4]_{12} \mapsto ([0]_2, [4]_6) \\ [5]_{12} \mapsto ([1]_2, [5]_6) \\ [6]_{12} \mapsto ([0]_2, [0]_6) \\ [7]_{12} \mapsto ([1]_2, [1]_6) \\ [8]_{12} \mapsto ([0]_2, [2]_6) \\ [9]_{12} \mapsto ([1]_2, [3]_6) \\ [10]_{12} \mapsto ([0]_2, [4]_6) \\ [11]_{12} \mapsto ([1]_2, [5]_6) \end{cases}$$

In this case we do not get a bijection. As always, this ought just to make us more curious . . .

LEMMA 11.11. *If m and n are relatively prime, then $\rho : \mathbb{Z}_{mn} \to \mathbb{Z}_m \times \mathbb{Z}_n$ is a bijection.*

PROOF. Since $|\mathbb{Z}_{mn}| = |Z_m \times \mathbb{Z}_n| = mn$, it suffices to show that ρ is injective.

If $\rho(x) = \rho(y)$, then n and m divide $x - y$. Hence the least common multiple of m and n divides $x - y$ (see Exercise 4.42). Because m and n are relatively prime, $\text{LCM}(m, n) = mn$ divides $x - y$. So $x \equiv y \ (\text{MOD } mn)$. □

Exercise 11.45 Show that the remainder map gives a bijection:

$$\rho : \mathcal{U}_{mn} \to \mathcal{U}_m \times \mathcal{U}_n$$

when m and n are relatively prime.

Exercise 11.46 Let the prime decomposition of n be $n = p_1^{k_1} p_2^{k_2} \cdots p_m^{k_m}$, with each $k_i \in \mathbb{N}$. Prove that

$$\phi(n) = \phi(p_1^{k_1})\phi(p_2^{k_2}) \cdots \phi(p_m^{k_m}) = \prod_{i=1}^{m} (p_i - 1)p_i^{k_i - 1}.$$

At this point we have an effective method for computing Euler's totient function. For example, $\phi(135) = \phi(3^3 \cdot 5) = \phi(3^3) \cdot \phi(5) = \left(2 \cdot 3^2\right)(4) = 72$.

Exercise 11.47 Compute $\phi(700)$.

Finally, we are in a position to present an alternate formula for ϕ that is often preferable:

THEOREM 11.12. *Let $n \in \mathbb{N}$. Euler's totient function ϕ is given by*

$$\phi(n) = n \prod_{p|n} \left(1 - \frac{1}{p}\right),$$

where $p|n$ stands for the set of primes p that divide n.

For example,

$$\phi(700) = 700 \prod_{p|700} \left(1 - \frac{1}{p}\right) = 700 \left(1 - \frac{1}{2}\right)\left(1 - \frac{1}{5}\right)\left(1 - \frac{1}{7}\right)$$
$$= 700 \left(\frac{1}{2}\right)\left(\frac{4}{5}\right)\left(\frac{6}{7}\right) = 240\,.$$

Exercise 11.48 Prove Theorem 11.12.

If you are ambitious, we recommend that you explore bounds on Euler's totient function in the following exercise.

Exercise 11.49 Figure 11 shows the graph of Euler's totient function $\phi : \mathbb{N} \to \mathbb{N}$, up to $n = 250$.

(a) Prove that this graph is bounded above by the straight line $y = x - 1$, as Figure 11 suggests.
(b) Prove that no line of slope $m < 1$ could be an upper bound on the graph of ϕ.
(c) In Figure 11 we have drawn in the line connecting the origin to $(210, \phi(210))$. It appears that this line gives a lower bound on ϕ. Show that this is in fact false.
(d) Prove that $\phi(n) \geq \sqrt{n/2}$ for all integers $n \geq 2$.

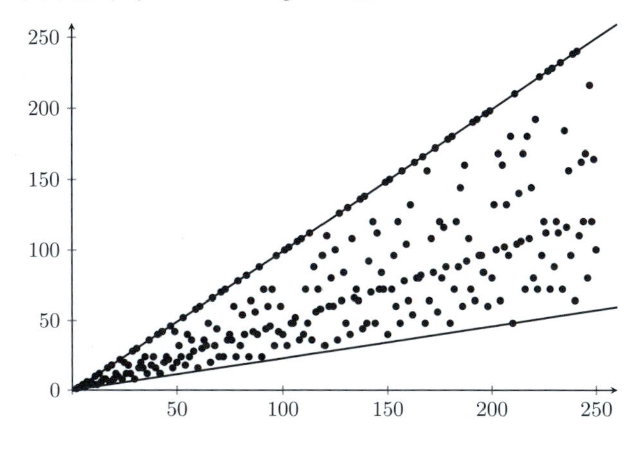

Figure 11. The Euler totient function for $n = 2$ to $n = 250$.

11.8. Proving the Schröder–Bernstein Theorem

In Section 7.6 we demonstrated the tremendous utility of the Schröder–Bernstein Theorem, but we did not provide a proof of this theorem. The goal of this project is for you to construct a proof, and before working on this project you should review Exercise 7.18.

THEOREM 11.13 (The Schröder–Bernstein Theorem). *Let A and B be sets, and assume there are injections $f : A \to B$ and $g : B \to A$. Then A and B have the same cardinality.*

Every element $a \in A$ determines a sequence of elements, alternating between the sets A and B. The sequence starts with a and then alternately applies f and g:

$$\{a, f(a), g(f(a)), f(g(f(a))), \ldots\}.$$

It is cleaner to visualize this sequence as the filling in of blanks in the following diagram:

$$a \xrightarrow{f} \square \xrightarrow{g} \square \xrightarrow{f} \square \cdots$$

You can also attempt to build out this sequence to the left, essentially applying the inverse of g and f. However, this is only possible if you are working with elements in the range of g and f, respectively. Thus, while the sequence will continue to the right indefinitely, it may not extend indefinitely to the left.

$$\cdots \xrightarrow{g} ? \xrightarrow{f} ? \xrightarrow{g} a \xrightarrow{f} \square \xrightarrow{g} \square \xrightarrow{f} \square \cdots$$

We will call these sequences *back-and-forth* sequences, as they bounce between A and B. When discussing a back-and-forth sequence, we always assume that it has been extended as far as possible to the left.

EXAMPLE 11.14. The Schröder–Bernstein Theorem can be easily applied to show that $[0, 1] \asymp (0, 3)$. For example, there are injections $f : [0, 1] \to (0, 3)$ and $g : (0, 3) \to [0, 1]$, where $f(x) = x + 1$ and $g(x) = \frac{1}{3}x$. These functions lead to the following sequences:

$$0 \xrightarrow{f} 1 \xrightarrow{g} \frac{1}{3} \xrightarrow{f} \frac{4}{3} \xrightarrow{g} \frac{4}{9} \xrightarrow{f} \cdots,$$
$$\frac{1}{8} \xrightarrow{f} \frac{9}{8} \xrightarrow{g} \frac{3}{8} \xrightarrow{f} \frac{11}{8} \xrightarrow{g} \frac{11}{24} \xrightarrow{f} \cdots,$$
$$\frac{1}{2} \xrightarrow{f} \frac{3}{2} \xrightarrow{g} \frac{1}{2} \xrightarrow{f} \frac{3}{2} \xrightarrow{g} \frac{1}{2} \xrightarrow{f} \cdots$$

The top back-and-forth sequence cannot be extended to the left, because the value 0 is not in the range of g. The middle sequence can be extended one step to the left, since $g(3/8) = 1/8$. However, $3/8 \notin \text{Range}(f)$, so this sequence extends no further to the left. Finally, it is clear that the bottom sequence extends indefinitely to the left. Thus we have:

$$0 \xrightarrow{f} 1 \xrightarrow{g} \frac{1}{3} \xrightarrow{f} \frac{4}{3} \xrightarrow{g} \frac{4}{9} \xrightarrow{f} \cdots,$$
$$\frac{3}{8} \xrightarrow{g} \frac{1}{8} \xrightarrow{f} \frac{9}{8} \xrightarrow{g} \frac{3}{8} \xrightarrow{f} \frac{11}{8} \xrightarrow{g} \frac{11}{24} \xrightarrow{f} \cdots,$$
$$\cdots \xrightarrow{g} \frac{1}{2} \xrightarrow{f} \frac{3}{2} \xrightarrow{g} \frac{1}{2} \xrightarrow{f} \frac{3}{2} \xrightarrow{g} \frac{1}{2} \xrightarrow{f} \frac{3}{2} \xrightarrow{g} \frac{1}{2} \xrightarrow{f} \cdots$$

Exercise 11.50 Let $f : A \to B$ and $g : B \to A$ be injections. Define a relation $\sim$ on the elements of A by saying $a \sim a'$ when a and a' are both in a shared back-and-forth sequence.

(a) Prove that $\sim$ is an equivalence relation, and therefore the equivalence classes partition A.
(b) State and prove the same result for the elements of B.

DEFINITION 11.15. A back-and-forth sequence that extends indefinitely to the left is a *bi-infinite* sequence. Those that cannot be extended indefinitely to the left are called A-sequences or B-sequences, depending on whether the left-most element is in A or B.

Exercise 11.51 This exercise refers back to the situation described in Example 11.14.

(a) Show that there are only two A-sequences for this example.
(b) Show that there are infinitely many B-sequences.

Exercise 11.52 Let $f : A \to B$ and $g : B \to A$ be injections.

(a) Prove that f is a bijection when restricted to the elements occurring in an A-sequence.
(b) Prove that g^{-1} is a bijection when restricted to the elements occurring in a B-sequence.
(c) Prove that f and g^{-1} are both bijections when restricted to the elements in a bi-infinite sequence.

Exercise 11.53 Prove the Schröder–Bernstein Theorem.

Exercise 11.54 Apply your proof of the Schröder–Bernstein Theorem to construct an explicit bijection between $[0, 1]$ and $(0, 3)$.

Exercise 11.55 Show that the bijection h given in Exercise 7.18 comes directly from the injections f and g discussed earlier in Section 7.6.

11.9. Cauchy Sequences and the Real Numbers

This project is in fact two projects. We begin with an elementary discussion of Cauchy sequences. The first sub-project proves that a real-valued sequence is Cauchy if and only if it converges. The second sub-project, which can be done independently of the first, outlines the construction of the real numbers via Cauchy sequences of rational numbers. We make frequent use of material in Chapter 8, up to and including Section 8.5, throughout this project.

DEFINITION 11.16. A sequence (a_i) of real numbers is a *Cauchy sequence* if, given any real number $\epsilon > 0$, there is an $N \in \mathbb{N}$ such that

$$|a_m - a_n| < \epsilon$$

whenever m and n are both greater than N.

The first exercise provides a good test case for practicing with this definition.

Exercise 11.56 Let $r = n.d_1d_2d_3d_4\ldots$ be a decimal expansion of some real number r. Here $n \in \mathbb{Z}$ and each $d_i \in \{0, 1, \ldots, 9\}$. For each $k \in \mathbb{N}$ define

$$r_k = n + \frac{d_1}{10} + \frac{d_2}{10^2} + \cdots + \frac{d_k}{10^k}.$$

Prove that the sequence of rational numbers (r_k) is a Cauchy sequence.

Exercise 11.57 Construct an example of a Cauchy sequence that is not monotone.

11.9.1. Real-valued Cauchy Sequences. The goal of this sub-project is to prove that a real-valued sequence (a_n) converges if and only if it is Cauchy.

Exercise 11.58 Let (a_i) be a real-valued Cauchy sequence. Because (a_i) is a Cauchy sequence, by its definition there is an $N \in \mathbb{N}$ such that $|a_m - a_n| < 1$ for all m and n greater than N. Consider then the set of initial terms $\{a_1, a_2, \ldots, a_N\}$ and the set of values that appear in the tail $\{a_{N+1}, a_{N+2}, a_{N+3}, \ldots\}$.

(a) Explain why the set of initial terms, $\{a_1, a_2, \ldots, a_N\}$, is bounded.
(b) Show that the terms in the tail, $\{a_{N+1}, a_{N+2}, a_{N+3}, \ldots\}$, are all contained in the closed interval $[a_{N+1} - 1, a_{N+1} + 1]$.
(c) Combine the facts above to prove that Cauchy sequences are bounded.

Because Cauchy sequences are bounded, given a Cauchy sequence (a_n) we may define an associated sequence of upper bounds (μ_n). We let μ_1 be any upper bound for all the terms in (a_n). We define μ_2 to be the average of a_2 and μ_1, if this average is an upper bound for $\{a_2, a_3, a_4, \ldots\}$, and otherwise we set $\mu_2 = \mu_1$. And in general

$$\mu_{n+1} = \begin{cases} \frac{a_{n+1}+\mu_n}{2} & \text{if this average is an upper bound for } \{a_i \mid i \geq n+1\}, \\ \mu_n & \text{otherwise.} \end{cases}$$

This sequence of upper bounds is illustrated in Figure 12, where the original Cauchy sequence is graphed using black circles, and the associated sequence of upper bounds (with μ_1 chosen to be 4) is displayed with crosses. The figure indicates, correctly, that both sequences converge to a limiting value of 2.

Exercise 11.59 Let (a_n) be a Cauchy sequence and let μ_1 be any upper bound on the set of terms $\{a_i \mid i \in \mathbb{N}\}$.

(a) Prove that the associated sequence of upper bounds (μ_i) defined as above is a bounded, monotone decreasing sequence. Conclude that it must converge to some real number μ.
(b) Prove that the Cauchy sequence (a_i) converges to μ as well.

Exercise 11.59 gives a very important method of demonstrating that a real-valued sequence converges. Constructing a proof of convergence using the original definition requires that you also already know the limit of the sequence. Proving that a sequence is Cauchy, however, requires only an argument that uses the terms of the sequence.

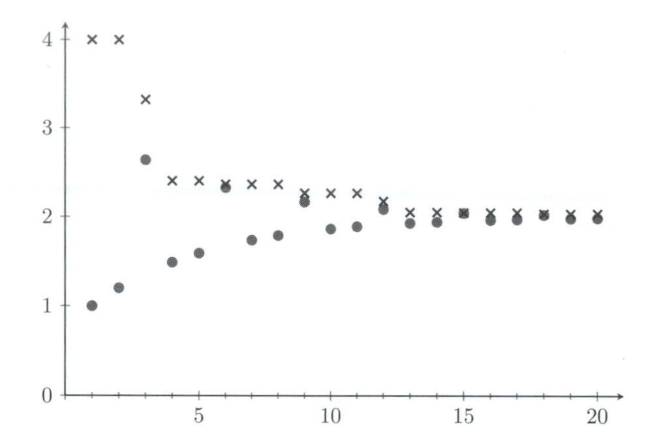

Figure 12. A Cauchy sequence is shown with black circles, and an associated sequence of upper bounds is displayed with crosses.

You are now in a position to establish the main result of this subsection. We anticipate that you will want to use the following approach: If (a_i) converges to a limiting value α, then for any $m, n \in \mathbb{N}$ you know that

$$|a_m - a_n| \leq |a_m - \alpha| + |\alpha - a_n|.$$

This is an application of the triangle inequality: $|a + b| \leq |a| + |b|$ for all $a, b \in \mathbb{R}$.

Exercise 11.60 Prove that a real-valued sequence (a_i) is Cauchy if and only if it converges.

11.9.2. Constructing the Real Numbers. Given the natural numbers, $\mathbb{N}$, you can construct the set of integers, $\mathbb{Z}$, by adding 0 and defining negative numbers. Given the integers, you can construct the rationals, $\mathbb{Q}$; one method for doing this is presented in Exercise 6.75. Here we describe a way to construct the real numbers, $\mathbb{R}$, from the rationals.

There are various definitions that lead to the real number system, and the following set of axioms is one of the most common. This definition begins by assuming there is a set of elements denoted by $\mathbb{R}$, two binary operations $+$ and $\cdot$ that map $\mathbb{R} \times \mathbb{R}$ to $\mathbb{R}$, and a relation $>$ on $\mathbb{R}$. This collection defines the real numbers if it satisfies the following 12 axioms.

- The Algebraic Axioms
  (1) For every a, b, and $c \in \mathbb{R}$, $(a + b) + c = a + (b + c)$ and $(a \cdot b) \cdot c = a \cdot (b \cdot c)$.
  (2) For every a and $b \in \mathbb{R}$, $a + b = b + a$ and $a \cdot b = b \cdot a$.
  (3) For every a, b, and $c \in \mathbb{R}$, $a \cdot (b + c) = a \cdot b + a \cdot c$.
  (4) There is an element called 0 such that $0 + a = a$ for all $a \in \mathbb{R}$.
  (5) Given any $a \in \mathbb{R}$, there is an element $b \in \mathbb{R}$ such that $a + b = 0$.
  (6) There is an element called 1 such that $1 \cdot a = a$ for all $a \in \mathbb{R}$.
  (7) For every $a \in \mathbb{R} \setminus \{0\}$, there is a $b \in \mathbb{R}$ such that $a \cdot b = 1$.

- The Order Axioms
  - (8) For every a, b, and $c \in \mathbb{R}$, if $a > b$ then $a + c > b + c$.
  - (9) For every a, b, and $c \in \mathbb{R}$, if $a > b$ and $b > c$ then $a > c$.
  - (10) For every a and $b \in \mathbb{R}$, exactly one of the following is true: $a > b$, $a = b$, or $b < a$.
  - (11) For every a and $b \in \mathbb{R}$ and every $c > 0$, if $a > b$ then $c \cdot a > c \cdot b$.

- The Completeness Axiom
  - (12) If S is a non-empty subset of $\mathbb{R}$ that is bounded above, then S has a least upper bound.

Our goal then is to start with the rational numbers, construct a set $\mathbb{R}$ along with $+, \cdot$, and $>$, and verify that the axioms above hold. In order to do so, we limit our discussion in this section to *rational* sequences, that is, sequences (a_i) where every $a_i \in \mathbb{Q}$. If q is a rational number, we let (q) be the constant sequence,

$$(q) = (q, q, q, q, q, \ldots).$$

DEFINITION 11.17. Two Cauchy sequences (a_n) and (b_n) are defined to be *equivalent* if the sequence of differences $(a_n - b_n)$ converges to 0.

Exercise 11.61 Prove that the sequence $(1) = (1, 1, 1, 1, \ldots)$ and the sequence $(1 - 1/10^i) = (0.9, 0.99, 0.999, \ldots)$ are equivalent Cauchy sequences.

Exercise 11.62 Prove that the "equivalent" relation defined above lives up to its name and is indeed an equivalence relation on the set of Cauchy sequences.

We define $\mathbb{R}$ to be the set of equivalence classes of rational Cauchy sequences. We define addition and multiplication of rational Cauchy sequences via

$$(a_i) + (b_i) = (a_i + b_i), \text{ and}$$
$$(a_i) \cdot (b_i) = (a_i \cdot b_i).$$

For example,

$$\left(\frac{1}{2^i}\right) + \left(\frac{1}{3^i}\right) = \left(\frac{1}{2^i} + \frac{1}{3^i}\right) = \left(\frac{2^i + 3^i}{6^i}\right) = \left(\frac{5}{6}, \frac{13}{36}, \frac{35}{216}, \ldots\right).$$

Exercise 11.63 Addition and multiplication are defined for specific sequences, not for equivalence classes. And our definition of $\mathbb{R}$ states the the elements of $\mathbb{R}$ are not sequences, but rather equivalence classes of sequences. This means you have some work to do! We need to establish that the sum and product of rational, Cauchy sequences are rational Cauchy sequences. We also need to verify that these operations are well defined on equivalence classes.

(a) Prove that if (a_i) and (b_i) are rational Cauchy sequences, then so is $(a_i + b_i)$.
(b) Prove that if (a_i) and (b_i) are rational Cauchy sequences, then so is $(a_i \cdot b_i)$.
(c) Let $(a_i) \sim (c_i)$ and $(b_i) \sim (d_i)$ be equivalent sequences. Prove that

$$(a_i + b_i) \sim (c_i + d_i).$$

(d) Let $(a_i) \sim (c_i)$ and $(b_i) \sim (d_i)$ be equivalent sequences. Prove that

$$(a_i \cdot b_i) \sim (c_i \cdot d_i).$$

Exercise 11.64 Verify that the Algebraic Axioms hold, with the constant sequence (0) functioning as the additive identity 0 and (1) functioning as the multiplicative identity 1.

DEFINITION 11.18. For two rational Cauchy sequences (a_i) and (b_i), define $(a_i) > (b_i)$ if they are not equivalent and if there is an $N \in \mathbb{N}$ such that $a_n > b_n$ for all $n \geq N$. We say $(a_i) \geq (b_i)$ if either the two sequences are equivalent or if $(a_i) > (b_i)$.

Exercise 11.65 We have defined $>$ for specific sequences, not on equivalence classes of sequences. We need to verify that $>$ makes sense on equivalence classes, and then verify that the order axioms hold.

(a) Let $(a_i) \sim (c_i)$ and $(b_i) \sim (d_i)$ be equivalent sequences. Prove that

$$(a_i) > (b_i) \Leftrightarrow (c_i) > (d_i).$$

Thus our definition is well defined on equivalence classes.

(b) Verify that the order axioms hold.

Exercise 11.66 Let S be a set of rational Cauchy sequences which is bounded above. Prove that S has a least upper bound by mimicking the "sequence of upper bounds" approach from the previous subsection. If you are working with representatives from the equivalence classes, be sure to verify that your arguments hold regardless of the chosen representatives.

Finally, having constructed the real numbers, it is a good idea for us to verify that the usual description in terms of decimal expansions is indeed accurate. Define a *standard decimal expansion sequence* to be a Cauchy sequence of the form discussed in Exercise 11.56, with the added requirement that given any $m \in \mathbb{N}$ there is an $n > m$ such that $d_n \neq 9$, so that decimal expansions are not allowed to have a tail of repeating 9s. This final exercise justifies the use of decimal expressions to represent the real numbers.

Exercise 11.67 Prove that every rational Cauchy sequence is equivalent to a standard decimal expansion sequence.

GOING BEYOND THIS BOOK. There are other ways of constructing the real numbers that do not involve equivalence classes of Cauchy sequences. The method of Dedekind cuts is nicely presented in Abbott's *Understanding Analysis* [**Abb15**]. Both the approach via Cauchy sequences and the approach of Dedekind date to the early 1870s. It should be noted, however, that the history of constructions of $\mathbb{R}$ is complicated, with credit certainly being due to Cantor, Heine, Dedekind, Méray, and Weierstrass (at the least!). Some would also include work of Simon Steven on numbers with unbounded decimal expansions in the 1600s, and perhaps even work of Eudoxus in ancient Greece. While it is not particularly focused on constructions of $\mathbb{R}$, Grabiner's *The Origins of Cauchy's Rigorous Calculus* [**Gra81**] is an excellent history of the development of analysis.

11.10. The Cantor Set

In this project we introduce the Cantor set, a subset of the closed unit interval $[0, 1] \subset \mathbb{R}$ that is formed by repeatedly removing open intervals. The Cantor set has a number of interesting and often counterintuitive properties, the most accessible being the fact that it is both a small subset of $[0, 1]$ and a large collection of points in $[0, 1]$. In this project we assume you have worked through Chapters 7 and 8, or at the least have a good intuition about the real numbers.

The process for defining the Cantor set $\mathcal{C}$ begins with the interval $\mathcal{C}_0 = [0, 1]$, and then creates

$$\mathcal{C}_1 = \mathcal{C}_0 - (\tfrac{1}{3}, \tfrac{2}{3}) = [0, \tfrac{1}{3}] \cup [\tfrac{2}{3}, 1].$$

The subset $\mathcal{C}_1$ consists of two closed intervals, and we remove the "middle thirds" from both of them to form $\mathcal{C}_2$. That is, we remove $(1/9, 2/9)$ and $(7/9, 8/9)$ from $\mathcal{C}_1$ to form $\mathcal{C}_2$:

$$\mathcal{C}_2 = \mathcal{C}_1 - \left((\tfrac{1}{9}, \tfrac{2}{9}) \cup (\tfrac{7}{9}, \tfrac{8}{9})\right) = [0, \tfrac{1}{9}] \cup [\tfrac{2}{9}, \tfrac{3}{9}] \cup [\tfrac{6}{9}, \tfrac{7}{9}] \cup [\tfrac{8}{9}, 1].$$

The subset $\mathcal{C}_2$ consists of four closed intervals. We then remove the middle thirds of each of them to form $\mathcal{C}_3$, consisting of eight closed intervals, and we continue this process to create a subset $\mathcal{C}_n$ for each $n \in \mathbb{N}$. The process is visualized in Figure 13.

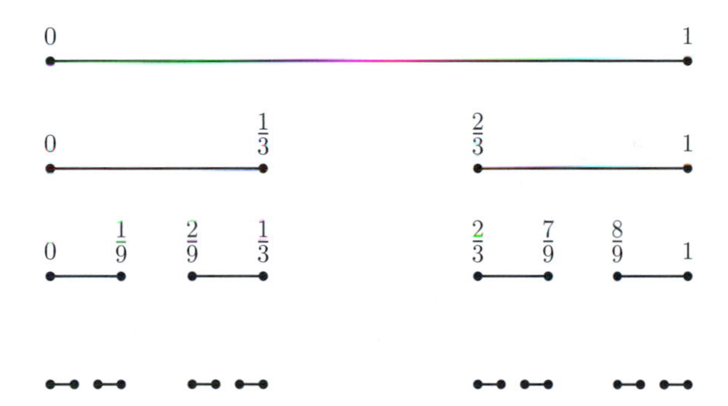

Figure 13. The first steps in constructing the Cantor set by removing middle thirds.

The Cantor set $\mathcal{C}$ is the end result of this process; that is, $\mathcal{C}$ is the set of points in $[0, 1]$ that are never removed in forming any of the sets $\mathcal{C}_i$.

You can also develop the Cantor set using rescaling and translating. If $X \subseteq \mathbb{R}$ and $a, b \in \mathbb{R}$, define

$$aX + b = \{ax + b \in \mathbb{R} \mid x \in X\}.$$

For example, if $X = [0, 1]$ then $\frac{1}{3}X + \frac{2}{3}$ is the interval $[\frac{2}{3}, 1]$. And, directly related to our discussion of the Cantor set,

$$\mathcal{C}_1 = \left\{\frac{1}{3}\mathcal{C}_0\right\} \cup \left\{\frac{1}{3}\mathcal{C}_0 + \frac{2}{3}\right\} = \left[0, \frac{1}{3}\right] \cup \left[\frac{2}{3}, 1\right].$$

Exercise 11.68 Verify that the next step of forming $\mathcal{C}_2$ can be expressed in the same manner:

$$\mathcal{C}_2 = \left\{\frac{1}{3}\mathcal{C}_1\right\} \cup \left\{\frac{1}{3}\mathcal{C}_1 + \frac{2}{3}\right\}.$$

DEFINITION 11.19. Let $\mathcal{C}_0$ and $\mathcal{C}_1$ be as above. Define the set $\mathcal{C}_{n+1}$ to be

$$\mathcal{C}_{n+1} = \left\{\frac{1}{3}\mathcal{C}_n\right\} \cup \left\{\frac{1}{3}\mathcal{C}_n + \frac{2}{3}\right\}$$

and the *Cantor set* $\mathcal{C}$ is the intersection

$$\mathcal{C} = \bigcap_{i \in \mathbb{N}} \mathcal{C}_i$$

of all of these sets of closed intervals.

Exercise 11.69 Prove that the set $\mathcal{C}_i$ is composed of 2^i closed intervals, each of length $(1/3)^i$, for a total combined length of $(2/3)^i$. Then prove that the open intervals removed in forming the Cantor set are pairwise disjoint and have combined length equal to 1.

Exercise 11.69 indicates that $\mathcal{C}$ is small, but it is not empty. The numbers 0 and 1 are in every $\mathcal{C}_i$, hence 0 and 1 are elements in the Cantor set $\mathcal{C}$. And there are quite a few more points in $\mathcal{C}$ in addition to these two.

Exercise 11.70 We pointed out that 0 and 1 must be in $\mathcal{C}$. Find two other real numbers that are in $\mathcal{C}$ and explain how you know that they must be in the Cantor set. Then describe an infinite set of real numbers in $\mathcal{C}$.

Our discussion of the cardinality of the Cantor set will use the *Nested Intervals Property* of the real numbers.

THEOREM 11.20 (Nested Intervals Property). *Let* $\{[a_i, b_i] \mid i \in \mathbb{N}\}$ *be a collection of closed, bounded intervals that are nested; that is,*

$$\mathbb{R} \supset [a_1, b_1] \supseteq [a_2, b_2] \supseteq [a_3, b_3] \supseteq \cdots.$$

Then the intersection of all of these intervals is non-empty:

$$\bigcap_{i \in \mathbb{N}} [a_i, b_i] \neq \emptyset.$$

Further, if the widths of the intervals decrease to zero, that is

$$\lim_{i \to \infty} (b_i - a_i) = 0,$$

then the intersection is a set containing a single real number.

Exercise 11.71 Exercise 8.39 asked you to prove the main part of this result: the intersection is non-empty. Please complete this exercise if you haven't done so already, and then prove that the additional claim that having the widths go to 0 implies that the intersection is a single real number.

In order to apply the Nested Intervals Property to the Cantor set we need a bit of additional notation. The first approximation to the Cantor set, $\mathcal{C}_1$, consists of two intervals,

$$\mathcal{C}_1 = [0, \tfrac{1}{3}] \cup [\tfrac{2}{3}, 1].$$

We will refer to $[0, 1/3]$ as the left interval, which we denote $\mathcal{C}_L$, and $[2/3, 1]$ as the right interval, $\mathcal{C}_R$. Thus $\mathcal{C}_1 = \mathcal{C}_L \cup \mathcal{C}_R$.

In passing to $\mathcal{C}_2$ we need to divide $\mathcal{C}_L$ into its left and right sub-intervals, $\mathcal{C}_{LL}$ and $\mathcal{C}_{LR}$, and we need to divide $\mathcal{C}_R$ into its left and right sub-intervals, $\mathcal{C}_{RL}$ and $\mathcal{C}_{RR}$. Thus

$$\mathcal{C}_2 = \underset{\mathcal{C}_{LL}}{[0, \tfrac{1}{9}]} \cup \underset{\mathcal{C}_{LR}}{[\tfrac{2}{9}, \tfrac{3}{9}]} \cup \underset{\mathcal{C}_{RL}}{[\tfrac{6}{9}, \tfrac{7}{9}]} \cup \underset{\mathcal{C}_{RR}}{[\tfrac{8}{9}, 1]}.$$

Similarly listing the eight intervals that comprise $\mathcal{C}_3$ from left to right, we have

$$\mathcal{C}_3 = \mathcal{C}_{LLL} \cup \mathcal{C}_{LLR} \cup \mathcal{C}_{LRL} \cup \mathcal{C}_{LRR} \cup \mathcal{C}_{RLL} \cup \mathcal{C}_{RLR} \cup \mathcal{C}_{RRL} \cup \mathcal{C}_{RRR}.$$

Define a *properly nested sequence of Cantor sub-intervals* to be a nested sequence where the first interval comes from $\mathcal{C}_1$, the second from $\mathcal{C}_2$, and so on. One example would be

$$\mathcal{C}_L \supset \mathcal{C}_{LR} \supset \mathcal{C}_{LRR} \supset \mathcal{C}_{LRRR} \supset \mathcal{C}_{LRRRL} \supset \cdots.$$

Exercise 11.72 Find a one-to-one correspondence between the set of properly nested sequences of Cantor sub-intervals and the Cantor set, and prove that your function is a bijection.

Exercise 11.73 Prove that the Cantor set is uncountable. (We recommend using the idea behind Cantor's Diagonal Argument, applied to properly nested sequences of Cantor sub-intervals.)

In Exercise 11.69 you showed that the Cantor set is formed by removing open intervals of total length 1, starting from the closed interval $[0, 1]$. This indicates that $\mathcal{C}$ must be rather small. Yet in Exercise 11.73 you proved that $\mathcal{C}$ has the same cardinality as $\mathbb{R}$. That is a good stopping place. However, if you feel like pushing forward just a bit more, feel free to work on a final exercise.

Exercise 11.74 If A and B are two sets of real numbers, we can define the sum of A and B as

$$A + B = \{x \in \mathbb{R} \mid x = a + b \text{ where } a \in A \text{ and } b \in B\}.$$

For instance, if $A = \{-4, 3\}$ and $B = [-1, 2]$, then

$$A + B = [-5, -2] \cup [2, 5].$$

As another example, if A and B are both equal to the closed interval $[0, 1]$, then $A + B = [0, 2]$. What's remarkable is that, despite how small $\mathcal{C}$ seems to be as a geometric subset of $[0, 1]$, you can show that $\mathcal{C} + \mathcal{C}$ is equal to the entire closed interval $[0, 2]$ as well.

(a) Recall that $\mathcal{C}_1 = [0, 1/3] \cup [2/3, 1]$. Thus $\mathcal{C}_1 \times \mathcal{C}_1 \subset \mathbb{R}^2$ consists of four solid squares in the first quadrant, as shown in Figure 14. Prove that $\mathcal{C}_1 + \mathcal{C}_1$ consists of those $r \in \mathbb{R}$ such that the line $y = r - x$ intersects $\mathcal{C}_1 + \mathcal{C}_1$. For example, Figure 14 shows that $3/2 \in \mathcal{C}_1 + \mathcal{C}_1$.

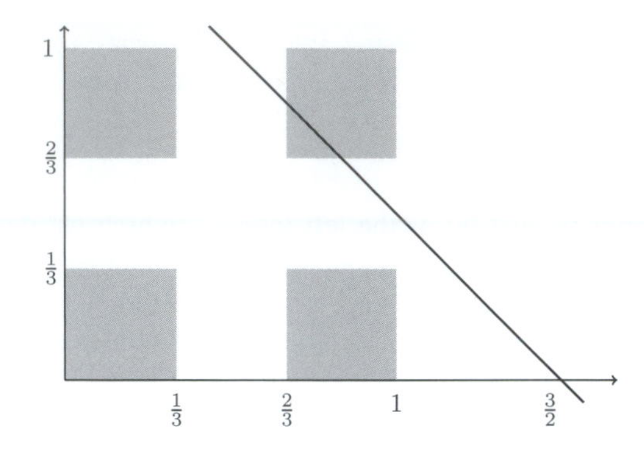

Figure 14. The Cartesian product $\mathcal{C}_1 \times \mathcal{C}_1$ for Exercise 11.74.

(b) Conclude that $\mathcal{C}_1 + \mathcal{C}_1 = [0, 2]$.
(c) Use induction to prove that $\mathcal{C}_n + \mathcal{C}_n = [0, 2]$.
(d) Prove that $\mathcal{C} + \mathcal{C} = [0, 2]$. Warning: This does not follow immediately from the previous statement!

11.11. Five Groups of Order 8

In Section 10.8 we explored ways you can distinguish some groups of order 12. In this project you will come up with arguments proving that five groups of order 8 are indeed distinct, that is non-isomorphic. It is also the case that these five groups are the only groups of order 8, meaning that any group of order 8 must be isomorphic to one of these five groups. Proving that they are the only groups of order 8 is, however, beyond what we can do given our limited introduction to group theory.

Exercise 11.75 Prove that the three Abelian groups

$$\mathbb{Z}_8, \quad \mathbb{Z}_4 \times \mathbb{Z}_2, \quad \text{and} \quad \mathbb{Z}_2 \times \mathbb{Z}_2 \times \mathbb{Z}_2$$

are distinct. That is, no two are isomorphic.

Exercise 11.76 Show that the symmetries of a square – $\mathrm{SYM}(\mathcal{Q})$ from Section 10.2 – is a non-Abelian group of order 8. What result proves that this group cannot be isomorphic to the three Abelian groups of order 8 in Exercise 11.75?

Let's look at one more group of order 8, the unit quaternions. Let

$$Q_8 = \{\pm 1, \pm i, \pm j, \pm k\},$$

where all products can be determined from the following equalities: $i^2 = j^2 = k^2 = -1$, $ij = -ji = k$, $jk = -kj = i$ and $ki = -ik = j$. These rules can be remembered by using the circle shown in Figure 15: the product of two consecutive elements is the third one, going clockwise. The same is true going counterclockwise, except that you get the negative of the third element.

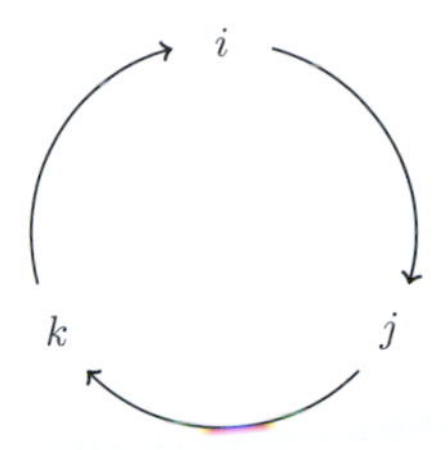

Figure 15. A mnemonic to help remember multiplication in Q_8.

Exercise 11.77 Make a Cayley table for Q_8, and illustrate via examples that Q_8 is not Abelian.

Exercise 11.78 If you have had exposure to matrices and matrix multiplication, then you can also view Q_8 as a group of eight matrices. As examples, the elements i and j are described by the following 4×4 matrices:

$$\mathbf{i} = \begin{bmatrix} 0 & 1 & 0 & 0 \\ -1 & 0 & 0 & 0 \\ 0 & 0 & 0 & -1 \\ 0 & 0 & 1 & 0 \end{bmatrix}, \quad \mathbf{j} = \begin{bmatrix} 0 & 0 & 1 & 0 \\ 0 & 0 & 0 & 1 \\ -1 & 0 & 0 & 0 \\ 0 & -1 & 0 & 0 \end{bmatrix}.$$

Write down all eight matrices that form Q_8.

Exercise 11.79 In order to prove that we now have five non-isomorphic groups of order 8, you need to show that Q_8 and $\text{SYM}(\mathcal{Q})$ are not isomorphic. One approach would be to determine how many elements in each group have order 2.

If you would prefer a bigger challenge, where we have not given as many guiding hints, then you might explore groups of order 18 as an alternative to this project. There are five non-isomorphic groups of order 18. Find them, describe them, and prove that it is the case that no two of them are isomorphic.

Solutions, Answers, or Hints to In-Text Exercises

As we mentioned in the Preface, solving the in-text exercises as you encounter them in the text is an essential task. Make earnest attempts on your own to solve each problem that shows up in the text before reading any further or consulting the material in this appendix. Use the solutions, answers, and comments below to check your work or assist your progress if you get stuck.

Chapter 1

Exercise 1.1 Here's one way to show that if a and b are two non-negative real numbers, then $(a+b)/2 \geq \sqrt{ab}$.

First,

$$(a-b)^2 \geq 0,$$

because the square of any real number is non-negative. Expanding the left-hand side yields

$$a^2 - 2ab + b^2 \geq 0.$$

Adding $4ab$ to both sides gives

$$a^2 + 2ab + b^2 \geq 4ab.$$

Since ab is non-negative, we know further that

$$a^2 + 2ab + b^2 \geq 4ab \geq 0.$$

This can be rewritten as

$$\frac{(a+b)^2}{4} \geq ab \geq 0.$$

The conclusion we desire now follows from taking square roots. □

Take a look at your scratch work and attempts on this problem. You might have started with $(a+b)/2 \geq \sqrt{ab}$, squared both sides, and then followed a series of calculations and deductions that are similar to the steps just given, but *in reverse*. That is certainly a correct approach to first understand the problem. However, the proof should *end* with $(a+b)/2 \geq \sqrt{ab}$ as the conclusion, not the first step. "If ..., then ..." statements and their proofs are studied in Chapter 2.

Exercise 1.2 In the sixth month, the original pair and their first and second offspring each create offspring, bringing the total number of pairs to 8.

In the seventh month, the original pair; their first, second, and third offspring; and their first offspring's offspring each create offspring, bringing the total number of pairs to 13.

Now things are getting quite complicated – there must be a better way to keep track of things . . .

Exercise 1.3 They are

$$f_8 = f_7 + f_6 = 13 + 8 = 21,$$
$$f_9 = f_8 + f_7 = 21 + 13 = 34,$$
$$f_{10} = f_9 + f_8 = 34 + 21 = 55.$$

Exercise 1.4 The key here is to make sure that f_{n+1} counts all of the rabbit pairs that exist in month $n + 1$ in a way that doesn't count any pair twice. Every rabbit pair in month $n + 1$ is exactly one of two types:

(a) It already existed in month n.
(b) It did not exist in month n.

How many pairs are in case (a)? That's f_n. How many pairs are in case (b)? That's f_{n-1}, because we can count their *parents*, pairs that must have existed two months prior to month $n + 1$. Add these two numbers together to get f_{n+1}.

Notice that, in order to justify that f_{n+1} is the correct count for the number of rabbit pairs in month $n + 1$, we are tacitly assuming that f_n is the correct count for month n and f_{n-1} is the correct count for month $n - 1$. This type of reasoning lies at the heart of proofs by mathematical induction, a topic we preview in this chapter and study in Section 2.6.

Exercise 1.5 Let $s(n)$ give the number of different ways to make sums of 1s and 2s that equal n. Then the displayed equalities show that $s(1) = 1$, $s(2) = 2$, $s(3) = 3$, and $s(4) = 5$. You might now guess that $s(n) = f_{n+1}$ for $n > 4$ as well. How could you make yourself feel more confident about this guess? And if the equality indeed holds in general, how might you prove it?

Exercise 1.6 The correct value for P should certainly be greater than $1/2$. If $P \le 1/2$, then $AB/AP \ge 2$, while $AP \le PB$ causes $AP/PB \le 1$.

Exercise 1.7 The calculations are

$$1 - \phi = 1 - \left(\frac{1+\sqrt{5}}{2}\right) = \frac{2 - 1 - \sqrt{5}}{2} = \frac{1-\sqrt{5}}{2}$$

and

$$-1/\phi = \frac{-1}{(\frac{1+\sqrt{5}}{2})} = \frac{-2}{1+\sqrt{5}} = \left(\frac{-2}{1+\sqrt{5}}\right)\left(\frac{1-\sqrt{5}}{1-\sqrt{5}}\right) = \frac{-2(1-\sqrt{5})}{-4} = \frac{1-\sqrt{5}}{2}.$$

Exercise 1.8 The first verification is the calculation

$$\frac{\phi^2 - (1-\phi)^2}{\sqrt{5}} = \frac{\phi^2 - (\phi^2 - 2\phi + 1)}{\sqrt{5}} = \frac{2\phi - 1}{\sqrt{5}} = \frac{\sqrt{5}}{\sqrt{5}} = 1 = f_2.$$

For the second one, first rewrite

$$\frac{\phi^3 - (1-\phi)^3}{\sqrt{5}}$$

as

$$\frac{(\phi^2+\phi)-((1-\phi)^2+(1-\phi))}{\sqrt{5}}.$$

While it would be easy enough to continue from here by substituting the numerical formulas for ϕ and $1-\phi$, let's instead foreshadow the proof technique of Proposition 1.4 by rearranging the terms in the numerator to give

$$\frac{\phi^2-(1-\phi)^2}{\sqrt{5}}+\frac{\phi-(1-\phi)}{\sqrt{5}}.$$

Since you already know that

$$\frac{\phi^2-(1-\phi)^2}{\sqrt{5}}=f_2$$

and

$$\frac{\phi-(1-\phi)}{\sqrt{5}}=f_1,$$

it must be the case that

$$\frac{\phi^3-(1-\phi)^3}{\sqrt{5}}=f_2+f_1=f_3.$$

Exercise 1.9 First use Lemma 1.5 to rewrite

$$\frac{\phi^4-(1-\phi)^4}{\sqrt{5}}$$

as

$$\frac{(\phi^3+\phi^2)-((1-\phi)^3+(1-\phi)^2)}{\sqrt{5}}.$$

Then rearrange the terms in the numerator to give

$$\frac{\phi^3-(1-\phi)^3}{\sqrt{5}}+\frac{\phi^2-(1-\phi)^2}{\sqrt{5}}=f_3+f_2=f_4.$$

Exercise 1.10 (a) If $a=2$, $b=2$, and $c=2$, then $a|c$ and $b|c$, but $ab=4$ and so $ab\nmid c$. Thus, the statement is false.

This is only one of many different counterexample sets of values for a, b, and c; another one is $a=10$, $b=15$, and $c=30$. You only need one counterexample to prove that the statement is false.

(b) All of your experimentation should have led you to guess that the statement must be true. For instance, if you tried $a=3$, $b=6$, and $c=8$, then that particular instance is true because $3|48$.

Here's a proof that the statement is true in general.

PROOF. Since you are assuming that $a|b$, you know that there is a natural number m such that $b=am$. Multiplying both sides of this equation by c gives

$$bc=(am)c=a(mc).$$

But now look: bc is a times the natural number mc. This means that $a|bc$ by the definition of divisibility. □

(c) This statement is true and is called the "transitive" property of divisibility. In Chapter 6 you will study the transitive law in several different contexts.

PROOF. Since $a|b$, there is some natural number m such that $b = am$. Since $b|c$, there is some natural number n such that $c = bn$. Since $c = bn = (am)n = a(mn)$ and mn is a natural number, we conclude that $a|c$. □

(d) This proof will follow the proof of Proposition 1.7 very closely. Be sure that you understand how the numbers 2 and 3 can be inserted into that proof with little difficulty.

PROOF. Since $a|b$, there is a natural number m such that $b = am$. Similarly, since $a|c$, there is a natural number n such that $c = an$. This means that

$$2b + 3c = 2am + 3an = a(2m + 3n),$$

where $2m + 3n$ is a natural number, which shows that $a|(2b + 3c)$. □

(e) Look ahead to Lemma 1.10!

Exercise 1.11 (a) Here's one way to prove that this statement is true.

PROOF. If $2|n$, then there exists a natural number c such that $n = 2c$. Multiplying both sides of this equation by n gives

$$n^2 = (2c)n = 2(nc),$$

which shows that $2|n^2$ (because nc is a natural number). □

Another way is to use the result of part (b) in Exercise 1.10. Can you see how to do this?

(b) This statement is not the same as the one in (a) because the hypothesis and conclusion have switched places, so don't expect a proof of (b) to be similar to a proof of (a).

PROOF. Natural numbers are either odd or even. If n were odd, then n^2 would also be odd, which would imply that $2 \nmid n^2$. So the only cases when $2|n^2$ must have n even, which means that $2|n$. □

(c) This statement is false: for a counterexample, simply let $n = 2$.

Exercise 1.12 Since $a|b$, there is a natural number c such that $b = ac$. Since

$$b - a = ac - a = a(c - 1)$$

and both a and $c - 1$ are non-negative, we know that $b - a \geq 0$. Thus $b \geq a$.

Exercise 1.13 The only way to factor 1 is $1 = 1 \cdot 1$, and those two factors aren't distinct.

Exercise 1.14 There are 25 primes less than 100:

$$2, 3, 5, 7, 11, 13, 17, 19, 23, 29, 31, 37, 41, 43,$$
$$47, 53, 59, 61, 67, 71, 73, 79, 83, 89, \text{ and } 97.$$

Notice that the proportion of natural numbers between 1 and n that are prime seems to decrease as n gets larger. If this fascinates you, research the "prime number theorem."

Exercise 1.15 They are

$$\begin{aligned}84 &= 2\cdot 2\cdot 3\cdot 7 = 2\cdot 3\cdot 2\cdot 7 = 3\cdot 2\cdot 2\cdot 7 = 2\cdot 2\cdot 7\cdot 3\\ &= 2\cdot 3\cdot 7\cdot 2 = 3\cdot 2\cdot 7\cdot 2 = 2\cdot 7\cdot 2\cdot 3 = 2\cdot 7\cdot 3\cdot 2\\ &= 3\cdot 7\cdot 2\cdot 2 = 7\cdot 2\cdot 2\cdot 3 = 7\cdot 2\cdot 3\cdot 2 = 7\cdot 3\cdot 2\cdot 2.\end{aligned}$$

Do you see the systematic way we enumerated them? Maybe you found a better way to do this.

Exercise 1.16 If a is any integer, then $0 = a \cdot 0$. This means that $a|0$.

If a is any non-zero integer, then there's no $c \in \mathbb{Z}$ such that $a = 0 \cdot c$. This means that $0 \nmid a$. However, given our definition of divisibility, $0|0$ seems to be true, since *any* integer c satisfies $0 = 0 \cdot c$. But you've likely heard that it's never good to divide by 0, so we should really modify our definition of divisibility for $\mathbb{Z}$ to exclude this possibility. If you need convincing, see *Zero: The Biography of a Dangerous Idea* by Charles Seife [**Sei00**].

Exercise 1.17

Proposition 1.7 *If a, b, and c are any integers such that both $a|b$ and $a|c$, then $a|(b+c)$.*

PROOF. Since $a|b$, there is an integer m such that $b = am$. Similarly, since $a|c$, there is an integer n such that $c = an$. This means that

$$b + c = am + an = a(m+n),$$

which shows that $b + c$ is an integer multiple of a. In other words, $a|(b+c)$. □

Lemma 1.10 *If a, b, and c are any integers such that both $a|b$ and $a|(b+c)$, then $a|c$.*

PROOF. Since $a|b$, there is an integer m such that $b = am$. Similarly, since $a|(b+c)$, there is an integer n such that $b + c = an$. Thus

$$c = (b+c) - b = an - am = a(n-m),$$

which shows that c is an integer multiple of a, so we conclude that $a|c$. □

Exercise 1.18 In view of

$$\begin{aligned}&\vdots\\ 83 - 5\cdot 11 &= 28,\\ 83 - 6\cdot 11 &= 17,\\ 83 - 7\cdot 11 &= 6,\\ 83 - 8\cdot 11 &= -5,\\ 83 - 9\cdot 11 &= -16,\\ &\vdots\end{aligned}$$

we see that $r = 6$.

Similarly,

$$\begin{aligned}
&\vdots \\
-11 - (-5)4 &= 9, \\
-11 - (-4)4 &= 5, \\
-11 - (-3)4 &= 1, \\
-11 - (-2)4 &= -3, \\
-11 - (-1)4 &= -7, \\
&\vdots
\end{aligned}$$

shows that $r = 1$.

Exercise 1.19 The Division Algorithm says that, given a natural number b, every integer can be written in the form $mb + r$, where r is an integer between 0 and $b - 1$ (inclusive). So if $b = 4$, every number can be written as $m \cdot 4 + r$, which is just a small variation on the statement that every integer can be expressed as $4k, 4k + 1, 4k + 2$, or $4k + 3$.

Exercise 1.20

(a) $0 \in \mathbb{N}$ is not true. Natural numbers are positive.
(b) $0 \in \mathbb{Z}$ is true from the definition of $\mathbb{Z}$.
(c) $\sqrt{2} \in \mathbb{Q}$ is not true, as was proved in Proposition 1.14.
(d) Since $-\sqrt{2} + \sqrt{2} = 0 \in \mathbb{Q}$, the statement is true.
(e) $1 + \sqrt{2} \in \mathbb{Q}$ is not true. We can prove this with a short proof by contradiction.

PROOF. Assume for the moment that the statement $1 + \sqrt{2} \in \mathbb{Q}$ is true. Then $1 + \sqrt{2} = r$, where $r = m/n$ is a rational number. That implies that $\sqrt{2} = r - 1$, where $r - 1$ is another rational number, namely $(m - n)/n$. This contradicts the fact that $\sqrt{2}$ is not a rational number. □

(f) $2\sqrt{2} \in \mathbb{Q}$ is also not true. Try a short proof by contradiction, like the one just given above.
(g) $\sqrt{2} + \sqrt{2} = 2\sqrt{2}$, so the previous result applies.
(h) $\sqrt{2}\sqrt{2} \in \mathbb{Q}$ is true, because $\sqrt{2}\sqrt{2} = 2$.

One of the take-aways from this exercise is that while the sum (or product, or difference) of two rational numbers is rational, the same can't be said for irrational numbers.

Chapter 2

Exercise 2.1 As in Example 2.1, let C be the proposition "Cecelia is a knight" and let D be "Desmond is a knight." Then the truth of Cecelia's statement is determined by the fourth column of the table at the top of the next page.

Since Cecelia is making the statement, we are looking for any rows where the first and final entries agree. This only happens in the final row, so both Cecelia and Desmond are knaves.

C	D	$\sim C$	$\sim C \wedge D$
T	T	F	F
T	F	F	F
F	T	T	T
F	F	T	F

Exercise 2.2 P is true and Q is false, hence $\sim Q$ is true. This means $P \vee \sim Q$ is true. One way to express this statement is "4 is a divisor of 6^2 or 4 is not a divisor of 6."

Exercise 2.3 They are

P	Q	$P \vee Q$
T	T	T
T	F	T
F	T	T
F	F	F

and

P	Q	$\sim P$	$\sim Q$	$\sim P \wedge \sim Q$	$\sim(\sim P \wedge \sim Q)$
T	T	F	F	F	T
T	F	F	T	F	T
F	T	T	F	F	T
F	F	T	T	T	F

Notice that the first, second, and last columns of the two tables agree. This means that whatever statements you might use to replace P and Q, the truth values of $P \vee Q$ and $\sim(\sim P \wedge \sim Q)$ will be the same. Later we will say that these two statements are "logically equivalent."

Exercise 2.4 There will be $8 = 2^3$ rows for each table because there are 3 statements (P, Q, and R) and 2 possible values for each variable (T and F). Here are the tables:

P	Q	R	$P \wedge Q$	$(P \wedge Q) \Rightarrow R$
T	T	T	T	T
T	T	F	T	F
T	F	T	F	T
T	F	F	F	T
F	T	T	F	T
F	T	F	F	T
F	F	T	F	T
F	F	F	F	T

and

P	Q	R	$P \vee Q$	$(P \vee Q) \Leftrightarrow R$
T	T	T	T	T
T	T	F	T	F
T	F	T	T	T
T	F	F	T	F
F	T	T	T	T
F	T	F	T	F
F	F	T	F	F
F	F	F	F	T

Exercise 2.5 Observe that the final column of

P	Q	$Q \Rightarrow P$	$P \Rightarrow (Q \Rightarrow P)$
T	T	T	T
T	F	T	T
F	T	F	T
F	F	T	T

is all Ts.

Exercise 2.6 Truth tables show that the statements in Exercise 2.3 are logically equivalent, while the statements in Exercise 2.4 are not.

Exercise 2.7 The third and sixth columns are the same in this truth table:

P	Q	$P \Rightarrow Q$	$\sim Q$	$\sim P$	$\sim Q \Rightarrow \sim P$
T	T	T	F	F	T
T	F	F	T	F	F
F	T	T	F	T	T
F	F	T	T	T	T

Exercise 2.8 You can verify this claim by comparing the truth tables for $P \Leftrightarrow Q$ and $(P \Rightarrow Q) \wedge (Q \Rightarrow P)$.

Exercise 2.9 You can visualize the sum $1 + 2 + \cdots + n$ as a sequence of stacked squares, the first stack being of height 1, the second of height 2, and so on. Add n squares to the top of the first stack, and then $n - 1$ squares to the second stack, and so forth. The result is a rectangular array of boxes that is $(n + 1) \times n$. Thus, as before, $2 \cdot (1 + 2 + \cdots n) = (n + 1)n$. Can you think of a way to draw a diagram illustrating the general case?

Exercise 2.10

(a) $1/11 = 0.\overline{09} = 0.090909\ldots$

(b) This might require some insight. Since

$$10 \cdot 56.1\overline{67} = 561.\overline{67}$$

and

$$1000 \cdot 56.1\overline{67} = 56167.\overline{67},$$

we have

$$(1000 - 10)56.1\overline{67} = 55606.$$

Thus $56.1\overline{67} = 55606/990 = 27803/495$.

Exercise 2.11 Let a be a natural number and let x be an irrational number. Assume to the contrary that the product ax is rational, so that $ax = m/n$ for some integer m and some natural number n. Then $x = m/an$, where an is a natural number, contradicting the fact that x is irrational.

What happens if you try the argument above with $a = 0$?

Exercise 2.12 No, because $x^3 + x^2 = 2$ has a rational solution of $x = 1$.

Exercise 2.13

(a) *Original implication*: "If n^2 is even, then n is even."
Contrapositive: "If n is not even, then n^2 is not even." Because the contrapositive is logically equivalent to the original implication, either both statements are true or both statements are false. In this case, the contrapositive is probably in the best form to prove that they are true: if n is not even, then $n = 2k + 1$ for some integer k, and so $n^2 = 4k^2 + 4k + 1 = 2(2k^2 + 2k) + 1$, which is certainly not even.
Converse: "If n is even, then n^2 is even." This is also true, but for a reason that is completely independent of the other two cases (see Exercise 1.11).

(b) *Original implication*: "If x^2 is rational, then x is rational." This is false: a counterexample is $\sqrt{2}$.
Contrapositive: "If x is not rational, then x^2 is not rational."
Converse: "If x is rational, then x^2 is rational." This is true. The proof is direct: if x is rational, then $x = m/n$ for some $m \in \mathbb{Z}$ and $n \in \mathbb{N}$, and so $x^2 = m^2/n^2$, which is also a rational number.

(c) *Original implication*: "If $|a + b| < 1$, then $|a| + |b| < 1$." This is false: a counterexample has $a = -1$ and $b = 1$.
Contrapositive: "If $|a| + |b| \geq 1$, then $|a + b| \geq 1$." Notice how negating $<$ leads to $\geq$.
Converse: "If $|a| + |b| < 1$, then $|a + b| < 1$." This is true. Can you prove it?

(d) *Original implication*: "If the professor is 5 minutes late, then class is cancelled." This is likely false; consult your instructor for the correct answer.
Contrapositive: "If class is not cancelled, then the professor is not 5 minutes late." Be sure you understand why this statement must have the same truth status as the original implication.
Converse: "If class is cancelled, then the professor is 5 minutes late." This is almost certainly false. Class could be cancelled for lots of reasons that have nothing to do with your instructor being late. For example, if she happens to walk by some pygmy marmosets just before class, she will probably not make it to class that day: she'll just keep petting them.

Exercise 2.14 We hope you can do these types of problems quite well by now ...

Exercise 2.15 ...including this one!

Exercise 2.16 After rewriting the inequality as $0 \leq n^2 - n$, we can factor the right-hand side as $n(n-1)$. If n is greater than 1, both n and $n-1$ are positive, so $0 \leq n(n-1)$. Also, if n is less than 0, both n and $n-1$ are negative, so $0 \le n(n-1)$. The only integer values of n we haven't yet considered are $n = 0$ and $n = 1$, and each one makes $n(n-1) = 0$. Thus, $0 \le n(n-1)$ for all integers n.

Exercise 2.17 Replace n with q in the preceding proof, and all is fine until you get to the sentence that starts with "The only . . . ". For a specific counterexample, try $q = 1/2$.

Exercise 2.18 Two of the four statements are true – which ones?

Exercise 2.19 Part (a) shows that a solution exists, while part (b) shows that the solution is unique. Proving the truth of a ∃! statement usually requires two arguments like these.

Exercise 2.20

(a) $\forall h \in H$, h is mortal.
(b) $\exists h \in H$, h scored higher than me on the test.
(c) $\sim (\exists t \in T$, t is better than apple pie).

For practice, rewrite part (c) starting with $\forall t \in T$.

Exercise 2.21 The first one can be "For every real number y there is a real number x such that $y < x$." The second can be "There is a real number x such that, for all real numbers y, $y < x$."

Exercise 2.22 She said ∀ persons, ∃! true soul mate; in this order, the soul mate can depend on the person. Dilbert stupidly thought ∃! true soul mate for all persons. His question about the monkey is a good one.

Exercise 2.23 Only one of the statements is false – which one?

Exercise 2.24

(a) $\forall r, s \in \mathbb{R}, r \neq s, \exists q \in \mathbb{Q}, (r < q < s) \vee (s < q < r)$.
(b) $\sim (\forall g \in G, g = \text{gold})$. Now try to write this with an ∃ but no ∼.
(c) $\forall t \in T$, t can't stop the Sandman!
(d) $\forall d \in D$, d will not open. One wonders why they bother to make the stop! What is a more accurate statement as the train approaches the station?

Notice that in part (a), the statement accommodates either $r < s$ or $s < r$. Another possible statement would be

(a) $\forall r, s \in \mathbb{R}, r < s, \exists q \in \mathbb{Q}, r < q < s$.

The reason this statement also works is that for any two distinct real numbers, one of them must be less than the other, and so we might as well call the smaller one r.

Exercise 2.25 $7^2 - 1 = 48 = 8 \cdot 6$ and $7^3 - 1 = 342 = 57 \cdot 6$.

Exercise 2.26 The inductive assumption lets us know that expression B is a multiple of 6.

Exercise 2.27 The only modification necessary to the template is replacing "Assuming that $P(k)$ is true for all values of k with $n_0 \le k \le n$" with "Assuming that $P(n)$ is true." Now revisit Theorem 2.12 with 9 and 10 in place of 6 and 7.

Exercise 2.28

(a) $1000 = 2^3 5^3$
(b) $1001 = 7 \cdot 11 \cdot 13$
(c) $1002 = 2 \cdot 3 \cdot 167$
(d) $1003 = 17 \cdot 59$

It is pretty clear that simple induction will not be helpful, as there is no apparent way to relate the factorization of $n + 1$ to the factorization of n.

Exercise 2.29 In the final presentation of your proof, five base cases might be acceptable but are certainly unnecessary and possibly distracting: only two bases are needed. In your initial work on the problem you might have directly verified that the statement holds for five cases like these, and that's totally fine.

Exercise 2.30 The proof uses strong induction because both $P(n)$ and $P(n-1)$, which are the respective formulas

$$f_n = \frac{\phi^n - (1-\phi)^n}{\sqrt{5}}$$

and

$$f_{n-1} = \frac{\phi^{n-1} - (1-\phi)^{n-1}}{\sqrt{5}},$$

are assumed to be true to prove that $P(n + 1)$ is true.

Exercise 2.31 Despite the fact that Exercises 1.45 and 1.46 involve the Fibonacci numbers, neither requires consideration of only the two previous cases: Exercise 1.45 can be proved using simple induction and a single base case, while Exercise 1.46 is probably best proved with a single base case and a strong induction step that assumes the truth of the statement over a wide range of previous cases.

Exercise 1.47 might best be proved directly: consider the expression

$$n - (a_k + a_{k-1} + \cdots + a_1 + a_0)$$

after writing n in terms of powers of 10.

Exercise 1.48 is probably best proved using strong induction, by letting 2^k be the highest power of 2 less than or equal to $(n + 1)$ and using the inductive assumption to tell you something about $(n + 1) - 2^k$. The argument is similar to one you might use for Exercise 1.46.

Exercise 2.32 Start by declaring at the outset that x is a real number such that $x \geq -1$. The proof uses simple induction on the natural number n.

Base Case: If $n = 1$, $(1 + x)^1$ and $1 + 1 \cdot x$ both equal $1 + x$, so the inequality holds (as an equality).

Inductive Step: We assume that the inequality is true for $n \geq 1$. Then to prove that the inequality is true for $n + 1$, consider

$$\begin{aligned}(1+x)^{n+1} &= (1+x)^n(1+x) \\ &\geq (1+nx)(1+x) \\ &= 1 + x + nx + nx^2\end{aligned}$$

$$\begin{aligned} &= 1 + (n+1)x + nx^2 \\ &\geq 1 + (n+1)x, \end{aligned}$$

where the first inequality in the derivation uses the inductive assumption, and the final inequality holds because nx^2 can't be negative. Once you have also said where it is important in the derivation that $x \geq -1$, you will have a complete proof.

Exercise 2.33 Look for them!

Exercise 2.34 There is no absolute minimum number of examples that are needed to provide the basis for a conjecture, and it is sometimes the case that more examples are better. Using data for only the first four positive integers might not be enough for this problem. On the other hand, you can waste a lot of time checking cases when an attempted proof might highlight the sort of cases that could be problematic.

Exercise 2.35 Before you play, be kind and wish your friend good luck.

Exercise 2.36 Among other things, be sure that you explain why the second player always has a move to make.

Exercise 2.37 The case of $3/7$ is handled in the paragraph that follows this exercise.

Exercise 2.38 If four cases aren't enough, try eight or more.

Exercise 2.39 For example,

$$\begin{aligned} \frac{5}{6} &= \frac{1}{2} + \frac{1}{3} \\ &= \frac{1}{2} + \frac{1}{3}\left(\frac{1}{2} + \frac{1}{3} + \frac{1}{6}\right) \\ &= \frac{1}{2} + \frac{1}{6} + \frac{1}{9} + \frac{1}{18}. \end{aligned}$$

Exercise 2.40 Here's an easy question to ask: "Can you always avoid using $1/2$?"

Exercise 2.41 In part (b), you need to point out that $g \geq 2$ so that you avoid dividing by 0.

Exercise 2.42 Using no lines leaves the plane whole, so $p(0) = 1$ seems best.

Chapter 3

Exercise 3.1 Try "$z > 0$" or "$z \geq 1$."

Exercise 3.2 When $k < 0$, the element $q = m/n^k$ is an integer, so it's already in the set in the form $q = q/n^0$.

Exercise 3.3 We know that $4/9 \in \mathbb{Q}_3$ because setting $m = 4$ and $k = 2$ in the definition of $\mathbb{Q}_3$ gives $4/3^2 = 4/9$.

We also know that $5/12 \in \mathbb{Q}_6$ because taking $m = 15$ and $k = 2$ in the definition of $\mathbb{Q}_6$ gives $15/6^2 = 15/36 = 5/12$. Note that there is no requirement in the definition of $\mathbb{Q}_n$ that m/n^k be in lowest terms when n^k is an integer.

Now let's show that $4/9 \notin \mathbb{Q}_2$, using a proof by contradiction. Suppose for the moment that $4/9 \in \mathbb{Q}_2$. Then Exercise 3.2 implies that

$$4/9 = m/2^k \quad \text{for some integers } m \text{ and } k \text{ with } k \geq 0.$$

Cross-multiplying gives $4 \cdot 2^k = 9m$, but the right-hand side is divisible by 3 while the left-hand side is not. This is a contradiction, so we conclude that $4/9 \notin \mathbb{Q}$.

Exercise 3.4 Regarding the existential quantifier, the assertion that $\{a \in A \mid P(a)\}$ is non-empty means that there is at least one element x in that set; and, by the definition of the set, $P(x)$ is true, so $\exists a \in A, P(a)$ is true as well. Conversely, if $\exists a \in A, P(a)$ is true, then $P(x)$ is true for some $x \in A$. This means that x is in the set $\{a \in A \mid P(a)\}$, so the set is non-empty.

Exercise 3.5 If $x \in \mathbb{Z}$, then $x = x/n^0$ for integers x and 0, so $x \in \mathbb{Q}_n$.

Here's one reason $\mathbb{Z}$ is a proper subset of $\mathbb{Q}_n$ if $n \geq 2$: $1/n \notin \mathbb{Z}$, but $1/n = 1/n^1 \in \mathbb{Q}_n$.

Exercise 3.6

(a) 4 is not equal to 1, 2, 3, or $\{1, 4\}$, so $4 \notin A$.
(b) Since $4 \notin A$, $\{1, 4\} \not\subseteq A$.
(c) $\{1, 2\}$ is not equal to 1, 2, 3, or $\{1, 4\}$, so $\{1, 2\} \notin A$.
(d) $\{1, 4\} \in A$, but $\{1, 4\} \notin \mathbb{N}$.

Exercise 3.7 We can put them in a line: $\mathbb{N} \subset \mathbb{Z} \subset \mathbb{Q} \subset \mathbb{R}$. They are proper subsets owing to numbers like -1, $1/2$, and $\sqrt{2}$, respectively.

Exercise 3.8 They are $\{1, 2\}$, $\{1, 3\}$, $\{1, 4\}$, $\{2, 3\}$, $\{2, 4\}$, and $\{3, 4\}$.

Exercise 3.9 This requires two arguments: one to show that $6\mathbb{Z} \subseteq 3\mathbb{Z}$, and one to show that $6\mathbb{Z} \neq 3\mathbb{Z}$.

For the first, let $x \in 6\mathbb{Z}$. Then $x = 6y$ for some $y \in \mathbb{Z}$, so that $x = 3(2y)$ with $2y \in \mathbb{Z}$. This means that $3|x$, so $x \in 3\mathbb{Z}$.

The second argument is easy: $3 \in 3\mathbb{Z}$, but $3 \notin 6\mathbb{Z}$.

Exercise 3.10 The sets are as follows.

(a) $C \cap D = \{2, 3, 5\}$
(b) $C \cup D = \{1, 2, 3, 4, 5, 7, 11, 13, 17, 19, \ldots\}$
(c) $C \setminus D = \{1, 4\}$
(d) $D - C = \{7, 11, 13, 17, 19, \ldots\}$
(e) $C \triangle D = \{1, 4, 7, 11, 13, 17, 19, \ldots\}$

Exercise 3.11 Both statements are logically equivalent to $A \cap B = \emptyset$.

For the first statement, $A \setminus B = A$ implies that B contains no elements of A, and $B \setminus A = B$ implies that A contains no elements of B; thus A and B have no common elements. Conversely, if A and B have no common elements, then we must also have $A \setminus B = A$ and $B \setminus A = B$.

Exercise 3.12 Remember that $\sqrt{2} \notin \mathbb{Q}$.

Exercise 3.13 (a) In Figure 2 on page 71, both $A - B$ and $A - (A \cap B)$ can be seen as the crescent-shaped region on the left.

(b) First, let $x \in A - B$. This means that $x \in A$ and $x \notin B$. Since x is not an element of B, it can't be an element of both A and B, so $x \notin A \cap B$. Thus, $x \in A$ and $x \notin A \cap B$, so $x \in A - (A \cap B)$.

Second, let $x \in A - (A \cap B)$. This means that $x \in A$ and $x \notin A \cap B$. Since $x \in A$, the only way that x is not in $A \cap B$ is if $x \notin B$. Since $x \in A$ and $x \notin B$, $x \in A - B$.

Exercise 3.14 The proof is similar to the first half. Let $x \in A \cup (B \cup C)$. By the definition of the union of two sets, this implies that $x \in A$ or $x \in B \cup C$, which (again by the definition of union) means that $x \in A$ or $x \in B$ or $x \in C$. This in turn means that $x \in A \cup B$ or $x \in C$, which implies that $x \in (A \cup B) \cup C$. Thus, we have shown that $A \cup (B \cup C) \subseteq (A \cup B) \cup C$.

Exercise 3.15 Let $x \in (A \cap B) \cup (A \cap C)$. Then it must be the case that $x \in A \cap B$ or $x \in A \cap C$. Either way, we know $x \in A$. We also know that $x \in B$ or $x \in C$, and so $x \in B \cup C$. Thus, $x \in A \cap (B \cup C)$.

Exercise 3.16 Begin by assuming that $x \in A$, and use the assumption that $B^c \subseteq A^c$ to prove that $x \in B$.

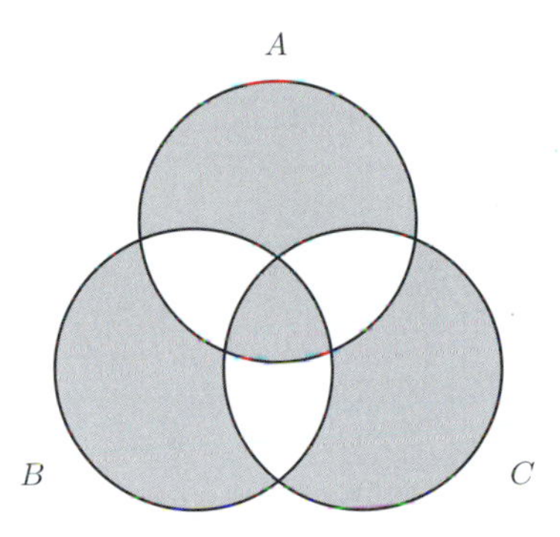

Figure 1. The set $(A \triangle B) \triangle C = A \triangle (B \triangle C)$.

Exercise 3.17 Both expressions describe the shaded region in Figure 1. This should be a sufficient hint for leading you to a proof (or a radioactive safe room).

Exercise 3.18 Looking at a 2-set Venn diagram, we see that enumerating all of the elements of A plus all of the elements of B double-counts the elements in $A \cap B$. So, to get the correct size of $A \cup B$, we must subtract the number of elements in $A \cap B$ from $|A| + |B|$.

A similar formula for $|A \cup B \cup C|$ accounts for double- and triple-counting certain elements:

$$|A \cup B \cup C| = |A| + |B| + |C| - (|A \cap B| + |A \cap C| + |B \cap C|) + |A \cap B \cap C|.$$

Exercise 3.19 There are many good ways to do this. Our attempt at a picture is shown in Figure 2. The intervals are nested, as each sits inside the previous interval, like a matryoshka doll.

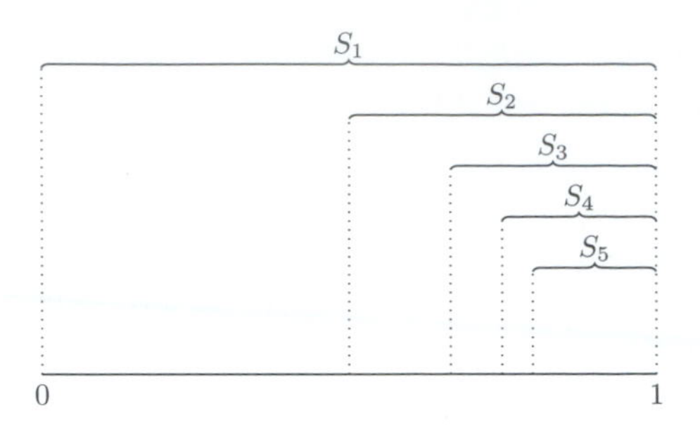

Figure 2. Our attempt for Exercise 3.19.

Exercise 3.20 The key is that "for some" expresses the same concept as "or," and "for all" corresponds to "and."

Exercise 3.21 The union consists of all of the points on the curves in Figure 3. The intersection contains four points:

$$\bigcap_{i \in I} S_i = \{(0, 1),\ (2\pi/3, -1/2),\ (4\pi/3, -1/2),\ (2\pi, 1)\}.$$

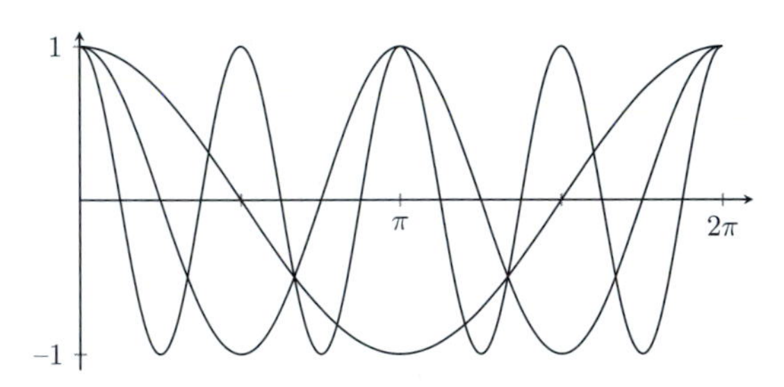

Figure 3. The curves $\cos(x)$, $\cos(2x)$, and $\cos(4x)$.

Exercise 3.22 There are no ordered pairs because there is no element available for the first entry, so $A \times B = \emptyset$.

Exercise 3.23 Suppose that A and B are both finite. Since an ordered pair is just a short sequence of length 2, Proposition 2.13 tells us that

$$|A \times B| = |A||B|$$

because there are $|A|$ choices for the first term and $|B|$ choices for the second term. If at least one of A and B is empty, then either Exercise 3.22 or Proposition 2.13 shows that the formula still holds. The only other case is if at least one of A and B is infinite and the other is non-empty, and it shouldn't be too hard to see that $A \times B$ must be infinite in this case.

Exercise 3.24 $(1, 2) = \{\{1\}, \{1, 2\}\}$, whereas $(2, 1) = \{\{2\}, \{1, 2\}\}$.

Exercise 3.25

(a) Any set A with $|A| = 1$ can be expressed as $A = \{a\}$ for some element a, so $|\mathcal{P}(A)| = |\{\emptyset, \{a\}\}| = 2$.
(b) Similarly, if $|A| = 2$, then $A = \{a, b\}$ for some distinct elements a and b, so $|\mathcal{P}(A)| = |\{\emptyset, \{a\}, \{b\}, \{a, b\}\}| = 4$.
(c) $|\mathcal{P}(A)| = 8$ if $|A| = 3$. Can you write down all eight elements of $\mathcal{P}(A)$?

Exercise 3.26 Since $\mathcal{P}(\mathcal{P}(\emptyset)) = \{\emptyset, \{\emptyset\}\}$, $\mathcal{P}(\mathcal{P}(\mathcal{P}(\emptyset)))$ is equal to the 4-element set

$$\{\emptyset,\ \{\emptyset\},\ \{\{\emptyset\}\},\ \{\emptyset, \{\emptyset\}\}\}.$$

$\mathcal{P}(\mathcal{P}(\mathcal{P}(\mathcal{P}(\emptyset))))$ contains 16 elements – can you find all of them?

Exercise 3.27 The first case requires $d \notin B$, so the eight possible subsets B are

$$\emptyset,\ \{a\},\ \{b\},\ \{c\},\ \{a, b\},\ \{a, c\},\ \{b, c\},\ \{a, b, c\}.$$

The second case requires $d \in B$. The subsets B are listed below, and the corresponding sets B' are obtained by simply removing the element d:

$$\{d\},\ \{a, d\},\ \{b, d\},\ \{c, d\},\ \{a, b, d\},\ \{a, c, d\},\ \{b, c, d\},\ \{a, b, c, d\}.$$

Exercise 3.28 In no particular order, they are

$$\begin{aligned} S_1 &= \{\{a\}, \{b\}, \{c\}\}, \\ S_2 &= \{\{a, b\}, \{c\}\}, \\ S_3 &= \{\{a, c\}, \{b\}\}, \\ S_4 &= \{\{a\}, \{b, c\}\}, \\ S_5 &= \{\{a, b, c\}\}, \end{aligned}$$

Exercise 3.29 For the partition into three blocks, note that

$$\mathbb{N} = \{1, 4, 7, \ldots\} \cup \{2, 5, 8, \ldots\} \cup \{3, 6, 9, \ldots\}.$$

Exercise 3.30 Since there are n options for the ith term of the sample, the total number of samples is

$$\underbrace{n \cdot n \cdots n}_{k \text{ terms}} = n^k.$$

Exercise 3.31 $(4)_2 = 4 \cdot 3 = 4!/2! = 12$ counts the number of ordered samples of size 2, without replacement:

$$\begin{array}{cccccc} (a, b) & (a, c) & (a, d) & (b, c) & (b, d) & (c, d) \\ (b, a) & (c, a) & (d, a) & (c, b) & (d, b) & (d, c) \end{array}$$

Exercise 3.32 The empty set is the first set in any path. Regardless of the choices made along the way, it is clear from the number of edges directly above each subset in the diagram that there are 4 options for the second set on the path, 3 options for the third, 2 options for the fourth, and the path ends with $\{a, b, c, d\}$. By Proposition 2.13, there are $4 \cdot 3 \cdot 2 = 24 = 4!$ such paths.

Now see if you can arrive at 4! in a different way, by having a permutation of $\{a, b, c, d\}$ determine the four edges of the path.

Exercise 3.33 Notice that each 2-combination of $\{a, b, c, d\}$ is represented by exactly $2! = 2$ of the $(4)_2 = 12$ ordered samples in Exercise 3.31, so the number of 2-combinations is

$$\binom{4}{2} = (4)_2/2! = 6.$$

Exercise 3.34 Every subset of an n-element set contains either 0 elements, or 1 element, or 2 elements, ...

Exercise 3.35 You will get the same partition by choosing S_1 to be a subset of S or its complement in S. And you had better not let S_1 equal either S or its complement, $\emptyset$.

Exercise 3.36 The number of blocks can be $n - 1$ only if all of the blocks contain a single element of S, except for one block containing two elements. In how many ways can you choose these two elements?

Chapter 4

Exercise 4.1 No. If you thought q was it, consider $q/5$.

Exercise 4.2 After some experimentation, and deciding that "0 cents postage" doesn't make much sense, we know that we can create the following values using only 10 and 4 cent stamps.

$$\begin{aligned}\text{Possible postage} &= \{4, 8, 10, 12, 14, 16, 18, \ldots\} = \{4, 8\} \cup \{8 + 2n \mid n \in \mathbb{N}\} \\ &= \{4, 8\} \cup \{2n \mid n \in \mathbb{N} \text{ and } n \geq 5\}.\end{aligned}$$

We can prove that all of these values are possible by considering numbers of the form $4a$ and $4a + 10$.

We conjecture that this is a complete list. To prove this conjecture, assume to the contrary that there are counterexamples. By the Well-Ordering Principle there must be a minimal counterexample, m, which we know by our earlier computations must be at least 20. We also know that m must be even, since 4 and 10 are even. Because $m \geq 20$, it must be the case that $m - 4$ is an even integer that is at least 16. Thus $m - 4 = 4a + 10b$ for some non-negative integers a and b. But then $m = 4(a + 1) + 10b$, which contradicts the claim that m is the minimal counterexample.

Exercise 4.3 Finding an integer combination that equals 1 isn't hard to do: $(-1) \cdot 8 + 3 \cdot 3$ is one, as is $2 \cdot 8 + (-5) \cdot 3$. How many different possibilities are there? Finding a combination of 8 and 3 that equals 0 might help.

On the other hand, 8 and 6 are both even, so any integer combination of them must be even, too.

Exercise 4.4 The pairs are $(1, 1)$, $(1, 2)$, $(2, 3)$, $(3, 5)$, $(5, 8)$, $(8, 13)$, $(13, 21)$, $(21, 34)$; no pair shares a common integer divisor greater than 1. But $2|2$ and $2|8$, as you might also recall from Section 2.9.2.

Exercise 4.5 There are only two integers relatively prime to 0, and 0 isn't one of them.

Exercise 4.6

(a) False: For example, let $a = 3$ and $b = 1$.
(b) True: A proof very similar to the one given in Lemma 4.7 works here, although you'll need to tweak Exercise 1.17.
(c) True: Again, a proof with the same overall logic as the one given in Lemma 4.7 works here. A few tweaks will be needed; and remember that if $d|a$, then $d|ac$ for any integer c.
(d) False: For example, let $a = 2$, $b = 1$, and $c = 2$.

Exercise 4.7 Yes. The "only if" part of the theorem doesn't depend on whether a and b are positive, negative, or 0. And by Exercise 4.5, 0 is only relatively prime with 1 and -1. Since $0 \cdot 0 + 1 \cdot 1$ and $0 \cdot 0 + (-1) \cdot (-1)$ both equal 1, the "if" part of the theorem is true as well.

Exercise 4.8 Yes. Again, the "only if" part of the theorem doesn't depend on whether a and b are positive, negative, or 0; and Exercise 4.7 handles the situations when at least one of them equals 0.

We are thus left with two cases not yet handled by Theorem 4.8:

- Both a and b are negative.
- One of a and b is negative, and the other is positive.

One possible way to handle each of these cases is to use induction on the quantity $|a| + |b|$ (which of course equals $a + b$ when both a and b are positive). You will need to rearrange and alter some of the inequalities given in the proof, and refer to $|a|$ and $|b|$ instead of a and b; and for one of the cases, you may need to use $a' = a + b$ instead of $a' = a - b$.

Another way to handle these cases is just to use Theorem 4.8 directly applied to $|a|$ and $|b|$, which are both positive. In any resulting linear combination, you can replace $|a|$ with a and $|b|$ with b, so long as you negate certain coefficients. You would then need only a lemma that says a and b are relatively prime if and only if $\pm a$ and $\pm b$ are relatively prime, regardless of their signs.

Good luck putting the pieces together!

Exercise 4.9 For certain values of a and b, it may be challenging to find integer combinations that equal 1 by trial and error. There is a method, called the Euclidean Algorithm, that works for any pair of integers; we hope that these exercises motivate you to research it!

(a) Since $(-3) \cdot 9 + 2 \cdot 14 = 1$, we have $(-9) \cdot 9 + 6 \cdot 14 = 3$.
(b) Since $7 \cdot 11 + (-4) \cdot 19 = 1$, we have $(-14) \cdot (-11) + (-8) \cdot 19 = 2$.
(c) Notice that $(-21) \cdot 21 + 13 \cdot 34 = 1$. Do you think it's a coincidence that 13 is another Fibonacci number?

Exercise 4.10 Recall the structure of the proof of Proposition 2.13.

Exercise 4.11 There will be a quiz later!

Exercise 4.12 For the first part, you can simply let one of a and b be equal to 15 and the other 225; there are other possibilities as well. For the second part, notice that $15 \nmid 220$; explain why this is problematic.

Exercise 4.13 For one approach, note that the product ab is a multiple of both a and b, so the smallest common multiple of a and b is in the finite set of natural numbers $S = \{1, 2, \ldots, ab\}$.

For a different approach, consider the set M of all common multiples of a and b, and notice that $ab \in M$ means that it is a non-empty subset of $\mathbb{N}$.

Exercise 4.14 Let $d = \text{GCD}(a, b)$. Since d must divide both a and b, its prime divisors can only be among the primes p_i, so $d = p_1^{c_1} p_2^{c_2} \cdots p_n^{c_n}$, where each c_i is a non-negative integer. If $c_i = \min(a_i, b_i)$ for each i, then certainly $d|a$ and $d|b$. On the other hand, if some $c_i > \min(a_i, b_i)$, then at least one of $p \nmid a$ and $p \nmid b$ is true because c_i would be greater than one of a_i and b_i.

The argument for LCM is similar.

Exercise 4.15 Using the notation in the preceding paragraph, the formulas for GCD and LCM imply

$$\text{GCD}(a, b) \cdot \text{LCM}(a, b) = p_1^{\min(a_1,b_1)+\max(a_1,b_1)} p_2^{\min(a_2,b_2)+\max(a_2,b_2)} \cdots p_n^{\min(a_n,b_n)+\max(a_n,b_n)}.$$

Once you have explained why $\min(x, y) + \max(x, y) = x + y$, you will know that

$$\text{GCD}(a, b) \cdot \text{LCM}(a, b) = p_1^{a_1+b_1} p_2^{a_2+b_2} \cdots p_n^{a_n+b_n} = (p_1^{a_1} p_2^{a_2} \cdots p_n^{a_n})(p_1^{b_1} p_2^{b_2} \cdots p_n^{b_n}) = ab.$$

Exercise 4.16 Since $(-3) \cdot 7 + 2 \cdot 11 = 1$, we have $(-3) \cdot 21 + 2 \cdot 33 = 3$.

Exercise 4.17 Good luck!

Exercise 4.18 There's only one such set.

Exercise 4.19 Computations like

$$\frac{m_1}{n^{k_1}} + \frac{m_2}{n^{k_2}} = \frac{n^{k_2} m_1 + n^{k_1} m_2}{n^{k_1+k_2}}$$

show that $\mathbb{Q}_n$ is closed under addition, subtraction, and multiplication. For division, consider $1/p$, where p is a prime that doesn't divide n.

Exercise 4.20 The hardest step involves the computation

$$(a + b\sqrt{2})(c + d\sqrt{2}) = (ac + 2bd) + (ad + bc)\sqrt{2}.$$

Chapter 5

Exercise 5.1 There are a lot of choices, including the floor and ceiling functions, which are defined later in this section if you have never seen them before. Or you could be boring and just do something like set $h(x) = -2$ for all $x \in \mathbb{R}$.

Exercise 5.2

(a) The number of characters in ω is the sum of the numbers of characters in ω' and ω''.
(b) The alphabet for Σ^* consists of only the characters 0 and 1, so $\|\omega\|$ is the number of 0s plus the number of 1s.

Exercise 5.3 One possibility for (d) would be

$$g((n,p)) = \begin{cases} 1 + x^2 + \cdots + x^{2|n|} & \text{if } p = \text{even}, \\ x + x^3 + \cdots + x^{2|n|+1} & \text{if } p = \text{odd}. \end{cases}$$

As examples, $g((3, \text{even})) = 1 + x^2 + x^4 + x^6$ and $g((-1, \text{odd})) = x + x^3$. What is important for your function is that *every* ordered pair in $\mathbb{Z} \times$ Parity is sent to a unique and *unambiguous* polynomial in $\mathbb{Z}[x]$. Thus the following would not define a function for (d), for a couple of reasons:

$$g((n,p)) = \begin{cases} x^n & \text{if } p = \text{even}, \\ x^{|n|+1} & \text{if } p \in \text{Parity}. \end{cases}$$

Exercise 5.4 The range of $\|\cdot\|$ is $\mathbb{N} \cup \{0\}$, and the range of ϕ is $\{0, 1\}$. As for Range(κ), does any polynomial map to the string $00\ldots0$ consisting of n zeros for any $n > 1$?

Exercise 5.5 Just two of the seven are not functions. In those cases, there exist elements of the domain whose images are not in the given codomains.

Exercise 5.6

(a) Let x_1 and x_2 be any elements of (the domain) $\mathbb{R}^+$. If $f(x_1) = f(x_2)$, then $\sqrt{x_1} = \sqrt{x_2}$, and squaring both sides gives $x_1 = x_2$. Thus, f is injective.
(b) The proof of injectivity is essentially the same as the one given in part (a), with reciprocating in place of squaring.
(c) Note that $h(-2) = h(2) = 1/4$, so h is not injective.
(d) Note that $m(10) = m(01) = 1 + x^2$, so m is not injective.

Exercise 5.7

(a) If a is any element in (the codomain) A, then that same a (which is also in the domain A) satisfies $1_A(a) = a$, and thus 1_A is surjective.

If a_1 and a_2 are any elements of (the domain) A such that $1_A(a_1) = 1_A(a_2)$, then

$$a_1 = 1_A(a_1) = 1_A(a_2) = a_2,$$

so 1_A is injective.
(b) The floor function is surjective because if n is an integer, then $\lfloor n \rfloor = n$. It is not an injective function because $\lfloor 2 \rfloor = \lfloor 2.1 \rfloor = 2$. Similar reasonings imply that the same results hold for the ceiling function.

(c) If a_1 and a_2 are any elements of A such that the ordered pairs $f(a_1) = (a_1, a_1)$ and $f(a_2) = (a_2, a_2)$ are equal, then their first coordinates, a_1 and a_2, must be equal, so f is injective.

If $|A| \geq 2$, then there are distinct elements $a_1 \neq a_2$ in A, and $f(a) = (a_1, a_2)$ is not satisfied for any $a \in A$, so f is not surjective. However, if $|A| = 1$ or $|A| = 0\ldots$ you can take it from here!

Exercise 5.8 For (b), perhaps show that $f(x) = 8x - 3$ is both injective and surjective. Does this example suggest a formula for a bijection from $(0, 1)$ to (a, b) for any real numbers $a < b$?

The same type of formula probably won't work for (c). Maybe incorporate $1/x$?

Exercise 5.9 Since f is an injection, so is $\hat{f}$:

$$\hat{f}(a_1) = \hat{f}(a_2) \text{ implies } f(a_1) = f(a_2), \text{ so } a_1 = a_2.$$

And Range(f) = Range($\hat{f}$), so $\hat{f}$ is surjective.

Exercise 5.10

(a) Each of the 3 elements in A can be paired with one of the 5 elements in B, so you can think of this as creating a sequence of length 3 with 5 options for each term. Perhaps Proposition 2.13 would be helpful?
(b) After an element of B has been paired with an element of A, it can not be paired with another element of A. Proposition 2.13 still seems like a good approach, or consult Section 3.8.
(c) None. Why?
(d) The answer to part (a) simply changes in value; the answer to part (b) changes entirely!

Exercise 5.11 It is meaningless, because the codomain of g and the domain of f are different.

Exercise 5.12

(a) g is not injective because, for example, $g(-1) = g(1) = 1$. But $(g \circ f)(x) = (e^x)^2$ is injective: if $(e^{x_1})^2 = (e^{x_2})^2$, then e^{x_1} and e^{x_2} both being positive means that $e^{x_1} = e^{x_2}$, which implies $x_1 = x_2$. Plotting a graph of $g \circ f$ will confirm this result.
(b) An example is shown in Figure 4.

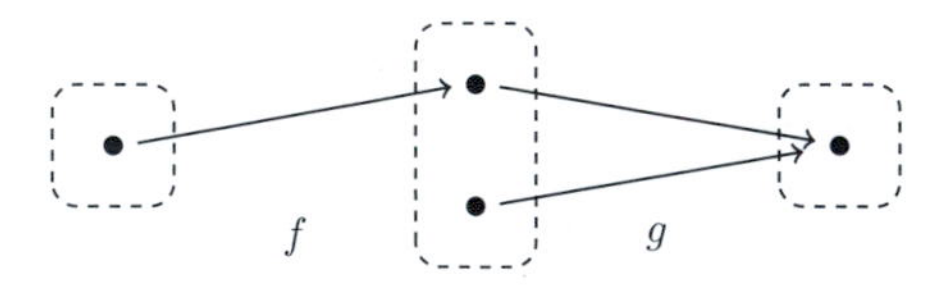

Figure 4. A simple example for Exercise 5.12.

Exercise 5.13 Continuing the hints given in the text ...

(a) ... there is a $b \in B$ such that $g(b) = c$. Since f is surjective, there is an $a \in A$ such that $f(a) = b$. Thus, $g(f(a)) = g(b) = c$, so $g \circ f$ is surjective.

(b) ... we know that $f(a_1) = f(a_2)$. Since f is injective, we know that $a_1 = a_2$. Thus, $(g \circ f)$ is injective.
(c) There's at most one sentence to write here!

Exercise 5.14 Yes: f must contain an ordered pair of the form $(3, b)$, and nothing else.

Exercise 5.15 The function $\|\cdot\|$ contains eight such elements: ω can be any element of the set

$$\{000, 001, 010, 011, 100, 101, 110, 111\}.$$

Exercise 5.16 A function is injective if $(x_1, y) \in f$ and $(x_2, y) \in f$ imply that $x_1 = x_2$. A function is surjective if for each $y \in Y$ there is an $x \in X$ such that $(x, y) \in f$.

Exercise 5.17 For the second step, if $\hat{f}$ is not injective, there exist distinct elements $x, y \in A - \{a\}$ and an element $z \in B - \{b\}$ such that both (x, z) and (y, z) are in $\hat{f}$. But $\hat{f}$ is a subset of f, so (x, z) and (y, z) are in f, too.

Exercise 5.18 To find all points of intersection, solve $x^3 = \sqrt[3]{x}$. This leads to

$$0 = x^9 - x = x(x^8 - 1) = x(x^4 + 1)(x^4 - 1) = x(x^4 + 1)(x^2 + 1)(x + 1)(x - 1),$$

so the x-values are -1, 0, and 1; and the y-values are the same because these points must lie on the line $y = x$. Be sure that your two graphs are reflections of each other across this line.

Exercise 5.19 Using the logical equivalences discussed in Exercises 2.7 and 2.8 and the paragraph that follows them, we just need to show that $Q \Rightarrow (P_1 \wedge P_2)$ is logically equivalent to $\underbrace{(Q \Rightarrow P_1) \wedge (Q \Rightarrow P_2)}_{\equiv \text{ to second paragraph}}$. Try a truth table!

Exercise 5.20 To show that f is injective, suppose that $f(a_1) = f(a_2)$ for elements $a_1, a_2 \in A$. Then $g(f(a_1)) = g(f(a_2))$, and so $1_A(a_1) = 1_A(a_2)$, implying that $a_1 = a_2$. To show that f is surjective, consider any $b \in B$. Since $g(b) = a$ for some $a \in A$, we have

$$f(a) = f(g(b)) = 1_B(b) = b.$$

A similar argument implies that g is bijective as well: just swap the symbols f and g, A and B, and a and b!

Finally, since f is bijective and thus the function f^{-1} exists, let's show that $g = f^{-1}$. Let $b \in B$, and suppose that $g(b) = a_1$ and $f^{-1}(b) = a_2$. Then

$$f(a_1) = f(g(b)) = b$$

from the given properties of f and g, and

$$f(a_2) = f(f^{-1}(b)) = b$$

from the definition of f^{-1}. Since f is injective, we must have $a_1 = a_2$. Thus, g and f^{-1} agree for all b in their common domain, B.

Exercise 5.21

(a) Since $f(x) = x^3 - x = (x+1)(x-1)x = 0$ exactly when $x \in \{-1, 0, 1\}$, the pre-image of 0 is $\{-1, 0, 1\}$.
(b) From the graph (or calculus) it is clear that the pre-image of 6 contains exactly one element. Since $2^3 - 2 = 6$, the pre-image of 6 is $\{2\}$.
(c) We have $f'(x) = 3x^2 - 1 = 0$ when $x = \pm\sqrt{1/3}$; these are the x-values of the points where the local minima and maxima occur. Since $f(\sqrt{1/3}) = -2/3\sqrt{3}$ and $f(-\sqrt{1/3}) = 2/3\sqrt{3}$, each of the pre-images of $2/3\sqrt{3}$ and $-2/3\sqrt{3}$ are 2-element sets.

Exercise 5.22

(a) Use the fact that $3.14 < \pi < 3.15$, so $9.8596 < \pi^2 < 9.9225$.
(b) $\{0, 1, 4\}$
(c) $\emptyset$
(d) $\{4, 5, 6, 7, 8, 9\}$
(e) Since the range of $f_{\mathcal{P}}$ can only consist of non-negative integers, and since $f_{\mathcal{P}}(S) = \mathbb{N} \cup \{0\}$ for $S = \{0, \sqrt{1}, \sqrt{2}, \ldots\}$, $f_{\mathcal{P}}(\mathbb{R}) = \mathbb{N} \cup \{0\}$ as well.

Exercise 5.23 $\sin([0, \pi]) = [0, 1]$, and $\cos_{\mathcal{P}}^{-1}(\{1\}) = \{2\pi n \mid n \in \mathbb{Z}\}$.

Exercise 5.24 The graphs are best suggested using dotted lines with *very* fine spacing. We will see in Chapter 7 that some of the lines should have much finer spacing than others!

(a) $f(7/3) = 1$ and $g(7/3) = 7/3$.
(b) $f(\sqrt{2}) = g(\sqrt{2}) = 0$.
(c) $f_{\mathcal{P}}([-1, 2]) = \{0, 1\}$, because there are rational and irrational numbers in the interval $[-1, 2]$.
(d) $g_{\mathcal{P}}(\{\sqrt{2}, \sqrt{3}, \sqrt{4}\}) = \{0, 2\}$.
(e) Notice that $g_{\mathcal{P}}^{-1}(\mathbb{Q}) = \mathbb{R}$, because $g(x) \in \mathbb{Q}$ for each $x \in \mathbb{R}$. Thus, $f_{\mathcal{P}}(g_{\mathcal{P}}^{-1}(\mathbb{Q})) = f_{\mathcal{P}}(\mathbb{R}) = \{0, 1\}$.
(f) Since $S \subset \mathbb{Q}$, we know that $g_{\mathcal{P}}(S) = S$, so $f_{\mathcal{P}}^{-1}(g_{\mathcal{P}}(S)) = f_{\mathcal{P}}^{-1}(S)$. Since S contains neither 0 nor 1, the answer is $\emptyset$.

Exercise 5.25

(a) Try finding a counterexample with $A = \{1\}$, $B = \{x, y\}$, and $T = B$.
(b) Let $b \in T$. Since f is surjective, there is an $a \in A$ with $f(a) = b$. So we know that $a \in f^{-1}(T)$, and this implies that $b \in f(f^{-1}(T))$. Thus $T \subseteq f(f^{-1}(T))$, and with part (b) of Proposition 5.24 we conclude that $f(f^{-1}(T)) = T$.
(c) Which is most helpful: part (a) or part (b)?

Exercise 5.26 Try letting X and Y be disjoint.

Chapter 6

Exercise 6.1 The graph of A is a line, the graph of B is a closed half-plane, and the graph of C is shown in Figure 5.

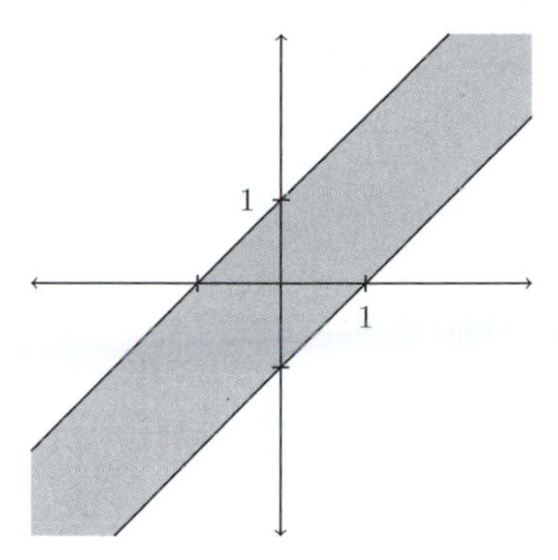

Figure 5. The graph of the relation $C = \{(x, y) \mid |x - y| \leq 1\}$.

Exercise 6.2 C is reflexive since $|x - x| = 0 \leq 1$ for all $x \in \mathbb{R}$, and it is symmetric because $|x - y| = |y - x|$ for all $x, y \in \mathbb{R}$. But it is not transitive: for example, $|3 - 2| \leq 1$ and $|2 - 1| \leq 1$, but $|3 - 1| \not\leq 1$.

B is reflexive and transitive; A is all three.

Exercise 6.3 Our coin flips asked us to find a relation that is reflexive and symmetric but not transitive. An example of such a relation is C of Example 6.2 .

Exercise 6.4 By the discussion in Section 5.5, any function f is a subset of $S \times S$, so it's a relation! One such g is $g(x) = 2x$.

Exercise 6.5 If $a|b$ and $b|a$, then we know there are natural numbers m and n such that $a = bm$ and $b = an$. By substitution this implies $a = (an)m = a(nm)$ and so $nm = 1$. Thus $m = n = 1$, so $a = b$.

Exercise 6.6 Yes – for example, move 5 and 10 to the left of the rows they are on.

Exercise 6.7 It is a poset because $\preceq$ is reflexive, antisymmetric, and transitive. For example, it is antisymmetric since any two rabbit pairs can't both be descendants of each other, so $x \preceq y$ and $y \preceq x$ implies that x and y are the same pair.

The Hasse diagram for $(S_5, \preceq)$ contains five rabbit pairs, with three minimal elements and the original pair at the top as the maximal element.

Exercise 6.8 This is an interesting infinite poset.

(a) $1/2 < 3/4$, but $1/3 \not< 1/4$.
(b) The most complicated argument should be the one establishing antisymmetry. Try a proof by contradiction, assuming that $[a, b] \neq [c, d]$ and considering what is implied by $[a, b] \preceq [c, d]$ and $[c, d] \preceq [a, b]$.
(c) Assume to the contrary that $[a, b]$ is a maximal element with $b < 1$. Then consider $[(b + 1)/2, 1]$.
(d) See the suggestion for part (c).
(e) $[0, 1]$ is both maximal and minimal.

Exercise 6.9 The only one that isn't is (c).

Exercise 6.10 The posets (b) and (d) are well ordered.

Exercise 6.11 We need to establish that $\equiv_e$ has three properties.

Reflexive: Since $n + n = 2n$ is even for all $n \in \mathbb{Z}$, $\equiv_e$ is reflexive.

Symmetric: If $m \equiv_e n$, then $m + n = 2k$ for some $k \in \mathbb{Z}$. But then $n + m$ also equals $2k$, so $n \equiv_e m$.

Transitive: If $k \equiv_e m$ and $m \equiv_e n$, then

$$k + m = 2i \text{ and } m + n = 2j$$

for some $i, j \in \mathbb{Z}$. Thus $k + n = 2i + 2j - 2m = 2(i + j - m)$, which is even. Therefore $k \equiv_e n$ as well.

Exercise 6.12 The proof of Proposition 6.11 can be adapted to prove this is an equivalence relation, and again we point you to the more general result in Exercise 6.57.

Exercise 6.13 To answer part (c), notice that an equivalence class corresponds to the set of x-values where a horizontal line intersects the graph of sine. If two x-values correspond to the same horizontal line, their equivalence classes are the same; and if they correspond to different horizontal lines, the two equivalence classes have no x-values in common (because sine is a function). The dotted horizontal line in Figure 6 intersects the graph of sine in the equivalence class of $\pi/6$, which may be helpful for part (b).

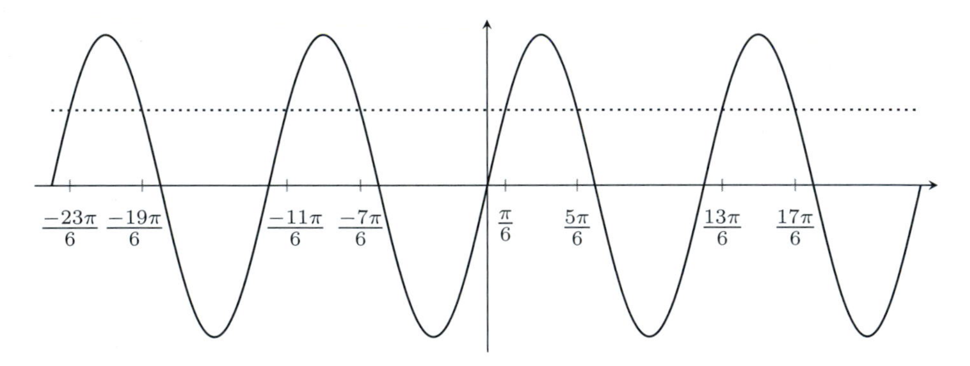

Figure 6. Equivalence classes for $\sim_s$ correspond to x-values that yield the same value in $y = \sin(x)$.

Exercise 6.14 It's equality.

Exercise 6.15 Look at Theorem 6.14 and Exercise 3.28.

Exercise 6.16 The equivalence classes modulo 2 are the evens and odds. The equivalence classes modulo 3 are:

$$[0] = \{3n \mid n \in \mathbb{Z}\} = \{\ldots, -3, 0, 3, 6, \ldots\},$$
$$[1] = \{3n + 1 \mid n \in \mathbb{Z}\} = \{\ldots, -2, 1, 4, 7, \ldots\},$$
$$[2] = \{3n + 2 \mid n \in \mathbb{Z}\} = \{\ldots, -1, 2, 5, 8, \ldots\}.$$

Exercise 6.17 We ask you ...

Exercise 6.18 ... to complete these three exercises ...

Exercise 6.19 ... without any additional hints from us.

Exercise 6.20 Each element of $\mathbb{Z}_4$ is an equivalence class that must be mapped to a unique element of $\mathbb{Z}_3$.

Exercise 6.21 The formulas follow from the fact that $(0+a)-a=0$ and $(1\cdot a)-a=0$ in $\mathbb{Z}$.

Exercise 6.22 The multiplication table for $\mathbb{Z}_6$ is shown in Figure 7. Notice that we have $[2]\cdot[3]=[0]$ in $\mathbb{Z}_6$, which is directly related to the factorization $6=2\cdot 3$.

·	[0]	[1]	[2]	[3]	[4]	[5]
[0]	[0]	[0]	[0]	[0]	[0]	[0]
[1]	[0]	[1]	[2]	[3]	[4]	[5]
[2]	[0]	[2]	[4]	[0]	[2]	[4]
[3]	[0]	[3]	[0]	[3]	[0]	[3]
[4]	[0]	[4]	[2]	[0]	[4]	[2]
[5]	[0]	[5]	[4]	[3]	[2]	[1]

Figure 7. Multiplication in $\mathbb{Z}_6$.

Exercise 6.23 Recall the proof of Proposition 2.16 and the discussion preceding it.

Exercise 6.24 The last digit is a 9. Here's a slightly more difficult computation: What's the last digit of 7^{312}?

Exercise 6.25 The argument given for $f:\mathbb{Z}_6\to\mathbb{Z}_3$ would fail when applied to the proposed squaring function from $\mathbb{Z}_4$ to $\mathbb{Z}_3$. If $a\equiv b\ (\text{MOD }4)$, then $b=a+4k$ for some $k\in\mathbb{Z}$. So

$$f([a+4k]_4)=[(a+4k)^2]_3=[a^2+8ak+16k^2]_3\,.$$

Since $8\equiv 2\ (\text{MOD }3)$ and $16\equiv 1\ (\text{MOD }3)$, we can't reduce $8ak+16k^2$ to zero as we could when we were defining the function from $\mathbb{Z}_6$ to $\mathbb{Z}_3$. In particular, we cannot make a statement similar to "Since 3 divides 12 and 36 we know $12ak\equiv 0\ (\text{MOD }3)$ and $36k^2\equiv 0\ (\text{MOD }3)$."

Exercise 6.26 Here is a proof of part (d).

If $[a]=[b]$ in $\mathbb{Z}_{mn}$, then $a=b+k(mn)$ for some $k\in\mathbb{Z}$. Therefore

$$a^2=b^2+2bk(mn)+k^2(mn)^2\,.$$

Thus $a^2\equiv b^2\ (\text{MOD }n)$, and so $[a^2]_n=[b^2]_n$ if $[a]_{mn}=[b]_{mn}$.

If you would like an additional challenge, find values of m and n where this squaring function from $\mathbb{Z}_{mn}$ to $\mathbb{Z}_n$ is not surjective.

Exercise 6.27 You can prove this by mimicking the proof of Theorem 6.19, or you can use an argument inspired by $[a]_m=[a-b+b]_m=[a-b]_m+[b]_m$.

Exercise 6.28 $4^2 \equiv 1 \ (\text{MOD } 15)$, so 4 plays the role of "1/4."

Exercise 6.29 Construct the rows corresponding to multiplication by 2 and 10 in order to verify that [2] and [10] do not have inverses. On the other hand, $3 \cdot 5 \equiv 1 \ (\text{MOD } 14)$, so [3] is the multiplicative inverse of [5] and [5] is the multiplicative inverse of [3].

Exercise 6.30 If $ab_1 \equiv 1 \ (\text{MOD } m)$ and $ab_2 \equiv 1 \ (\text{MOD } m)$, then $a(b_1 - b_2) \equiv 0 \ (\text{MOD } m)$. This means that $m | a(b_1 - b_2)$, with $\text{GCD}(a, m) = 1$. What can you conclude about m and $b_1 - b_2$?

Exercise 6.31 Figure 8 should get you started, if you are truly stuck.

·	1	3	5	9	11	13
1	1	3	5	9	11	13
3	3	9	1	13	5	11
5	5	1	-	-	13	-
9	9	13	-	-	-	-
11	11	5	13	-	9	-
13	13	11	-	-	-	-

Figure 8. A hint to help construct the multiplication table for $\mathcal{U}_{14}$.

Exercise 6.32 Search for hints among the various facts we have proven about relatively prime integers.

Chapter 7

Exercise 7.1 Suppose that n new guests arrive. Hilbert should announce over the intercom to everyone who already has a room: "I again apologize for the inconvenience, but in order for us to accommodate some new guests, we need you to change rooms. If you are in room #k, please move to room #$(k + n)$ for the night. Thank you!" He can then send the new arrivals to rooms #1 through #n.

Exercise 7.2 Hilbert can first move every current guest into an even-numbered room by announcing: "I once again apologize for the inconvenience, but in order for us to accommodate some new guests, we need you to change rooms. If you are in room #k, please move to room #$2k$ for the night. Thank you!" He then tells the new guests to move into the odd-numbered rooms, with new guest #n sent to room #$(2n - 1)$.

Exercise 7.3 None! Cookie #n will be eaten in round n, and that accounts for all of them.

Exercise 7.4 The answers are:

(a) Cookies #1 through #9 will be on his plate.
(b) The even-numbered cookies will be on his plate.

If you found this to be an interesting exercise, or if you found it difficult, then you should also work on Exercise 7.22 at the end of the chapter.

Exercise 7.5 If you have worked earnestly on this question for more than two hours and are stuck, then perhaps you should stop and begin reading the remainder of this section and the next, where almost all is revealed.

Exercise 7.6 We recommend using the fact that all of these functions have inverses: the natural logarithm $\ln(x)$, $\arctan(x)$, and $y = (x-a)/(b-a)$.

Exercise 7.7 Constructing an explicit function might help you prove that this is a bijection. Using $(-1)^n$ in your formula will allow you to change signs back and forth between positive and negative, and the function $\lfloor n/2 \rfloor$ is also handy, as for example $\lfloor 6/2 \rfloor = \lfloor 7/2 \rfloor = 3$.

Exercise 7.8 We already gave you a nice hint!

Exercise 7.9 For part (a), by shifting his current guests from room #n to room #$(n+1)$, Hilbert has opened up room #1 for a frog. It's quite generous of him to go to all of this trouble for a frog.

For part (b), you might wish to view the partial table presented in this section as one corner of a vast parking lot.

Exercise 7.10 Again, we already gave you a nice hint!

Exercise 7.11 Since Range($\hat{h}$) is an infinite subset of $\mathbb{N}$, use Lemma 7.7.

Exercise 7.12 $(876, 924) \notin \text{Range}(\tau)$ because $\frac{876}{924} = \frac{73}{77}$. There are surely simpler examples than this one.

Exercise 7.13 Randomly selecting the image of each $a \in A$, we created the injective function $f : A \to \mathcal{P}(A)$ defined by

$$\begin{array}{ccc} & f & \\ 1 & \to & \{2\} \\ 2 & \to & \{1,3\} \\ 3 & \to & \{1,3,4\} \\ 4 & \to & \{1\} \end{array}$$

The resulting set is $S_f = \{1, 2, 4\}$, which is not in the range of f.

Exercise 7.14 Just change "$\mathcal{P}(\mathbb{N})$ is countable" to "$A \asymp \mathcal{P}(A)$," and change all occurrences of $\mathbb{N}$ to A. And change n to x for style points, if you think n should only represent integers.

Exercise 7.15 We are not uncomfortable with the original argument, so we skipped this exercise.

Exercise 7.16 We know that $\mathbb{R} = \mathbb{Q} \cup \mathbb{Q}^c$. If we assume to the contrary that $\mathbb{Q}^c$ is a countable set, then we have expressed $\mathbb{R}$ as the union of two countable sets. Lemma 7.10 would then imply that $\mathbb{R}$ must be countable, which contradicts Theorem 7.15.

Exercise 7.17 The function $f : [0, 1) \to (0, 1]$ defined by

$$f(x) = \begin{cases} x & \text{if } x > 0, \\ 1 & \text{if } x = 0 \end{cases}$$

is a bijection, so $[0,1) \asymp (0,1]$. In the previous paragraph we applied Schröder–Bernstein to prove $[0,1) \asymp [0,1]$. Thus, by the transitivity of $\asymp$ we will be finished if we can show that $(0,1) \asymp [0,1]$. We can do this by considering containment as the injection $(0,1) \hookrightarrow [0,1]$ and $g(x) = \frac{1}{2}x + \frac{1}{4}$ as the injection $[0,1] \hookrightarrow (0,1)$, and then appealing to Schröder–Bernstein.

Exercise 7.18 First prove that h is a surjection. Then argue that it is sufficient to prove that h is injective when restricted to X and when restricted to $[0,1] \setminus X$.

Exercise 7.19 You could try to do this by mimicking the proof of Lemma 7.20, or you could instead quote results we have already established: $\mathbb{R}$ has the same cardinality as the open interval $(0,1)$, which leads to $\mathbb{R}^2$ having the same cardinality as $(0,1)^2$, and then you can apply Lemma 7.20.

Exercise 7.20 In fact we only need Lemmas 7.21 and 7.22 to prove

$$\mathbb{R} \asymp \mathbb{R}^2 \asymp [0,1]^2 .$$

Chapter 8

Exercise 8.1 One lower bound is -1. Any $\ell \leq -1$ is also a lower bound, because if $s \in (-1,2]$, then $\ell \leq -1 < s$. Similarly, an upper bound is 2, and any $m \geq 2$ will serve as an upper bound. Figure 9 might help to clarify things.

Figure 9. The half-open interval $(-1,2]$ for Exercise 8.1 is displayed in bold.

Exercise 8.2 Every real number r is an upper bound for the empty set, since $s \leq r$ for all $s \in \emptyset$: there are no such s! Similarly, every real number is a lower bound.

Since every element of $\mathbb{R}$ is an upper bound, there is certainly no least one.

Exercise 8.3 A lower bound is 2 and an upper bound is 3. The greatest lower bound is 2.7, and the least upper bound is e. For a proof that e is irrational, see Section 8.6.2.

Exercise 8.4 The set of all lower bounds for A is $(-\infty,-2]$, so the greatest lower bound is -2. The set of all upper bounds for A is $[1,\infty)$, so the least upper bound is 1.

Exercise 8.5 Corollary 8.4 guarantees the existence of an $n \in \mathbb{N}$ such that $1/n < r$. There is a largest power of 10 within the non-empty finite set $\{1,2,\ldots,n\}$; suppose it is 10^{k-1} for some $k \geq 1$. Then $n < 10^k$, so $1/10^k < 1/n < r$.

Exercise 8.6 Assume to the contrary that $\ell \in \mathbb{R}$ is a lower bound for $E(r)$. Since r is negative and $r^2 > 1$, we know that

$$\ell \leq r^{2n} r < r$$

for all $n \in \mathbb{N}$. Since r is negative, dividing by r changes the order of the inequalities, giving

$$1 < r^{2n} \leq \frac{\ell}{r}$$

for all $n \in \mathbb{N}$. Thus

$$0 < n \cdot \ln(r^2) \leq \ln\left(\frac{\ell}{r}\right)$$

for all $n \in \mathbb{N}$. Since $\ell < r < 0$, we know that $\ell/r > 1$ and thus $\ln(r^2)$ and $\ln(\ell/r)$ are positive real numbers, so we have a contradiction to Corollary 8.3.

Exercise 8.7 The terms of the first two sequences are heading toward a limit of 0. The terms of the third sequence aren't heading toward any finite value, so the limit doesn't exist, or you might say that it's "$+\infty$"; see Exercise 8.27. The fourth sequence does have a limit, but it's not rational. Write the first few terms of the sequence in decimal notation and see if that helps with your guess.

Exercise 8.8 Parts of the definition are matched with phrases of the sentence: "Regardless of how close [ϵ] you want the sequence terms [a_n] to be to the limiting value [L], you will get your wish [$|L - a_n| < \epsilon$] so long as you ignore an appropriate number [$N - 1$] of the initial terms of the sequence." And order is important here: the value of N depends on the tolerance ϵ. You might wish to write N_ϵ in place of N, if doing so helps you remember this order.

Exercise 8.9 The same argument that worked for $\epsilon = 0.1$ works for $\epsilon = 0.01$, requiring the solution of

$$\frac{3}{n} < \frac{1}{100}$$

at the end. Now, sketch a (wide!) extension of Figure 1 on page 190 showing that the terms of the sequence eventually stay inside of a thinner $\epsilon = 0.01$ band around $L = 1$.

Exercise 8.10 Since the limit is $L = 0$, we need to show that for every $\epsilon > 0$ there is an $N \in \mathbb{N}$ such that $|0 - r^n| < \epsilon$ for all natural numbers $n \geq N$. Since

$$|0 - r^n| = |r^n| = r^n$$

for $r \in (0, 1)$, we can appeal to the first proof given in Theorem 8.5 to conclude that there is an $N \in \mathbb{N}$ such that $r^N < \epsilon$. Since $0 < r < 1$, for any natural number $n \geq N$ we have

$$r^n \leq r^N < \epsilon.$$

Exercise 8.11 For the first, it's

$$\frac{1/2}{1 - 1/2} = 1.$$

For the second, it's

$$\frac{A}{1 - 1/4} = \frac{A}{3/4} = \frac{4A}{3}.$$

Exercise 8.12 If $a = 0$, then the sequence of partial sums is simply $(0, 0, 0, \ldots)$, so the series converges to 0. Otherwise, if $a \neq 0$ and $|r| \geq 1$, the series doesn't converge. For example, if $a = r = 1$, then the series is $1 + 1 + 1 + \cdots$ and the sequence of partial sums is $(1, 2, 3, \ldots)$, which has no finite limit.

Showing that a geometric series doesn't converge in other cases where $a \neq 0$ and $|r| \geq 1$ requires other arguments, some of which might be assisted by the Comparison Test of Section 8.5.

Exercise 8.13 It is a constant sequence: all of the terms must be equal.

Exercise 8.14

(a) The sequence $(1/n) = (1, 1/2, 1/3, \ldots)$ is monotone and bounded.
(b) The sequence $((-1)^{n-1}) = (1, -1, 1, -1, \ldots)$ is bounded but not monotone. The sequence presented in Figure 1 (page 190) is another example, and also shows that a sequence need not be monotone to have a limit.
(c) The sequence $(n) = (1, 2, 3, 4, \ldots)$ is monotone but not bounded.
(d) The sequence $((-1)^n n) = (-1, 2, -3, 4, \ldots)$ is neither monotone nor bounded.

Exercise 8.15 Change μ to λ, "increasing" to "decreasing," "least upper bound" to "greatest lower bound," and so on. Also reverse most of the inequalities and change some $-$ signs to $+$; but ϵ should remain positive!

Exercise 8.16 You can show that the series converges, and find crude upper and lower bounds for the value of the series, by first comparing it with the geometric series $\sum_{i=1}^{\infty} \frac{1}{3^i}$, and then with $\sum_{i=1}^{\infty} \frac{1}{3^{i+1}}$.

Exercise 8.17 First show that $n! \geq 2^{n-1}$ for all $n \in \mathbb{N}$; induction will work, as will a direct argument. This implies that

$$\frac{1}{n!} \leq \frac{1}{2^{n-1}}$$

for all $n \in \mathbb{N}$, and the 0th terms are both 1, so the Comparison Test applies.

Exercise 8.18 Integration by parts yields

$$\int_0^{\pi} f(x)\sin(x)\,dx = (f(x)(-\cos(x)))|_0^{\pi} - \int_0^{\pi} f^{(1)}(x)(-\cos(x))\,dx,$$

where the final term can be replaced via

$$\int_0^{\pi} f^{(1)}(x)\cos(x)\,dx = (f^{(1)}(x)(\sin(x)))\Big|_0^{\pi} - \int_0^{\pi} f^{(2)}(x)(\sin(x))\,dx,$$

whose final term can be replaced via

$$\int_0^{\pi} f^{(2)}(x)\sin(x)\,dx = (f^{(2)}(x)(-\cos(x)))\Big|_0^{\pi} - \int_0^{\pi} f^{(3)}(x)(-\cos(x))\,dx,$$

and so on, until a final term is 0 because $f^{(2k+1)}(x) = 0$.

When $k = 2$, the resulting expression is

$$-f(x)\cos(x) + f^{(1)}(x)\sin(x) + f^{(2)}(x)\cos(x) - f^{(3)}(x)\sin(x) - f^{(4)}(x)\cos(x)\Big|_0^{\pi},$$

which evaluates to

$$(f(\pi) - f^{(2)}(\pi) + f^{(4)}(\pi)) - (-f(0) + f^{(2)}(0) - f^{(4)}(0)) = F(\pi) + F(0),$$

as desired.

Exercise 8.19 Among the results you might need are that

$$\int_a^b f(x)\,dx \leq (b-a)M$$

when $f(x)$ is non-negative and bounded above by M, and that

$$\lim_{k\to\infty} \frac{c^k}{k!} = 0$$

for any real number c.

Chapter 9

Exercise 9.1 There are eight equally likely outcomes:

HHH HHT HTH HTT THH THT TTH TTT

Of these eight, three contain exactly two heads: HHT, HTH, and THH. So you might expect this result to occur about three-eighths of the time, if you repeat the activity many, many times.

Exercise 9.2 List the 16 equally likely outcomes and tally the ones you want, as in the previous exercise.

Exercise 9.3 Out of 250 flips, there are 124 heads and 126 tails. Out of all 249 2-flip strings, there are 59 HHs, 64 HTs, 64 THs, and 62 TTs. If you only count 125 2-flip strings starting from flips 1, 3, 5, ..., 249, you get 29 HHs, 34 HTs, 32 THs, and 30 TTs.

What does this evidence suggest, if anything?

Exercise 9.4 Only the denominator changes: $p(G) = 2/37$ and $p(R) = 18/37$. Since $18/37 > 18/38$, the ball is slightly more likely to land on red on a European wheel.

Exercise 9.5 You can determine the numerators of these fractions...

Exercise 9.6 ...by counting carefully in Figure 2 (page 212).

Exercise 9.7 Is it close to 1/8?

Exercise 9.8 Symmetry follows from

$$\binom{n}{k} = \frac{n!}{k!(n-k)!} = \frac{n!}{(n-k)!(n-(n-k))!} = \binom{n}{n-k}.$$

For unimodality, use the fact that

$$\frac{n!}{k!(n-k)!} = \frac{n!}{(k+1)!(n-(k+1))!} \cdot \frac{k+1}{n-k} \leq \frac{n!}{(k+1)!(n-(k+1))!}$$

is true so long as $(k+1)/(n-k) \leq 1$, and so when $k < n/2$.

Exercise 9.9 100! is a 158-digit integer beginning 93326... and ending with 24 0s.

The real number $\sqrt{2\pi}\ 100^{100+\frac{1}{2}}\, e^{-100}$ is approximately 9.32485×10^{157} in scientific notation.

Exercise 9.10 The general statement is that, for non-negative integers n,

$$(x+y)^n = \sum_{k=0}^{n} \binom{n}{k} x^n y^{n-k} = \binom{n}{0} x^n + \binom{n}{1} x^{n-1} y + \cdots + \binom{n}{n} y^n.$$

Mathematical induction seems like an appropriate way to prove this, using the identity $(x+y)^{n+1} = (x+y)^n(x+y)$ in the inductive step along with Theorem 9.8.

Exercise 9.11 Again, count carefully in Figure 2 (page 212).

Exercise 9.12 The values of the binomial coefficients are

$$76\,904\,685, 18\,643\,560, 3\,838\,380, 658\,008, 91\,390, 9880, 780, 40, \text{ and } 1.$$

The sum of these numbers is 100 146 724, which upon dividing by 2^{40} yields

$$P(X \geq 32) = 0.000\,091\,08\ldots$$

Now, what's your conclusion?

Exercise 9.13 If you're not close, keep rolling!

Exercise 9.14 If X counts the points you earn on a die roll, then

$$E(X) = 2 \cdot \frac{1}{6} + 0 \cdot \frac{1}{6} + 6 \cdot \frac{1}{6} + 0 \cdot \frac{1}{6} + 10 \cdot \frac{1}{6} + 0 \cdot \frac{1}{6} = \frac{18}{6} = 3.$$

Exercise 9.15 If A is the event that at least one 6 is seen, then A^c is the event that no 6s are seen. It is easier to compute $p(A^c)$: by Proposition 2.13, $|A^c| = 5^4$, so

$$p(A^c) = 5^4/6^4 \approx 0.4823,$$

and thus $p(A) = 1 - p(A^c) \approx 1 - 0.4823 = 0.5177$. This number is larger than $1/2$.

Exercise 9.16 We know that $D_5(2) = A_5(2) - A_5(1)$ and $A_5(1) = f_7 = 13$, so we are left to determine $A_5(2)$. Since

$$A_3(2) = A_2(2) + A_1(2) + A_0(2) = 4 + 2 + 1 = 7$$

and

$$A_4(2) = A_3(2) + A_2(2) + A_1(2) = 7 + 4 + 2 = 13,$$

we find that

$$A_5(2) = A_4(2) + A_3(2) + A_2(2) = 13 + 7 + 4 = 24.$$

Thus, $D_5(2) = 24 - 13 = 11$.

Exercise 9.17 It was, in fact, faked.

Chapter 10

Exercise 10.1 If (x_1, y_1) and (x_2, y_2) are two points in $\mathcal{Q}$, then the distance between them is

$$\sqrt{(x_1 - x_2)^2 + (y_1 - y_2)^2},$$

which is equal to

$$\sqrt{((-y_1) - (-y_2))^2 + (x_1 - x_2)^2},$$

the distance between $\rho[(x_1, y_1)]$ and $\rho[(x_2, y_2)]$.

Now just see where the two corners go.

Exercise 10.2 It is probably easiest to note that $\rho[(x, y)] = (-y, x)$, which can be applied twice for ρ^2 and three times for ρ^3:

$$(x, y) \mapsto (-y, x) \mapsto \underbrace{(-x, -y)}_{=\rho^2} \mapsto \underbrace{(y, -x)}_{=\rho^3}$$

Exercise 10.3 $\phi_D[(x, y)] = (y, x)$ and $\phi_O[(x, y)] = (-y, -x)$.

Exercise 10.4 You are interested in $\phi_X \circ \phi_D$, so you first perform ϕ_D and then ϕ_X. Since

$$(-1, 1) \overset{\phi_D}{\to} (1, -1) \overset{\phi_X}{\to} (1, 1)$$

and

$$(1, 1) \overset{\phi_D}{\to} (1, 1) \overset{\phi_X}{\to} (1, -1),$$

the composition must be ρ^3.

Exercise 10.5

(a) In order for a subset $A \subseteq S$ to be an identity element, it must be the case that $A \cup B = B \cup A = B$ for all $B \in \mathcal{P}(S)$. The empty set can function as an identity element, as

$$\emptyset \cup B = B \cup \emptyset = B$$

for all $B \in \mathcal{P}(S)$. Let A be a non-empty subset of S. The set A cannot serve as the identity element as we would need it to be true that $A \cup \emptyset = \emptyset$, but $A \cup \emptyset = A \neq \emptyset$.

(b) Since S is non-empty, the set S cannot have an inverse with respect to $\cup$. Such a subset A would have to satisfy $S \cup A = \emptyset$, but $S \cup A = S$ since $A \subseteq S$.

(c) We'll let you do these last two parts ...

(d) ... on your own.

Exercise 10.6 Here are three very strong hints. You can prove that associativity holds by appealing to the fact that it held for G and H; the identity element is the ordered pair of identity elements (e_G, e_H); and the inverse of (g, h) is (g^{-1}, h^{-1}).

Exercise 10.7 Table 1 on the next page is a partially completed Cayley table for $\mathbb{Z}_4 \times \mathbb{Z}_2$. In order to avoid cumbersome notation we express $([a]_4, [b]_2)$ as (a, b).

Exercise 10.8 You might do this by first proving that a symmetry of $\mathfrak{R}$ is determined by where it sends any three vertices on a face of $\mathfrak{R}$. Then prove that where these three vertices go is determined by where any one of them goes using the fact that the box is irregular.

Exercise 10.9 In Theorem 10.11 you proved that $\text{SYM}(\mathfrak{R})$ is closed under composition, so you know that $\alpha \in \text{SYM}(\mathfrak{R})$. By tracing out the destinations of the corners you will

Table 1 *Part of the Cayley table for* $\mathbb{Z}_4 \times \mathbb{Z}_2$.

	(0, 0)	(1, 0)	(2, 0)	(3, 0)	(0, 1)	(1, 1)	(2, 1)	(3, 1)
(0, 0)	(0, 0)	(1, 0)	(2, 0)	(3, 0)	(0, 1)	(1, 1)	(2, 1)	(3, 1)
(1, 0)	(1, 0)	(2, 0)	?	?	(1, 1)	?	?	?
(2, 0)	(2, 0)	?	(0, 0)	?	?	?	?	?
(3, 0)	(3, 0)	?	?	?	?	?	?	?
(0, 1)	(0, 1)	(1, 1)	?	?	?	?	?	?
(1, 1)	(1, 1)	?	?	?	?	?	(3, 0)	(0, 0)
(2, 1)	(2, 1)	?	?	?	?	(3, 0)	?	?
(3, 1)	(3, 1)	?	?	?	?	(0, 0)	?	?

see that α takes a corner to its opposite corner of the box, which the rotations and reflections cannot do. This map is known as the *antipodal* map, and it has a simple formula: $\alpha[(x, y, z)] = (-x, -y, -z)$.

Exercise 10.10 In doing our work we noticed that $\phi_X \circ \rho_X = \alpha$ for each $X \in \{L, W, H\}$.

Exercise 10.11

(a) To show that this is a group, you need to mention that multiplication is associative, 1 is the identity element, and as long as $m/n \neq 0$ then $\frac{m}{n} \cdot \frac{n}{m} = 1$. To show it is an Abelian group, you need to highlight that multiplication of rational numbers is a commutative operation.
(b) This would be similar to part (a), except that 2 does not have a multiplicative inverse within $\mathbb{Z} - \{0\}$, so it's not even a group!

Exercise 10.12 An examination of Table 2 (page 237) shows that $\text{SYM}(\mathfrak{R})$ is Abelian. Can you see why a certain form of symmetry in the table implies that the product is commutative? If so, write up your proof by first proving and then using this insight.

Exercise 10.13 The proof of Proposition 10.21 shows that [1] is a cyclic generator in $\mathbb{Z}_5$. The element [2] is as well, since

$$\begin{aligned} [2] + [2] &= [4], \\ [2] + [2] + [2] &= [1], \\ [2] + [2] + [2] + [2] &= [3], \\ [2] + [2] + [2] + [2] + [2] &= [0]. \end{aligned}$$

Similar computations will show that every non-zero element of $\mathbb{Z}_5$ is a cyclic generator.

Exercise 10.14 Gauss characterized when $\mathcal{U}_n$ is a cyclic group:

$\mathcal{U}_n$ is cyclic if and only if $n = 2, 4, p^k$, or $2p^k$, where p is an odd prime and $k \in \mathbb{N}$.

For example, $\mathcal{U}_9$ is cyclic while $\mathcal{U}_{12}$ is not.

Exercise 10.15

(a) $|\mathbb{Z}_3 \times \mathcal{U}_{15}| = 24$. (You may want to look up your answer to Exercise 3.23 if you think the answer is different.)
(b) $|(1,1)| = 3$.
(c) $|(1,7)| = 12$.

Exercise 10.16 If you are stuck after a couple of hours ...

Exercise 10.17 ... then you might want to pause until you get to Section 10.8.

Exercise 10.18 Here's a proof for part (a). Since we know that $f(a \cdot b) = f(a) \cdot f(b)$ for all $a, b \in G$, we know that $f(g^2) = f(g \cdot g) = f(g) \cdot f(g) = [f(g)]^2$. Assume by induction that $f(g^n) = [f(g)]^n$ for an arbitrary $n \in \mathbb{N}$. Then

$$f(g^{n+1}) = f(g^n \cdot g) = f(g^n) \cdot f(g) = [f(g)]^n \cdot f(g) = [f(g)]^{n+1} ,$$

so this part follows by induction.

Lemma 10.27 establishes (b). An argument like that given above for (a), combined with Lemma 10.28, can be used to prove (c).

Exercise 10.19 There might be some smaller value $j < k$ with $g^j = e$. For example, in $\mathcal{U}_{15}$ we have $2^8 \equiv 1 \pmod{15}$, so $|2| \leq 8$. However it is also the case that $2^4 \equiv 1 \pmod{15}$, so $|2| \leq 4$; and in fact $|2| = 4$ in $\mathcal{U}_{15}$.

Exercise 10.20 Let $f : G \to H$ be an isomorphism and assume that G is Abelian. Let h_1 and h_2 be any two elements of H. Since f is a bijection, there is a unique pair g_1 and g_2 in G such that $f(g_1) = h_1$ and $f(g_2) = h_2$. Thus $f(g_1)f(g_2) = h_1h_2$. By applying the definition of isomorphism twice and appealing to our assumption that G is Abelian, we find that

$$h_1h_2 = f(g_1)f(g_2) = f(g_1g_2) = f(g_2g_1) = f(g_2)f(g_1) = h_2h_1 .$$

Thus H is Abelian.

Similarly, if H is Abelian, we may apply the argument above using the isomorphism $f^{-1} : H \to G$ to prove that G is Abelian.

Exercise 10.21 Try two adjacent reflections, those whose axes of reflection form an angle of 30°. Their product will be a rotation through 60°, but the order in which you multiply them will determine if it is a clockwise or counterclockwise rotation.

Bibliography

[Abb15] Stephen Abbott, *Understanding Analysis*, second edition. Undergraduate Texts in Mathematics. Springer, New York, 2015.

[AF03] Titu Andreescu and Zuming Feng, *102 Combinatorial Problems: From the Training of the USA IMO Team*. Birkhäuser Boston, Inc., Boston, MA, 2003.

[Arc04] Archimedes, *The Works of Archimedes, Vol. I: The Two Books on the Sphere and the Cylinder*. Cambridge University Press, Cambridge, 2004. Translated into English, together with Eutocius' commentaries, with commentary, and critical edition of the diagrams by Reviel Netz.

[Asc91] Marcia Ascher, *Ethnomathematics: A Multicultural View of Mathematical Ideas*. Brooks/Cole Publishing Co., Pacific Grove, CA, 1991.

[BB61] Edwin Beckenbach and Richard Bellman, *An Introduction to Inequalities*, New Mathematical Library 3. Random House, New York-Toronto, 1961.

[BBJ07] George S. Boolos, John P. Burgess, and Richard C. Jeffrey. *Computability and Logic*, fifth edition. Cambridge University Press, Cambridge, 2007.

[BCG01] Elwyn R. Berlekamp, John H. Conway, and Richard K. Guy, *Winning Ways for your Mathematical Plays*, Vol. 1, second edition. A. K. Peters, Ltd., Natick, MA, 2001.

[BCG03a] Elwyn R. Berlekamp, John H. Conway, and Richard K. Guy, *Winning Ways for your Mathematical Plays*, Vol. 2, second edition. A. K. Peters, Ltd., Natick, MA, 2003.

[BCG03b] Elwyn R. Berlekamp, John H. Conway, and Richard K. Guy, *Winning Ways for your Mathematical Plays*, Vol. 3, second edition. A. K. Peters, Ltd., Natick, MA, 2003.

[BCG04] Elwyn R. Berlekamp, John H. Conway, and Richard K. Guy, *Winning Ways for your Mathematical Plays*, Vol. 4, second edition. A. K. Peters, Ltd., Wellesley, MA, 2004.

[Car09] Nathan C. Carter, *Visual Group Theory*. Classroom Resource Materials Series. Mathematical Association of America, Washington, DC, 2009.

[CG96] John H. Conway and Richard K. Guy, *The Book of Numbers*. Copernicus, New York, 1996.

[Con13] John H. Conway, On unsettleable arithmetical problems, *Amer. Math. Monthly*, 120(3):192–198, 2013.

[CS03] John H. Conway and Derek A. Smith, *On Quaternions and Octonions: Their Geometry, Arithmetic, and Symmetry*. A. K. Peters, Ltd., Natick, MA, 2003.

[CS13] John H. Conway and Joseph Shipman, Extreme proofs I: the irrationality of $\sqrt{2}$, *Math. Intelligencer*, 35(3):2–7, 2013.

[Cun16] D.W. Cunningham, *Set Theory: A First Course*. Cambridge Mathematical Textbooks. Cambridge University Press, Cambridge, 2016.

[Dun99] William Dunham. *Euler: The Master of Us All*, The Dolciani Mathematical Expositions 22. Mathematical Association of America, Washington, DC, 1999.

[ES35] P. Erdös and G. Szekeres, A combinatorial problem in geometry, *Compositio Math.*, 2:463–470, 1935.

[GG78] I. Grattan-Guinness, How Bertrand Russell discovered his paradox, *Historia Math.*, 5(2):127–137, 1978.

[Gil03] Rick Gillman (ed.), *A Friendly Mathematics Competition: 35 Years of Teamwork in Indiana*. MAA Problem Books Series. Mathematical Association of America, Washington, DC, 2003.

[GM64] Israel Grossman and Wilhelm Magnus, *Groups and their Graphs*. Random House, New Mathematical Library 14. New York; The L. W. Singer Co., New York, 1964.

[Gor11] Prakash Gorroochurn. Errors of probability in historical context, *Amer. Statist.*, 65(4):246–254, 2011.

[Gou11] Fernando Q. Gouvêa, Was Cantor surprised? *Amer. Math. Monthly*, 118(3):198–209, 2011.

[Gow02] Timothy Gowers, *Mathematics*, volume 66 of *Very Short Introductions*. Oxford University Press, Oxford, 2002. A very short introduction.

[Gow11] Timothy Gowers, Is mathematics discovered or invented? In John Polkinghorne (ed.), *Meaning in Mathematics*. Oxford University Press, Oxford, 2011, pp. 3–12.

[Gra81] Judith V. Grabiner, *The Origins of Cauchy's Rigorous Calculus*. MIT Press, Cambridge, 1981.

[Gra83] Judith V. Grabiner, Who gave you the epsilon? Cauchy and the origins of rigorous calculus, *Amer. Math. Monthly*, 90(3):185–194, 1983.

[Guy88] Richard K. Guy, The strong law of small numbers, *Amer. Math. Monthly*, 95(8):697–712, 1988.

[GWW98] Ellen Gethner, Stan Wagon, and Brian Wick, A stroll through the Gaussian primes. *Amer. Math. Monthly*, 105(4):327–337, 1998.

[HCV52] D. Hilbert and S. Cohn-Vossen, *Geometry and the Imagination*, trans. P. Neményi. Chelsea Publishing Company, New York, 1952.

[HH99] András Hajnal and Peter Hamburger, *Set Theory*, trans. Attila Máté. London Mathematical Society Student Texts 48. Cambridge University Press, Cambridge, 1999. Translated from the 1983 Hungarian original by Attila Máté

[Hin13] Arie Hinkis, *Proofs of the Cantor-Bernstein Theorem: A Mathematical Excursion*. Science Networks. Historical Studies 45. Birkhäuser/Springer, Heidelberg, 2013.

[Imh06] Annette Imhausen, Ancient Egyptian mathematics: new perspectives on old sources. *Math. Intelligencer*, 28(1):19–27, 2006.

[Imh16] Annette Imhausen, *Mathematics in Ancient Egypt: A Contextual History*. Princeton University Press, Princeton, NJ, 2016.

[JN54] R.D. James and Ivan Niven, Unique factorization in multiplicative systems. *Proc. Amer. Math. Soc.*, 5:834–838, 1954.

[Jon10] Timothy W. Jones, Discovering and proving that π is irrational, *Amer. Math. Monthly*, 117(6):553–557, 2010.

[Kle89] Israel Kleiner, Evolution of the function concept: A brief survey, *College Math. J.*, 20(4):282–300, 1989.

[Kli83] Morris Kline, Euler and infinite series, *Math. Mag.*, 56(5):307–314, 1983.

[Knu98] Donald E. Knuth, *The Art of Computer Programming, Vol. 2: Seminumerical Algorithms*, third edition. Addison-Wesley, Reading, MA, 1998.

[Lit86] J.E. Littlewood, *Littlewood's Miscellany*, ed. Béla Bollobás. Cambridge University Press, Cambridge, 1986.

[Loc07] Kari Lock, Mixing a night out with probability. . . & making a fortune, *Math Horizons*, 14(3):8–9, 2007.

[Loh07] Po-Ru Loh, Stepping to infinity along Gaussian primes, *Amer. Math. Monthly*, 114(2):142–151, 2007.

[Mei08] John Meier, *Groups, Graphs and Trees: An Introduction to the Geometry of Infinite Groups*. London Mathematical Society Student Texts 73. Cambridge University Press, Cambridge, 2008.

[Mir71] L. Mirsky, A dual of Dilworth's decomposition theorem, *Amer. Math. Monthly*, 78:876–877, 1971.

[Niv47] Ivan Niven, A simple proof that π is irrational, *Bull. Amer. Math. Soc.*, 53:509, 1947.

[Pis02] Leonardo Pisano, *Fibonacci's Liber Abaci: A Translation into Modern English of Leonardo Pisano's Book of Calculation*. Sources and Studies in the History of Mathematics and Physical Sciences. Springer-Verlag, New York, 2002. Translated from the Latin and with an introduction, notes and bibliography by L.E. Sigler.

[Pro13] James Propp, Real analysis in reverse, *Amer. Math. Monthly*, 120(5):392–408, 2013.

[RH14] Benoît Rittaud and Albrecht Heeffer, The pigeonhole principle, two centuries before Dirichlet, *Math. Intelligencer*, 36(2):27–29, 2014.

[RSW06] Frank Ruskey, Carla D. Savage, and Stan Wagon, The search for simple symmetric Venn diagrams, *Not. Amer. Math. Soc.*, 53(11):1304–1312, 2006.

[Sch90] Mark F. Schilling, The longest run of heads, *College Math. J.*, 21(3):196–207, 1990.

[Sei00] Charles Seife, *Zero: The Biography of a Dangerous Idea*. Penguin Books, New York, 2000.

[Smu11] Raymond M. Smullyan, *What is the Name of This Book? The Riddle of Dracula and Other Logical Problems*. Dover Publications, Inc., Mineola, NY, 2011.

[Ste99] Sherman Stein. *Archimedes: What Did He Do Besides Cry Eureka?* Mathematical Association of America, Washington, DC, 1999.

[Tao06] Terence Tao, *Solving Mathematical Problems: A Personal Perspective*. Oxford University Press, Oxford, 2006.

[Thu94] William P. Thurston, On proof and progress in mathematics, *Bull. Amer. Math. Soc. (N.S.)*, 30(2):161–177, 1994.

[Wey89] Hermann Weyl, *Symmetry*. Princeton Science Library. Princeton University Press, Princeton, NJ, 1989. Reprint of the 1952 original.

[Wil14] Robin Wilson, *Four Colors Suffice: How the Map Problem was Solved*. Princeton Science Library. Princeton University Press, Princeton, NJ, 2014. Revised color edition of the 2002 original, with a new foreword by Ian Stewart.

[Win04] Peter Winkler, *Mathematical Puzzles: A Connoisseur's Collection*. A. K. Peters, Ltd., Natick, MA, 2004.

Index